ULTRA LIBRIS

POLICY, TECHNOLOGY, AND THE CREATIVE ECONOMY OF BOOK PUBLISHING IN CANADA

ULTRA LIBRIS

ROWLAND LORIMER

ECW Press

Co-published with the Canadian Centre
for Studies in Publishing (CCSP) Press

Copyright © Rowland Lorimer, 2012

Co-published by ECW Press
2120 Queen Street East, Suite 200, Toronto, Ontario, Canada M4E 1E2
416-694-3348 / info@ecwpress.com

with Canadian Centre for Studies in Publishing Press

All rights reserved. No part of this publication may be reproduced, stored in a retrieval system, or transmitted in any form by any process — electronic, mechanical, photocopying, recording, or otherwise — without the prior written permission of the copyright owners and ECW Press. The scanning, uploading, and distribution of this book via the Internet or via any other means without the permission of the publisher is illegal and punishable by law. Please purchase only authorized electronic editions, and do not participate in or encourage electronic piracy of copyrighted materials. Your support of the author's rights is appreciated.

Library and Archives Canada Cataloguing in Publication

Lorimer, Rowland, 1944–
Ultra libris : policy, technology, and the creative economy of book publishing in Canada / Rowland Lorimer.
ISBN 978-1-77041-076-3
Also issued as 978-1-77090-289-3 (PDF) and 978-1-77090-290-9 (ePub)
1. Publishers and publishing—Government policy—Canada. 2. Book industries and trade—Government policy—Canada. 3. Publishers and publishing—Technological innovations—Canada. 4. Book industries and trade—Technological innovations—Canada. 5. Canada—Cultural policy. 6. Government aid to publishing—Ontario. I. Title.
Z481.L673 2012   070.50971   C2012-902719-7

Editor: Nadia Halim
Cover and text design: Tania Craan
Cover images: shelf © horiyan/iStockphoto; books © Luis Pedrosa/iStockphoto; glasses © homestudio/Veer
Indexing: Iva Cheung
Typesetting: Kendra Martin
Printing: BVG Arvato 1 2 3 4 5

We wish to thank Simon Fraser University's "University Publication Fund" for financial assistance in the publication of this book.

This book has been published with the help of a grant from the Canadian Federation for the Humanities and Social Science, through the Awards to Scholarly Publications Program, using funds provided by the Social Science and Humanities Research Council of Canada.

The publication of *Ultra Libris* has been generously supported by the Canada Council for the Arts which last year invested $20.1 million in writing and publishing throughout Canada, and by the Ontario Arts Council, an agency of the Government of Ontario. We also acknowledge the financial support of the Government of Canada through the Canada Book Fund for our publishing activities, and the contribution of the Government of Ontario through the Ontario Book Publishing Tax Credit. The marketing of this book was made possible with the support of the Ontario Media Development Corporation.

Printed and bound in the United States

To the authors and publishers, and the politicians and government officials of Canada who made this book worth writing. And also to my family: Anne, Stefan, Conor, and especially Julia, who had yet to enter the world when last I dedicated a brand new title.

TABLE OF CONTENTS

PREFACE AND ACKNOWLEDGEMENTS ix

INTRODUCTION Policy, Technology, and the Creative Economy of Book Publishing in Canada 1

CHAPTER 1 Of books, publishing, and social context 19

CHAPTER 2 Prelude to modernity: Some historical notes on Canadian book publishing and cultural development 53

CHAPTER 3 Establishing a book publishing industry: from the 1960s to the 1990s 77

CHAPTER 4 Reconceiving book publishing from the middle 1990s forward 121

CHAPTER 5 The current state of Canada's book industry, government policies, and cultural partnerships 157

CHAPTER 6 Operational and market realities in Canadian book publishing 205

CHAPTER 7 Trajectories of technology-incubated massive change 247

CHAPTER 8 Changing realities and the future of Canadian book publishing 291

APPENDIX 339

NOTES 351

BIBLIOGRAPHY 389

INDEX 409

Preface and acknowledgements

From 1977 until the late 1990s, I was involved in tracking policy and industry developments in book publishing and in the writing of reports for provincial and federal governments, their agencies, and publishers' organizations. Mainly, the reports assessed the state of the industry and individual firms, existing policy and support programs, and possible additional programs. In the middle '90s, I wrote a draft of Chapters 1 through 6 of this book, which I used to teach students in the Master of Publishing program at Simon Fraser University. But just before the century turned, I became more involved in scholarly journal editing and publishing and, with the help of a very valued colleague, Richard K. Smith, tested the limits of technology by, for example, making the back issues of the *Canadian Journal of Communication* available openly online. That activity, along with the hiring of a second invaluable colleague, John W. Maxwell, gave me the insight into emerging technology that formed the basis of my understanding of the influence of technology on book publishing. Inspired in part by John's and Richard's analyses and insights, I put forward, and received funding for, a proposal to develop software for the creation of what we called PExOD, Publishers Extensible Online Database, an ONIX-based information and communication tool that would transform small book publishers' websites and allow them to send and receive machine-based book metadata as they interacted with large business partners such as Chapters/Indigo. The software was a great help to at least two publishers, New Society Publishers and Dundurn Press, and it raised the level of awareness of digital bases, so that small Canadian-owned publishers were better prepared to purchase commercial products once they became available. This activity led to the writing of Chapter 7.

My continuing interest in policy, together with my evolving understanding of technology, provided the foundation for reviewing and renewing the policy analysis in the early chapters, extending it into current times, contextualizing the initial draft of the technology chapter, and integrating the two areas in Chapter 8. It has been an exciting journey, and it has been especially intriguing to project current developments into the future.

My purpose in writing this book was always to address, not only the very small academic community that might share my interests, but a general audience with an interest in books and book publishing. By casting the narrative with such an audience in mind, I thought this book would have value for industry members (especially those just joining the industry), students, government officials with responsibilities for publishing, and academics interested in publishing.

My motivation to focus on writing and books did not come from a childhood love of books — I was too enamoured with social contact for reading to establish itself as an early infatuation. Rather, it arose from my graduate student and early professorial days, where I came into contact with the ideas of Harold Innis and certain classicists, and with Marshall McLuhan, whom I actually managed to meet as a co-examiner of a Ph.D. dissertation. From these thinkers I acquired an understanding of "literate society" that I grafted on to the developmental psychology training I received mainly from David Ausubel and Edmund O'Sullivan. Partly with the help of David R. Olson, who helped me extend my understanding of Jean Piaget and Lawrence Kohlberg by introducing me to the writing of Jerome Bruner, I delved ever more deeply into the nature and boundaries of literate thought.

I also benefitted greatly from eight months in Oxford, England. I went with the idea of working informally with Bruner but quickly, through a landlady who was also a student, fell in with the social anthropologists there and at Cambridge, one of whom, Jack Goody, wrote directly on literacy.

My brother, James Lorimer, the publisher, led me out of a concern with developmental thought dynamics and into a concern with textbook content. He asked me to undertake some research into the nature of the content of school materials, as background to his attempts to publish culturally relevant school material that spoke to the lives of young Canadians of all social and ethnic backgrounds. The transition from examining content to addressing the need to examine the organization of publishing was obvious and easy. From there, my interest in examining policy, an interest my brother shared, arose partly as a result of his making me aware of the lobbying activities of the Association of Canadian Publishers.

In the early 1980s, a continuing education colleague, Ann Cowan, who happened to be a next-door neighbour, embarked on some developmental research into the training needs of book publishers and asked me, as the only professor at Simon Fraser teaching about publishing, to join her project. Out of that project was born the Canadian Centre for Studies in Publishing and the professional Master of Publishing program at Simon Fraser University. And out of those two structures, a colleague, Ron Woodward, brought to Simon Fraser the Summer Publishing Workshops, continuing education courses taught by members of industry. The workshops were born, or at least lived out their early years, at the Banff Centre.

In addition to the many policy and lobbying documents cited (and uncited) in this book, my continuing interaction with Canada's book publishers — in carrying out my research for commissioned reports, and in working with them when they serve as instructors in our MPub degree program and Summer Publishing workshops — has formed a rich foundation for the analysis I present here. As well, the requirement that students in the MPub degree program produce a project report, usually based on their industry internships and designed to be of value to both their internship hosts and future students (of the MPub and other publishing programs), has kept me abreast of industry activities and concerns, as time to take on publishing research contracts has become more scarce. Most recently, my membership on the Board of Directors of Access Copyright has given me the opportunity to listen to the perspectives of publishers and creators as they try to deal with the evolving dynamics of copyright.

While on the topic of students, I should say that not only have I benefitted from class discussions of most of the material presented here, I also had one class work on an earlier draft of this book as an editing exercise, and they provided me with 18 sets of suggestions on how it could be improved. Some of these were very valuable. The students were Vanessa Chan, Kelsey Everton, Cari Ferguson, Kathleen Fraser, Cynara Geisler, Kristin Gladiuk, Tamara Grominski, Tracy Hurren, Liz Kemp, Megan Lau, Chris Leblanc, Ann-Marie Metten,

Katarina Ortakova, Eva Quintana, Shannon Smart, Suzette Smith, Emma Tarswell, and Chelsea Theriault.

A life in academe has persuaded me that significant concrete realities are often set aside in social science in favour of conceptualizations. While this book is quite interdisciplinary in that it is informed by media theory, technology theory, classical economic theory, intellectual property theory, social theory, development theory, creative economy theory, book and other history, and marketing theory, I have eschewed theories of policy development in favour of dealing with the concrete realities.

For the record, I never intended for this book to be published by ECW. Thus I felt well able to express my admiration for what, for years, was Jack David's press. The book came into ECW's hands as a result of a casual mention in a phone call or email I made to Jack while attempting to find an internship for a student. When Jack said that yes, ECW would be interested in dipping back into its original mandate and publishing a book on book publishing, possibly to follow Doug Gibson's memoirs, I shipped the manuscript datafile off more or less immediately. Once Jack made the decision to publish, I felt an immediate swell of interest and effort from the ECW team, for which I am immensely grateful. Thanks to editor Nadia Halim for her professional acuity and sensitivity, and to the marvellous team at ECW, including Alexis Van Straten, Anya Oberdorf, Crissy Boylan, David Caron, Erin Creasey, Jen Hale, and Rachel Ironstone.

The foundations of knowledge described above are not the easiest to document, nor is it easy to remember, in anything like a complete fashion, the names of all of those I encountered over 35 years of interactions who deserve to be acknowledged. I suspect that I will be adding to this list until the very early morning prior to publication. So, in addition to those named above, here is an incomplete list of those in whose debt I am, mainly for their information and insight. In the order they came into my mind, they include: James J. Douglas, Don and Barbara Atkins, Rodger and Pat Touchie, Allan MacDougall, Jamie Broadhurst, Karl Siegler, Howard White, Stephen Osborne, Mary Schendlinger, Roberto Dosil, Carol Martin, Cynthia Good, David Kent, Kevin Williams, Kevin Hanson, Jack David, Roy

MacSkimming, David Godfrey, Peter Milroy, Paul Audley, Susan Renouf, Doug Gibson, Nancy Flight, Linda Cameron, Hamish Cameron, Brian Henderson, Brian Lam, Chris and Judith Plant, Suzanne Norman, Jo-Anne Ray, Shane Kennedy, David Caron, Brad Martin, Allan Reynolds, Margaret Reynolds, Karen Gilmore, Rob Sanders, Gordon Platt, Allan Clarke, Bill Clarke, Nadia Laham, Susan Bosse, Xiaoyan Huang, John Curtin, Paul Whitney, Diana Newton, Michael Levine, Coral Kennett, Paddy Scannell, Robert Hayashi, Rhonda Bailey, Marcel Ouellette, Bill Zerter, Nancy Gerrish, Greg Nordal, Jeff Miller, Mark Jamison, Carolyn Wood, Kate Walker, Bob Tyrell, Michael Tamblyn, Doug Plant, David B. Mitchell, David Moorman, Walter Hildebrandt, Errol Sharpe, Lynn Copeland, Brian Owen, Guylaine Beaudry, Gérard Boismenu, Peter Saunders, Antoine del Busso, Raym Crow, Nancy Duxbury, Ralph Hancox, Alan Twigg, Basil Stuart-Stubbs, Elizabeth Eve, Margaret Long, Peter Buitenhuis, Suzanne Williams, Avie Bennett, and Frits Pannekoek.

My greatest debt of gratitude I owe to my wife and partner, Anne Carscallen, who has given me the time to take on far more work than I ever should have. Such generosity would not have been possible were Anne a fully employed professional, as she could easily have been. Not only did she carry the bulk of the hours required for our preferred model of parenting, but also, on numerous occasions, she served as an initial editor. I have also benefitted greatly from the emergence of three fine people in my life, Stefan, Conor, and Julia, only one of whom I have been able to entice into providing reading/editorial services.

Having mentioned my debts and gratitude, I hope that a reasonable number of readers will appreciate my efforts. I offer this to the world with some trepidation. I did not review every available government document for this book. Nor did I delve into the ACP archives that are housed at Simon Fraser University for the detailed discussions of policy and the negotiations that took place between government and industry. I left the archives for a future historian and concentrated on the existing public record. Nor did I construct questionnaires and methodically sample opinion, mainly because my interactions with industry were often with leading members, and they were both formal and casual. Thus, to some degree, this is a personal story. May it please you well.

Introduction: Policy, Technology, and the Creative Economy of Book Publishing in Canada

Introduction 3

Books as cultural objects 3

Organization of this book 6

Publishing policy 8

Some quantitative measures of contemporary Canadian book publishing 10

    Size and ownership 12

    A note on book retailing 13

Some international comparisons of support for book publishing 13

Sectors and genres 16

Introduction

Beginning in earnest in the 1970s and carrying on to 2010, Canada's book market was in continuous evolution. It began the period as a market almost wholly dominated by imported authors, books, topics, and ideas. It ended the period as a market characterized by a presence of books written by Canadian authors, often on Canadian topics, alongside other authors and topics from literary communities that matter to Canadians. This book traces the policies and efforts that established the modern Canadian book publishing industry over the past forty-plus years. The various chapters explore the dynamics and nature of books and publishing within a history of Canadian book publishing policy and cultural history. *Ultra Libris* documents the actions of publishers, governments, and other institutions and organizations supportive of books and publishing, paying attention to a range of environmental variables that enabled the establishment and sustenance of a national book publishing industry. The final two chapters consider the likely directions of development and how Canadian book publishers might best address the future. The key themes of the narrative are the value and nature of books and publishing; the development of Canadian policy and its relationship to international ideas of cultural diversity, the contribution of publishing to social capital, and the creative economy; current operational and market realities; technological and market changes; and speculation about the future.

Books as cultural objects

While books are material objects, their defining identity is cultural. And as much as their cultural nature derives from the vision and writing of authors, the professional practices of publishers, which combine with authors' efforts to create the cultural objects we call books, are foundational to the creation of cultural meaning. This emphasis on cultural value and meaning making sets the stage for an exploration of the evolution of a national capacity to originate titles that reach readers and contribute to the mobilization of public understanding and opinion. Also, in focusing, as I do, on the publishing of books that address salient realities, this book documents the efforts of publishers, authors, and governments, in partnership with bookstores and other cultural actors, to create a Canadian literary voice in both fiction and non-fiction. The discussion concludes with some consideration of strategies for the future, given both technological and social change.

The creative, professional, and social surrounds of books, in both print and electronic form, are the cultural realities of publishing. Those cultural realities develop through the planning efforts of authors, publishers, or both, when an idea is transformed into a book proposal. They are furthered by authors bringing creativity and insight to their work. They continue to evolve within publishing firms, where a manuscript is shaped, edited, and fine-tuned visually as it is flowed onto pages, given a powerful title and cover, marketed, promoted, and sold at retail by booksellers. In turn, readers and the professional reception community — critics, librarians, awards givers, and book publishers of other countries who purchase rights — join the chorus of contributors. By following this chain of creative processes — which is most completely manifest in the origination of new titles, mainly by companies that have been founded, nurtured, and controlled by Canadians operating within Canada, most with a focus on Canadian readers — we can discern the key elements in the creation of both a literature and an industry of considerable value to Canadians.

As simple as books are as physical or quasi-physical objects, they are supported by two socio-technical processes, one intrinsic to their production, the other to shaping their content. The more technical of the two is production, a process that has evolved from scribal copying to mass paperback production and high-quality colour printing. Print forms have developed from wooden blocks, to movable metal type, to computer-generated inking, to images on a screen. Recently, ink itself has transformed, switching from a petroleum to a vegetable-oil base with few harmful volatiles. Even more recently, ink has become electronic energy interacting with a substrate. Typography has also increased in aesthetic sophistication, as the visual opaqueness of Gothic letters was replaced with easily read Roman script. The material form of the book has repeatedly been born anew, beginning with the folio and moving through the quarto and octavo to various modern printed and electronic forms. Binding (sewn, perfect, and saddle-stitched), protective covers, and paper manufacture have also undergone change and will continue to do so as new forms of electronic file formats and new e-readers enter the market.

Superimposed on these production technologies is a second set of professional socio-technical practices, focused on the preparation of content for publication and for the marketplace. These practices, which involve both authors and publishing professionals, encompass the development, marketing, and promotion of manuscripts as they turn into books. Most obvious are the various stages of editing, but cover design, page layout, marketing and publicity, and sales support are also included in the totality of publishing practices, given the current means of manuscript acquisition, development and reproduction, and entry into the marketplace. Certain elements of each of these systems are being carried forward into evolving electronic forms of publishing, but they are being reinvented, following in the grand tradition of the development of printing and publishing.

A focus on the cultural attributes of books pays heed to the collaboration inherent in their creation. Like other cultural objects, books emerge from a creative relationship between creators and society. Reflective engagement within a social milieu leads publishers and authors to produce original expressions. Publishers see opportunities for authors, and authors select and reconstruct elements, filter them through their thought and talent, and assemble them into meaningful and imaginative wholes. Copyright confers a set of rights on creators, protecting the author's expression of an idea that they can license to publishers for exploitation, leaving the ownership of ideas themselves in the hands of the commonwealth of humanity.

The author/publisher relationship, at its heart, is a creative partnership. Authors bring to this relationship their manuscripts and their ability to refine them; publishers, through their sense of the marketplace and of the flow of creativity and ideas, and their teams of editors, designers, production personnel, and marketers, bring mastery of production, content, and marketing practice. The publishing process goes mostly unrecognized by readers, and it is also not uncommon for authors to minimize the contribution of the publisher. When the essence of their vision is maintained, and marketing seems to them to be nothing more that what the book calls for, authors may believe the role played by the publisher is relatively minor.[1] But a creative partnership it is, as we shall see, even if an increasing number of self-published titles come to the market without the benefit of publishers' input.

Organization of this book

The central purpose of *Ultra Libris* is to trace how this creative partnership of authors, publishers, and society established itself as a domestic book publishing industry in Canada, beginning in the 1970s and carrying through to 2010. Following an opening chapter on the cultural nature of book publishing, including consideration of the social environment necessary for a national book publishing industry to thrive, Chapter 2 highlights Canadian publishing efforts prior to the 1960s and the beginnings of the cultural policies that laid the general groundwork for both the publishing of books and publishing policy. Chapters 3, 4, and 5 focus on the next forty years, emphasizing the actions of government, including the establishment of the Ontario Royal Commission on Book Publishing, which produced the founding document of book publishing policy. Those three chapters also review the industry's efforts to gain favourable policy and support programs, the results of the policies enacted, and the efforts of authors and firms.

Chapter 6 delves into the structure of today's marketplace and shows how economies of scale and the organization of global production affect, and continue to challenge, the Canadian industry. Currently, the forces of mass production and consumption in book markets are most readily seen in the increasing market share of bestsellers and in the retail practices of Canada's single national bookstore chain.

Chapters 7 and 8 explore the future of book publishing, and how it will be shaped by the rapidly developing information and communication technologies that are reorganizing production, markets, and the preferences of consumers. Information technology has already unleashed forces of such power that the ability to read text by itself and understand its meaning may become a passing desideratum of a bygone age. This would mean the eclipse of books containing nothing but text as such a dominant form. The component professions of the book community — printers, retailers, publishers, authors, and new entrants — are already adapting to a changing technological foundation, and in doing so they are creating new products for meaning making. These and many others are the issues of Chapter 7, in which technological change is discussed along three trajectories: the reorganization of firms around electronic database technology, the emergence of new media, and the increasing injection of interactivity into the "reading" experience. In turn, six business applications of these trajectories fill out some details of their dynamics.

Carla Hesse has argued that the book survived the aftermath of the French Revolution because the Marquis de Condorcet invented philosophical underpinnings and legal structures for books that legitimized authorial ownership and responsibility in post-revolutionary France.[2] Out of these conceptual inventions came a sufficiently orderly market for books to survive — and with them authors, publishers, and, arguably, the course of Western civilization, carried forward by books. As the twenty-first century unfolds, the challenge to books is not humanity's changed vision of itself (wrought by the Enlightenment and played out in the French Revolution), but a revolution in information and communication technology, the organization of markets, and an evolution in our understanding of the economic contribution of books and other cultural objects to society and the economy.

Two important tasks face those involved with the creation and commerce of books today. The first is adapting technological forms to address emerging patterns by which readers seek information and leisure. In the face of today's and tomorrow's information technology and the economy that derives from it, the current materiality of books will almost certainly disappear in order to allow their survival. In terms of its cultural character, what we think of as the book's definition — a long form of textual meaning making by an individual — may cease to be its overwhelmingly dominant form, given interactivity and its economics. The contribution that books and their descendants make to human imaginings, to society, and to the economy will decide their future.

The second task for book people of this time in history is to find the appropriate balance between author and publisher reward, on the one hand, and public access, on the other. At various points in this presentation of book realities, it becomes quite apparent that both writers and the employees of publishers earn less for their work than others with equivalent qualifications working in other industries. As detailed in Chapter 4 and the final chapter, while book publishing is currently receiving increasing attention as part of the macro-economic contribution of the creative economy to society, the unfortunate micro-economic underbelly of that contribution — the toil for small rewards by the vast majority of Canadian-owned publishers and Canadian authors — is too easily accepted as inevitable. Happily, the closing chapter argues, models exist that can restructure the micro-economics to increase opportunities for bright, creative people, and thereby stem the potential talent drain from book publishing. The spiritual devotion without sufficient material reward that book publishers now demand of their authors and employees is quite unnecessary.

Publishing policy

Two features of policy development form the background to this book's discussion of the policies that Canadian governments have put in place to help secure the foundations of Canada's book publishing industry and to assist in maintaining it. First, in the absence of direct, industry-specific policy, general policies form a framework within which social and business activity is carried on. For example, Canada's encouragement of foreign investment impinges on book publishing even though much of the policy does not speak directly of book publishing. Second, social policy is contested most commonly at its birth, when it has yet to fully emerge in social practice. Thus, the principle of freedom of speech was much contested at its initial appearance in Britain, in the Bill of Rights of 1689. Three hundred and twenty-plus years later, so ingrained is freedom of speech that it is uncontested in principle, even though its specific application is continuously debated. Gender equality is a good example of a principle at an earlier stage of development. Even though it is manifest in Canadian law and largely accepted as an ideal, its social acceptance has only begun to evolve; hence, social policy upholding the principle of gender equality is more contested.

Cultural industries policy in general, and Canadian book publishing policy in particular, are closer in age to gender equality than they are to freedom of speech. As young policies, their fundamental legitimacy continues to be contested, both domestically and internationally. Canada has played a leading role in the development of book publishing policy. First among its several achievements is the successful emergence of a Canadian writing community that has achieved worldwide recognition. The creation of a distribution right to control book imports is a second notable policy achievement, as is a third, the maintenance of a heterogeneous industry that continues to bring forward a wide range of titles appealing to readers of varying tastes, ages, and identities. The middle chapters explore these and other policy landmarks.

The policy analysis throughout this book foregrounds the conceptual foundations underpinning the policies that ultimately proved successful in establishing a substantial Canadian-owned book publishing industry. Without exception, each policy was challenged by free market and free trade ideology in various guises. Two recent concepts in social philosophy hold further promise for book publishing on the international stage. The first focuses on the value of cultural diversity and the right of each nation to foster domestic cultural expression within and among the diversity of human communities. It emerged from the Canadian federal government, in consultation with industry, as a reaction to an American challenge to Canadian magazine policy. The second concept was articulated by French social philosopher Pierre Bourdieu and applied with some imagination by American public policy analyst Robert Putnam. The notion of social capital redefines the dynamics of economic benefit, taking into account the spin-offs of social participation and meaning making.

Like other social policy innovations in the past, the concepts of cultural diversity and social capital are being challenged. The main reason for this challenge is that such policies foster domestic opportunities for cultural expression, production, and exchange in countries that have served as client markets for the entertainment industries of large, powerful, and influential nations, such as the United States and the United Kingdom, which have needed only free market and free trade policies to nurture cultural expression, production, and export to world markets. In the twentieth century, the cultural products of these nations became lucrative exports and established an overwhelming presence around the world. A policy that sets forth the right to national cultural expression, and a complementary policy that emphasizes the economic advantages of community participation, threaten the exports of the United States and United Kingdom as powerfully as freedom of speech threatened the rule of the kings and queens of yesteryear. Canada has much to gain from such policies, which also stand to benefit emerging powers such as India, China, and Brazil. The policies have the potential to be a powerful foundation for building creative economies worldwide.

Some quantitative measures of contemporary Canadian book publishing

The preceding paragraphs outline the framework and focus of this book, but make no reference to any discussion of the normal quantitative industrial measures of book publishing, such as the size, makeup, profitability, market share, and number of titles published. This is because the importance of book publishing in society stems from the content of books. Numbers of readers, sales figures, and the like are indications of the character and relative health of the industry overall, but one book can change the world — even one book read by a limited number of people. The measures that matter in describing the cultural impact of Canadian book publishing include: our books' cogency and persuasiveness in addressing important social realities, their demonstrable creativity, the nature of media discussion devoted to them and their authors, the nature of support structures put in place to encourage writing, the activities of cultural partners, the nature and extent of the attention books command in society, the many different publishing and writing awards, the Canadians who are inspired by writing workshops and literary festivals, and Canadians' general recognition of the importance of books. These factors are taken up in Chapter 5. Although most economists and public policy analysts would argue that standard quantitative measures of industry performance must, in the end, be the foundation for the allocation of scarce dollars, foregrounding industrial measures for an activity whose primary value is non-economic would tilt the discussion away from the very reason for the existence of book publishing. The main discussion of the quantitative elements is to be found in Chapter 5. For now, here are some basic orienting data, all of which are the latest available at the time of writing.

As of 2006, Statistics Canada reported that there were 282 Canadian-controlled and 11 foreign-controlled book publishing firms operating in Canada. Together, in 2008, they had earnings totalling $2.13 billion.[3] Book sales in Canada (also referred to as the domestic market) accounted for nearly 75 percent ($1.5 billion) of industry revenue. Firms operating in English, including both foreign-controlled and Canadian-controlled book publishers, attained approximately 80 percent of those sales.

Sales in 2008 of trade books in Canada — that is all general books, including children's books, sold in bookstores and online — accounted for about half the value of the domestic market. Canadian-controlled firms (French and English) account for just over half of trade book sales, and English-language publishers (Canadian-controlled and foreign-controlled) account for over three-quarters of all domestic trade sales.

Own titles — the books that publishers originate and bring to market — are the defining products of book publishers. Unfortunately, Statistics Canada includes as "own titles" titles for which the publisher purchases the rights from a foreign publisher that originated the title in its homeland. With that caveat, own-title domestic sales by all firms accounted for just over 60 percent of domestic market sales. The remaining were books imported and marketed by publishing firms based (and sometimes controlled) in Canada.[4] Of the 60 percent of own titles, Canadian-controlled firms accounted for two-thirds of all own-title sales in Canada.[5]

The eleven foreign-owned firms that operate in Canada are all multimillion-dollar enterprises connected by ownership to large foreign corporations. Their main business is bringing foreign-originated books into the Canadian market. They include three large trade publishers: Random House Canada (part of Bertelsmann), Penguin Canada (part of Pearson), and HarperCollins Canada (part of Rupert Murdoch's News Corp). Simon & Schuster is a fourth large operation that is restricted to importing and distributing by Investment Canada regulations. On the other hand, Canadian firms range in size from Harlequin (part of Torstar) and Nelson Education (formerly owned by the Thomson Corporation, but operating within an international partnership called Cengage and "owned" by the Ontario Municipal Employees pension fund) to firms doing $50,000 a year in business.[6] The main business of the Canadian-owned sector is originating titles for the Canadian market.

Size and ownership

The eleven foreign-owned firms hold just under half the domestic market share of the Canadian book publishing market. The 282 Canadian-owned firms reporting to Statistics Canada in 2006, as well as smaller firms publishing fewer than half a dozen titles, account for just over half the domestic market share. Canadian-owned firms can boast the publication of emerging outstanding authors and prize-winning titles, just as can foreign-owned firms. Large foreign firms predominate in the general educational market, but small Canadian-owned firms also serve certain parts of that market. In various professional markets — law, for example — there is a greater mixture of foreign and Canadian firms, large and small, especially when the focus is on provincial law. In the trade book sector, the large foreign-owned publishers serve the mainstream market and established authors, while large Canadian-owned firms (small in comparison to their foreign-owned counterparts) maintain a grip on a small portion of such titles. The many small trade publishers scattered across the country serve emerging authors.

A note on book retailing

Patterns of book retailing influence the nature of the market. National bookstore chains have existed throughout the period under study, and the Canadian-owned sector has been consistently leery of their influence. The concern derives from the tendency of larger businesses to serve the mainstream of the market; the more successfully they do so, the less that market is available to independent booksellers, many of whom are voracious and eclectic readers, a trait they encourage in their customers.

This concern notwithstanding, the chains — now a single chain, Chapters/Indigo, in English Canada — have increased their influence on the market through increased concentration of ownership and overall market share. In the 1970s, the market share of three national chains — Coles, W.H. Smith, and Classics — was about 25 percent. In 2007, one study estimated the market share of the single chain to be 44 percent.[7] Certain publishers report much higher percentage sales to Chapters/Indigo. Not surprisingly, some others, often those publishing for market niches, report that Chapters/Indigo is insignificant to the success of many of their titles. Indeed, some small publishers have increasingly been exploring non-traditional outlets — for example, wine stores for books about wine, or kitchen stores for cookbooks — while others have taken to selling online from their own websites. Online outlets, such as Amazon and Chapters/Indigo, have seen continued success in online sales and, according to several senior industry members, general retailers like Costco and Walmart have discovered that book sections can be profitable and are growing their share.[8]

Some international comparisons of support for book publishing

The support Canadian-owned publishers receive from Canadian governments can be compared with support book publishers receive in other countries. Canada's governmental direct-support system for book publishing is robust. Indeed, Canada leads the world in terms of direct governmental financial support of book publishing and other cultural industries. Some of Canadian governments' motivations for this are: the predominant role of imports in the Canadian market; Canada's smaller population; and the fact that the nation shares a language and cultural similarities with the United States. One might even claim that in the 1970s, thanks to a combination of prescience and happy coincidence, Canadians began to build the infrastructure for a creative economy, the value of which is now becoming apparent as the creative economy begins to be understood and to grow in earnest.

Yet, financially speaking, this direct support turns out to be often substantially less than the indirect support provided by value-added tax (VAT) reductions in European countries. Even if the zero rating of provincial sales taxes is factored in, Canada still is not the front-runner in subsidizing books. The Canadian federal government's support is approximately $50 to $60 million, but if the federal government were to reduce the goods and services tax (GST) charged on books to zero, the cost to the federal treasury would be close to $180 million,[9] approximately three times the value of direct federal subsidies. One reason the government does not zero-value the GST on books is that it would lose roughly $100 million in tax revenue from sales by foreign corporations.

In Europe, the main mechanism used to support book publishing is VAT reductions, which decrease the price of books and thereby, theoretically, increase consumption. In Austria, Belgium, the Czech Republic, Estonia, Finland, France, Germany, Greece, Hungary, Italy, Lithuania, Luxembourg, the Netherlands, Portugal, Romania, Slovakia, Slovenia, Spain, and Sweden, where VAT (roughly equivalent to Canada's GST and PST) in the range of 20 percent is charged on most consumer purchases, governments have reduced it to 5 percent on books. Ireland, Poland, the United Kingdom, and Norway reduced VAT on books to zero. In 2002, Sweden's reduction of VAT from 25 percent to 6 percent resulted in an increase in sales that was sustained over the following few years.[10]

The other mechanism used to support diversity in book publishing is price maintenance, a device some call price fixing. As of 2009, price maintenance was in effect in Austria, Denmark, France, Germany, Greece, Italy, the Netherlands, Norway, Portugal, Spain, and Hungary, where books were sold at prices set by publishers that could not be discounted. Price maintenance prevents competition based on price, which otherwise allows for the emergence of bestseller stores and inevitably favours large companies, as they can discount long enough to put smaller companies out of business. With price maintenance in place, small bookstores can use the lure and profits of bestsellers to bring in customers and, with the help of knowledgeable staff, introduce them to lesser-known titles. Various studies have demonstrated that price maintenance allows booksellers to stock a wide range of titles and makes it easier for independent booksellers to survive.[11]

As this book was nearing completion, U.S. publishers appeared to have found a method for bringing back something close to price maintenance. In response to Amazon's heavily discounting e-books, which the publishers saw as a way of forcing prices down to levels that would make them completely unprofitable for the publishers, most large publishers have set up an agency relationship with retailers. This relationship recasts the role of the retailer as an agent of the publisher, who agrees to sell the book for the publisher at a defined markup. At the time of writing, it was unclear whether such a relationship would be found to be legal in the American, and Canadian, courts.

One further type of indirect subsidy, common in some European countries, is a highly efficient book distribution system, which allows booksellers to capture a higher percentage of potential sales. In Germany, for example, an efficient train-based network ensures a book can be delivered to most bookstores within twenty-four hours of its being ordered.

No English-speaking country other than Canada provides direct financial support to publishers at the level the Canadian government does. England directs grants mainly to authors and events, to international organizations that involve English authors, and to a few small publishers for specific development projects. Scotland experimented with block grants to publishers in 2006 and 2007, but has since reverted to five types of grants: grants for individual titles, writing or storytelling fellowships, literature and storytelling development grants, translation grants for foreign publishers, and capital grants. Wales and Ireland focus solely on subsidies to writers and offer no other form of financial support to the book industry.

In Australia, printing subsidies used to predominate alongside support to writers, but in 2008 the Australia Council for the Arts "invested more than $5.7 million for literature overall, including more than $4.6 million through the literature board." This support took the form of "grants to individual literary creators, financial and operational support to organizations that provide infrastructure for the sector, market and audience development initiatives and strategic initiatives to build capacity across the sector."[12] The Australia Council's funds were spread among fourteen programs, with much more going to writers than to Australia's $1.5 billion (Australian currency) book industry. In 2009, book publishers received a mere $352,000 in publishing grants, with an additional $185,000 for promotion and $97,000 for author residencies.[13] (Throughout this book, monetary values are expressed in the currencies of the country or region.)

Sectors and genres

Given that the French-language publishing industry in Canada receives only passing mention, this book is by no means an encyclopedic analysis of Canadian book publishing. The French-language publishing industry is quite separate and distinct from English-language book publishing in Canada and is complex enough to deserve a book of its own. The focus of this study is on trade books, leaving aside the other major sectors — education, reference, imports, and, for the most part, scholarly book publishing. Indeed, within the trade-book category, this study focuses on titles originated by English-language Canadian-owned publishers, which account for $250 million of a $1.5 billion domestic bilingual industry. Nevertheless, that $250 million represents approximately one book for every two Canadians, and it is those books that generate attention, are talked about in book clubs and on TV, and win literary prizes. With their counterparts on the French-language side of the industry, they are the books that represent the backbone of cultural value for which books are revered.

The dynamics of Harlequin Enterprises and its titles, and what are typically termed genre fiction titles, are generally set aside, as their contents do not stretch the bounds of knowledge and literary creativity. This lack of attention is not a value judgement. In some ways, Harlequin is Canada's most successful book publisher in terms of income, reaching readers with appealing content, experimenting with genres and technologies, marketing, and so on. But Harlequin, the company, is not a repeatable phenomenon. It is without peer in the English-speaking world. It is certainly worthy of study, and I am aware of aspects of its operation.[14] But it does not fit into the narrative of this book, nor is it a counter-indicator of the thesis of this book.

This book is written for anyone interested in books, book publishing, and reading culture, particularly in Canada. It should appeal to those interested in cultural industries and policy and the nature of publishing and the creative economy. It offers a friendly, but not fawning, analysis for book publishers and other members of the book community, based on what I have learned rubbing shoulders with industry members, and the various studies I have undertaken for provincial associations of book publishers, for provincial governments, for the federal government, and for the Canada Council for the Arts. It draws on Canadian publishing history, current patterns and trends, and emerging technological, social, and economic dynamics. Scholars and students interested in book publishing as a cultural industry, an element of the creative sector, a medium of communication, and a beneficiary of cultural policy may also find value in its pages.

Chapter 1 lifts off with an examination of book writing and publishing practices, and a consideration of the social and policy foundations necessary to a national book industry. The information of the first part will be familiar to seasoned book publishers, but the main components of publishing practice are reviewed because few outside the industry understand the nature of the creative partnership of publishers with authors in the shaping of books and the shaping of a publishing house in the marketplace. Even within the industry, some publishers do not understand the full importance of their creative and market-strategic role: They just do it. A second reason for such a descriptive foundation is that it ensures that all readers are familiar with the basics.

The first chapter also examines the power of books and publishing in an international and historical context. The examples used are international because they are better known, and will likely be more successful than Canadian examples in setting the atmosphere for this study. The international and historical contexts are also relevant because, even though the focus of this book is on contemporary Canadian book publishing, there is little in the analysis that does not apply universally.

CHAPTER 1

Of books, publishing, and social context

What is a book? 21

Professional book publishing practice 22

Professional practices of cultural value 26

    Acquisition 26

    Editing 27

    Marketing 28

    Selling Rights 29

Building a list 29

Styles of publishing 31

Publishing and the history of ideas 32

Book publishing as cultural practice 36

    Of authors, readers, and books as media 37

    Books and literacy 39

    Books and social change 41

The necessary social context for book publishing 43

    The cultural environment 44

    The political environment 45

    Law, policy, and derivative support programs 47

    Economics and the market environment 48

    The educational environment 49

    The technological and organizational environment 50

Conclusion: Preconditions and formative roots 51

What is a book?

The material heart of contemporary book publishing is an organized display of textual symbols meant to convey information and ideas. In the first decade of the twenty-first century, the predominant medium of display was print on paper within a bound volume of no less than forty-nine pages.[15] The printed book is a manufactured physical object of some durability that can be purchased and owned by persons and institutions. One specific public institution, the public library, has as its primary mandate the purchase of books for lending to borrowers.

Current physical manifestations of the book have not been with us since the dawn of time, nor even since the dawn of writing in the seventh millennium BC. Following the universal development, from at least 3000 BC, of symbol systems and, in the West, phonetic alphabets — starting with the consonants of the Phoenician alphabet, to which were added the vowels of the Greek alphabet — impressionable media such as clay were replaced gradually with lighter media, such as papyrus and parchment in sheet and scroll form, followed by paper. Inks, and the manner in which ink was applied to paper, also changed. By Gutenberg's time (1398–1468), scrolls had already evolved into the codex — handwritten and hand-bound books with protective hard covers. Subsequently, there was a significant diminution in size, and soft covers re-emerged, protecting the octavo, a readily portable volume popularized by Aldus Manutius (1450–1515) that freed books from reading rooms. As the twenty-first century proceeds, the e-book and myriad other digital forms knock at the door of print on paper.

Fairly early in its history, in the midst of these physical changes, the book came to be seen as a repository of thought and human creativity. Books also developed into beacons of the communities, societies, and civilizations to which they belonged; they came to reflect the confluence of individual expression of knowledge and ideas and, through both authorship and readership, the social and cultural nature of society.

In Canada, a modern, domestically owned book publishing industry emerged over the past six decades, driven in part by Canada's determination to step outside the shadow of its colonizer, Great

Britain, and to resist the spillover exuberance of its adjacent southern neighbour, the United States of America. The Canadian-owned or domestic industry established itself in collaboration with Canadian governments. To some degree, as we shall see, that state involvement was an inspired move. Canadian governments admirably eschewed interference with the nature of the ideas being published and concentrated on providing support for business operations and cultural titles in general. The major achievement of the domestic industry has been the articulation of a distinct Canadian cultural voice in the world of ideas. During the same period, the foreign-owned sector was also transformed: from an industry focused almost totally on importation and distribution of foreign titles, to one that also engages in the publication of a certain percentage of Canadian authors, dealing with Canadian subject matter, primarily for Canadians to read.

Today, new challenges arise, brought forth by changing business forms, technological developments, shifting social realities and priorities in Canada, and environmental concerns. Against the background of a cultural awakening, this study examines the formative roots of the contemporary Canadian book publishing industry, its current functioning, the challenges it faces, and the potential for book publishing to remain a vibrant contributor to Canadian society, Canadian and world culture, and the Canadian economy. But first, some fundamentals, in the form of an outline of the elements of the book publishing process.

Professional book publishing practice

The first task of a book publisher is to define a reason for being — that is, to set up a business with a guiding purpose and mission that define its path of development and establish somewhat permeable boundaries of activity. With a purpose and mission in place, a book publisher can announce itself to the world and begin to acquire manuscripts, guided by particular objectives that translate into an emphasis on selected genres, subject areas, and, sometimes, targeted authors.

While established writers of creative works tend to keep in touch with their publishers or their agents about work they are contemplating to obtain initial reactions, first-time authors of works of fiction, poetry, or creative non-fiction are apt to submit letters of inquiry or full manuscripts to chosen publishers. In the case of non-fiction, publishers and authors seek each other out to discuss ideas and treatments, which develop into discussions of the organization and focus of the manuscript, as well as of the author's voice or orientation to the subject. Ideally, such discussions are followed by the author's submission of an outline, which gives the publisher a chance to engage in what is termed developmental editing. The outline often makes clear the author's intent, and, given that intent, the developmental editor works with the author to create a good book that will work in the marketplace. There are many potential problems that a developmental edit can allay. For instance, authors often want to accomplish far too much in a single manuscript, and an editor can help the author focus on the truly distinguishing features of the work.

After the author has drafted and polished the book, it moves on to a substantive edit, in which the editor assesses whether the author has created an appropriately lively, engaging, or authoritative book manuscript. Some publishers and editors use the terms developmental and substantive editing interchangeably, and substantive editing can be seen as part of the development of the manuscript. In fiction, where editors are often dealing with more skilled and language-sensitive authors, a substantive edit may involve asking questions about the characters, setting, plot, or such implicit elements as atmosphere. In non-fiction, the editor will focus on the logic and structure of the argument, narrative, and information. That said, in the same way that character and events come together to create a living fiction, a good piece of serious non-fiction relies on the author's ability to come up with an overarching structure that readers can understand, one that generates new insights even for authors as they explore various facets of that structure.

After substantive editing and author revisions come stylistic editing and copy editing, where words, phrases, sentences, paragraphs, and chapters are fine-tuned to allow the reader to follow the text unhindered by semantic and grammatical gaffes. Then comes fact-checking, where references are reviewed and facts are checked by an editor and/ or by the author, often at the behest of the editor. At the completion of this process, the copy editor or another editor gives the manuscript a complete read-through to catch any last-minute typos, repetitions, or inconsistencies.

As the manuscript moves into production, other pre-press elements are given attention. Once the font, page size, margins, line spacing, style for headers and footers, and so on are determined, the manuscript is flowed into the established format by means of a page-layout program, and any graphics being used are integrated with the copy. The result is then tweaked by the designer. At the same time, the cover design is commissioned — front cover, back cover, and spine, plus flaps in some cases. When the pages and cover are complete, a proofreader will check the page proofs or galleys (an image of the work to be printed), to insure that there have been no glitches and that no extraneous elements have been introduced that detract from the meaning on the page.

As the book comes together, the publisher solicits bids from printers, makes decisions on the weight and type of paper to be used, determines how many copies of the book will be printed, and sets up a schedule for delivering files to the printer. When those files are in hand, the printer will produce a set of printer's proofs, which are sent to the publisher to ensure the book corresponds to expectations. After these proofs are okayed, the print run is produced.

Marketing, which prepares the book for the public sphere, overlaps with production in that such elements as cover design and cover copy serve the dual purpose of presenting and marketing the book. While those responsible for sales and marketing are usually involved from the initial stages, advising acquiring editors and publishers on the market potential of a manuscript, they become more active as the book takes form. The marketing department will set a book's release date; decide how to present it in a catalogue; work with the cover designer to make the book stand out in a particular section of a bookstore; create shelf talkers, posters, and bookmarks; circulate review copies; and work with the publicity department, which carries out a specialized extension of marketing. The job of the publicist is to gain media attention for the book and its author beyond the attention being purchased by marketers. Marketers often speak of the four Ps of marketing: product, price, placement (distribution), and promotion (including sales and publicity).

The sales force also becomes increasingly involved as the book takes shape and a publication date is set. Publishers generally conduct biannual sales conferences, in which books are presented to the sales staff as they prepare to go out into the field to sell the new titles to bookstores. The essential tasks of sales reps are to maintain a relationship with bookstores, booksellers, and their clientele; to pre-select appropriate titles from many lists; and to convey enthusiasm in a memorable way, so that booksellers order an appropriate number of copies and bear the title in mind as worth mentioning to customers. Additionally, sales reps may make special sales to non-traditional retail outlets, or to organizations that will buy numerous copies of a book as gifts for members or clients.

Warehousing, fulfillment, and distribution begin as the books roll off the press. Some copies are sent directly to retailers, while others are sent to various warehouses belonging to the publisher or its distributor. The key element in distribution is getting the books out the door and to retailers in time to meet demand, whether this demand occurs as a result of marketing and publicity activity; for seasonal reasons, as with the Christmas shopping spree and the summer fiction boom; or perhaps because a title has won, or been nominated for, an award, and has begun to sell at levels beyond anyone's wildest expectations. Fulfillment of initial orders is usually not a problem, but fulfillment of re-orders — that is, the replenishment of stock to booksellers as they sell out and order more — can be challenging.

Once sold to retailers, publishers' books are in the hands of their business partners, the booksellers. This transfer of responsibility presumes that the retailer is not the publisher itself (as was Chapters/Indigo with its now inactive imprint Prospero Books). It also presumes that the major retail venue for the publisher's titles is not the publisher's own book sales operation, although selling directly to the public from a website is increasingly seen as a viable option. Books have traditionally sold to booksellers at a discount of 40 percent off the retail price, with an escalating discount as more books are ordered. Terms of sale normally allow unsold books to be returned after three months and not more than twelve months after the invoice date.

Beyond retail sales are rights sales. Copyright law provides for infinite divisibility of rights, of the whole manuscript or its parts, in one format or another, in one language or another, in one territory or another. The costs of making rights sales are relatively low, since, strictly speaking, these sales do not involve manufacture. Hence, publishers are keen to exploit their potential if there is any realistic expectation of a market.[16] For the publisher, a crucial element of the contract negotiations over a manuscript is establishing the rights it wishes to, or can afford to, acquire. Holding all rights is obviously most advantageous, in part because such rights can be partitioned and sold to others. However, it is a fairer contract if a publisher seeks only those rights it has the resources to exploit, either by itself or through sales to others. An author's agent will insist on this, and the tradition in the industry is to be flexible, at least with unestablished authors, should they be successful in attracting offers for rights as a result of their own efforts.

Professional practices of cultural value

While all the practices outlined above are necessary, several are key to the cultural value and market success of a book. As such, they distinguish one company from another and contribute directly to a company's survival and competitiveness in the marketplace.

Acquisition

The nature and quality of what a publisher acquires is the foundation of its identity and success in achieving its goals. These goals are at times market driven, especially for large publishers, but just as often, and particularly with smaller publishers, they are determined by the firm's founding purpose and mission. For example, university presses almost never seek out blockbuster titles. Regional presses do not look for international bestsellers. Few children's publishers accept crime fiction. Publishing houses that specialize in non-fiction may choose not to acquire fiction, and vice versa. In the end, the nature and quality of the titles published within a press's market niche bring recognition from booksellers, critics, and readers and attract new and established authors.

Editing

Acquiring quality manuscripts is an important starting point, but what a house does with its manuscripts is equally important. Here begins an intimate creative partnership between author and publisher that starts with developmental and substantive editing. Shyla Seller has documented the developmental editing process for *Gardens of Shame: The Tragedy of Martin Kruze and the Sexual Abuse at Maple Leaf Gardens* by Cathy Vine and Paul Challen, published in 2002 by Douglas & McIntyre, a trade-oriented house that produces mainly non-fiction titles.[17] The subject of *Gardens of Shame* is the childhood sexual abuse of NHL player Martin Kruze at one of Canada's hockey shrines, Maple Leaf Gardens. As an adult, Kruze brought his abusers to court, obtained a judgement against them, and subsequently committed suicide.

In their first outline, the authors proposed a book whose central focus was an analysis of the effects of sexual abuse on children, with the life story of Martin Kruze as an example. The authors proposed to undertake extensive research and to reveal the devastation of sexual abuse. The book would begin with Martin Kruze's suicide, followed by alternating chapters on general issues surrounding sexual abuse and Kruze's specific experiences.

The acquiring editor responded to this original proposal by saying that while the story was compelling, she believed the manuscript attempted to cover too much information. She suggested that instead of an academically tinged discussion of the effects of child abuse, the authors write a chronological, descriptive narrative focusing on Kruze, with stories of the other men abused at Maple Leaf Gardens as a secondary element.

The authors went back to the drawing board. Their revised outline began with a prologue describing Kruze's going public. Chapter 1 chronicled Kruze's childhood. Chapter 2 introduced the voices of other men abused at Maple Leaf Gardens. Chapters 3 and 4 showed Kruze's struggle to live with himself after the abuse, his decision to sue Maple Leaf Gardens, and the difficulties he experienced in going to the police and the media. Chapter 5 detailed the trial and included the reactions of other abuse survivors to the trial and sentence. Chapter 6 focused on Kruze's suicide, the reaction of his family, and the reactions of the other abused men. Chapter 7 described Kruze's funeral and the pledge of the other men to keep his memory alive. An epilogue included an apology from Ken Dryden on behalf of Maple Leaf Gardens, information about the Martin Kruze Memorial Fund, and an update on the other men.

In this final form, the power of the narrative replaced the authority of scholarship, giving the book a wider audience but the same credibility. Casting the story in the form of a chronological memoir allowed readers to understand the connections between Kruze's earlier and later life. The reactions of the other abused men gave voice to Kruze's struggles, and ending the book with Kruze's death underscored for readers the devastation produced by the sexual abuse of children. The editorial recast was a success, as was the book.

Marketing

The job of marketing, which gradually takes over as editing comes to an end, is to create an alluring discourse of interest and value that can be maintained long enough to garner sales. Ideally, that discourse is inherent in, and thus can be drawn from, the manuscript. Refined in editorial, the discourse is recognized and amplified in the marketing department, conveyed to the sales force, and transferred to retailers, the media, and a waiting public. The launch of *Seabiscuit: An American Legend* at a sales conference in the New York offices of Random House in 2001, and its subsequent marketing, tell the story well. The editor of *Seabiscuit* presented the book as the story of an undersized, crooked-legged colt who was recognized by an unlikely trainer, brought to the big time, ridden not by a jockey but by an ex-prizefighter, and set against the best horse America had to offer in a race run in 1938, just as the world was coming out of the depths of the Great Depression. It was a classic rags-to-riches tale, and when the New York editor outlined the story at the sales conference and screened the film of the actual race, he had his colleagues breaking into cheers for Seabiscuit's sixty-five-year-ago triumph.[18] The story of the little horse that made good — actually the third book on Seabiscuit — set a discourse in place that carried the book to the bestseller list and inspired a movie, two years later, that propelled books off bookstore shelves for a second time. Such a launch story is exceptional, as was the book; yet, to some degree, each title published seeks a similar reverberation in a particular audience, which can range from a few hundred to millions of readers.

Selling Rights

The fourth culturally significant professional practice of publishers is rights sales, an activity of growing importance. When sales of rights for different formats (i.e., trade paperback, mass-market paperback), languages, and countries are combined with the potential for digital repurposing (beginning with e-books, but expanding to extracts in a variety of contexts), and these opportunities are further combined with derivative adaptations, including turning books into movies, games, and television series, rights sales significantly expand the presence of a title in the marketplace and pad the bottom lines of publishers, agents, and authors.

These four key practices enhance both the cultural and market value of a book by casting it in a manner that is aimed to help a title readily reverberate with its intended audience. Moreover, they are supported by other professional practices necessary to the effective functioning of a publishing operation, such as design, copy editing, proofreading, publicity, distribution, and retailing.

Building a list

The practices described in the previous section are discrete tasks in the creation and promotion of one title. As publishers take on title after title, they build a list that gives them a presence in the minds of booksellers, avid readers, critics, and reviewers. The list is the public manifestation of a publisher's identity and its publishing strategy. Besides creating a public identity, the list influences each title, subtly or obviously, often affecting, if not determining, the treatment a topic receives. The books that a publisher publishes are also a signal to authors seeking an appropriate home for their manuscript — a signal that often goes unrecognized by new authors. The list indicates the kind of author or subject that will be of interest to the publishing house and provides a primary thematic — say, modern Canadian literature — which may be complemented in a second list, perhaps of literary biographies, once the primary list is established.

In its early incarnation as a publisher of inexpensive editions of literary classics, Penguin established itself in the minds of members of the public through its format, pricing, and list. Penguin has managed to carry its public identity as a publisher of note, though not necessarily of cheap classics, through to the present day. Other publishers have followed suit. Alfred Knopf made his list distinct by bringing European authors such as Conrad, Lawrence, Mann, Forster, Freud, de Beauvoir, and Camus to the United States, and by doing so with style, paying particular attention to outstanding printing, binding, and design.

In Canada, the modern identity of McClelland & Stewart first took shape in the imagination of a brash young man. Fresh home from the Second World War, Jack McClelland, son of company co-founder John McClelland, struck out west in 1947 at the age of twenty-five in search of (more) Canadian literature, resolving to make Canadian trade publishing the centrepiece of the family firm.[19] This vision was a long time coming to full fruition, but by 1960, Jack McClelland had expanded the M&S list to thirty titles, up from its initial nineteen, and the following year he added thirty-eight more. In 1962, he dropped all but the strongest agency lines and proclaimed his company "The Canadian Publishers." By 1967, M&S was producing eighty-one new titles a year.[20] This substantial list of Canadian fiction and non-fiction paved the way for, and responded to, a rapidly growing interest in things Canadian, which received a major boost in Canada's centennial year, 1967. McClelland became the darling of the writing community with his phrase, "I publish authors, not books" — not an entirely original sentiment, but a first in Canada. This declaration lent a moral superiority to his dealings with bankers.

The accumulation of titles into a list creates a house identity or brand. A publisher builds the list by remaining informed about issues and authors, and looking for opportunities to bring out titles that the trade recognizes as within that publisher's area of expertise — books that build the marketplace presence and the imprint. List building also involves the cultivation of authors, and of the public profile of significant authors and their titles. It takes place at the intersection of culture and commerce. As such, list building is a strategic activity, and publishers invest much thought and planning in the task. And even though readers may be blissfully unaware of publishers, other book professionals, such as booksellers and librarians, take note of the nature of the suppliers providing books to the market. Through their lists, and under the guidance of the dominant senior personalities within each firm, publishers create stable but evolving professional identities in the trade. Building on these foundations, they extend their mandates, all the while instilling in their employees an engagement with meaning making that is consistent with the established nature and commitments (purpose and mission) of the publishing house.

Styles of publishing

To some degree, a publisher's list is reflective of its overall style of publishing, as the examples in this section illustrate. ECW Press, which took its name from its initial publishing project, the journal *Essays on Canadian Writing*, changed its publishing strategy in the 1990s as a result of one book, a k.d. lang biography that had considerable success in the United States. As ECW co-proprietor Jack David notes, that biography "opened our eyes to the U.S. market. Then we decided that no one was preventing us from producing books about non-Canadians. Our first full-fledged effort to go after the world market [with a U.S. pop-culture-based title] was . . . *The Duchovny Files*, in 1995."[21] ECW then began to add popular culture titles in such areas as music, celebrity biography, TV series, and movies that transcended national boundaries. As ECW transformed itself from a literary publisher to a successful pop-culture publisher, the letters of its name came to stand for Entertainment (film and TV, music, wrestling, and sports and games), Culture (travel, humour and art, biography and memoir, popular science and education, and business), and Writing (fiction, poetry, mystery).[22]

Harbour Publishing took quite a different route. Founded as a magazine publisher in the 1970s, Harbour opened for business with *Raincoast Chronicles*, an occasional periodical containing interviews and articles dealing with the social history of the B.C. coastal region. Harbour went on to publish numerous books focused on British Columbia, many relying on first-hand accounts rather than archives. Particularly significant was its publishing of the *Encyclopedia of British Columbia*. Over the years, Harbour's titles encompassed forestry, First Nations, marine life, coastal places and communities, the North, nature, aviation, history, people, sports and leisure, as well as instructional/reference books, collections of photographs or art, humour, fiction, and poetry. The list's range, combined with its regional focus, provided the foundation for Harbour Publishing's professional identity in the trade.

Lone Pine Publishing, in contrast, publishes bird and plant guides for many different regions and cities in the United States and Canada. Lone Pine's genius is in emphasizing regional specificity in each book's title, while reusing information across books, as many birds or plants are common to several regions. This practice can also be seen in the firm's collections of ghost stories: there are ghost titles for thirty-seven states, regions, and provinces, in addition to eight thematic volumes, including *Haunted Battlefields*, *Haunted Cemeteries*, *Haunted Highways*, and *Haunted Hotels*. This is inspired marketing and does not exhaust the overlap possibilities of Lone Pine's list, by any means.

Publishing and the history of ideas

Occasionally, a list and publishing style come to have a broader cultural significance and historical import. Hogarth Press, a project led by Leonard and Virginia Woolf, core members of the Bloomsbury Group, is an example. The Bloomsbury Group was a collection of writers, artists, and intellectuals who first came together in 1905–6 at the home of Virginia Woolf and her sister Vanessa Bell. Its members created some of the central works of the modernist movement in Britain in the period 1905–30. They were in conscious revolt against the artistic, social, and sexual restrictions of the Victorian age.

In 1917, Virginia and Leonard Woolf founded Hogarth Press as a way to disseminate more widely ideas consonant with their world view. With the market success of Virginia's *Kew Gardens*, published by Hogarth, the press began to emerge as a significant cultural venture. Hogarth published Freud in English for the first time; it provided the general reading public with affordable booklets by such well-known writers and critics as Gertrude Stein and H.G. Wells; and it gave many artists, photographers, illustrators, and designers an avenue for expression.

An examination of the list by year reveals a gradual start. By 1920 the press had published eleven titles. Four years later, it boasted fifty-four titles. In 1925, twenty-seven books were published, while in 1927, the year of *Kew Gardens*, forty-two titles were produced. By 1938, when Virginia bowed out, the press had accumulated 440 titles, although its output had slowed beginning in 1933 (the Depression years). After Virginia left, title numbers diminished further. In the final three years, 1943 to 1946, its output was six titles.[23]

The essence of the ideas around which Bloomsbury organized and Hogarth Press published was a transition from the Victorian era to modernism. The seeds of modernism, brought forward in a social context by Bloomsbury, were planted prior to the beginning of the Victorian era in the scientific revolution of the seventeenth and eighteenth centuries, followed by the Industrial Revolution, at the root of which was the application of reason to observations of nature. Bloomsbury represented the imaginative and behavioural embrace of these same ideas that the elites of Britain and Europe had accepted intellectually, but which were not as influential socially. As much in their behaviour — which included conscientious objection and sexual openness — as in their writing, Bloomsbury, and the Hogarth Press, manifested a reconstruction of Victorian-inflected social and personal mores and the nature of art and expression.

Appealing to a wide number of readers in a variety of fields, the ideas present in the works of the Bloomsbury authors and artists — who in turn were influenced by others, such as philosophers G.E. Moore and Bertrand Russell — became the founding tenets of social change. While Bloomsbury's particular ideas were unique, the same interplay of thought and inventive creative expression can be seen in other creative groups in different times and places — for instance, in the art of the Impressionists, or the innovations of the group of "empirical philosophers" who laid the foundations for the social, organizational, scientific, technical, and commercial development that became the Industrial Revolution. In the latter case, ten members of the self-styled Lunatics,[24] including Erasmus Darwin (Charles's grandfather), Matthew Boulton, James Watt, Josiah Wedgwood, and Joseph Priestley, were recognized for their contributions to philosophical inquiry by being named fellows of the Royal Society.[25]

When creators coalesce into artistic movements, or around one or more dynamic imprints, they can rise, in interaction with each other and in counterpoint to prevailing norms, to reflect an emerging consciousness and become a point of reference for its development. Whether in painting, writing, or science, the dynamism of a group is conveyed to the outside world through publishing, in its literal meaning: that is, making expression public. In Bloomsbury's case, it was book publishing. In the case of the Impressionists, it was paintings on exhibit. With the Lunatics, it was announcements of scientific discovery. In each instance, the core members attracted a wider following, and the movement as a whole, in publishing its works, changed society forever. The Bloomsbury Group captured a *zeitgeist*, and Hogarth Press capitalized on the creativity of the group's members.

The history of ideas contains many more examples of groups such as Bloomsbury, the Impressionists, and the Lunatics changing the nature of human perception and existence. But there have also been individuals who have had a similar impact. Galileo, Voltaire, Darwin, Freud, and Einstein come to mind. The case of Voltaire is particularly instructive regarding what publishers must do to achieve the most consequential manifestation of their social role: bringing revolutionary ideas to market. Voltaire's story, in brief, runs as follows.

The transition from feudalism to democracy in eighteenth-century Europe took place against a backdrop of rivalries among European states, and book publishers played a key role. In France, the legally enfranchised book publishing industry was a privileged monopoly composed of the Paris Corporation of Printers and Publishers, a guild ruled by a few families based in Paris.[26] As book historian Robert Darnton points out, the Paris Corporation avidly protected its monopoly of the Parisian book trade, but it failed to satisfy the changing demands of the reading public. This failure to meet market demand created a potentially profitable opportunity for French provincial publishers. They "formed alliances with foreign publishers, who sent them unbound illegal works [by authors such as Voltaire] under cover of a customs permit that protected book shipments from all inspection between the border and their points of destination within France where they were to be examined by the nearest official bookseller."[27] At that point, with a certain amount of persuasion, books could be bound and distributed to the market with sufficient secrecy and speed that a waiting public could gain access to them before they were banned.

The Jesuit-educated Voltaire, born François-Marie Arouet in Paris on November 21, 1694, eventually wrote over 2,000 books and pamphlets and 14,000 letters. His social circle included intellectuals and the nobility in England, France, Germany, and Switzerland, and he was embraced by the many who read his works or attended his plays. Voltaire's writings were subject to severe censorship by the French government. Most were banned from distribution through legal channels, an action that caused Voltaire to write certain works anonymously and even to write critical essays condemning his own work. The banning of his books made them highly prized, but dangerous to possess.

Voltaire's publisher of record was Gabriel and Philibert Cramer of Geneva, where Voltaire took up residence after 1753. His books were distributed in France by the Société typographique de Neuchâtel (STN), a major publishing house and wholesaler that distributed illegal works but did not print them. Voltaire worked with other publishers beside the brothers Cramer. Jacques-Benjamin Teron, also of Geneva, ran a small publishing business that focused on publishing a few banned books that were in high demand. Teron specialized in the works of Voltaire, supplying them to the STN when the Cramers did not.[28] Roger Pearson describes the efforts of the publisher to avoid both censors and pirates in bringing Voltaire's banned *Candide* to market:

> The original publication of "Candide" was a carefully calculated coup. Unbound copies of the work were discreetly dispatched

from Geneva on 15 and 16 January 1759; 1000 to Paris, 200 to Amsterdam, and others to London and Brussels. They were then bound at their respective destinations and published on a previously agreed date, the idea being to circulate as many copies of the original editions as possible throughout Europe before pirated editions usurped and corrupted it. The aim, too, was to create the maximum stir in as many countries as possible before the authorities could suppress this subversive tale.[29]

American and French book historians have created a rich literature dealing with the dynamics of book publishing from this period to the aftermath of the French Revolution. As their work illustrates, just as a foment of ideas drove the Bloomsbury Group, and hence Hogarth Press, so Voltaire's activities drove his publishers, other French provincial and Swiss publishers, and a plethora of lesser authors of the time. Voltaire's ideas and his essays, plays, and letters were key to his influence, as was his socializing among the courts and intelligentsia of Europe. Had national boundaries not created the possibility of underground publishing, the established Parisian publishers of the *ancien regime* might easily have smothered his ideas.

Book publishing as cultural practice

As publishing missions take shape in the title and development choices publishers make, and as those choices are manifested in a list and a style of publishing, publishers catalyze the invention and flow of ideas in society by commissioning or accepting works from authors, questioning and challenging them, and giving them faith that the world wants to read what they are able to write. Far more often than authors and readers imagine, publishers seed the idea of a manuscript in the mind of a potential author. Like authors, publishers thrive on being at the centre of discourse. There are, of course, no barriers to prevent those who wish to pursue wealth through book publishing from doing so, but as many publishers' memoirs attest, ideas and literary innovation are the main motivators.

A press is not founded, and a list is not shaped, out of a desire to process manuscripts and bring them to market, even though such activities are necessary means to ultimate ends. Neither is it usual to found a press solely to produce bestsellers or to make pots of money. If it were, Faber & Faber would have shut up shop in the 1980s, when the millions of pounds in royalties from Andrew Lloyd Webber's musical *Cats*, based on T.S. Eliot's *Old Possum's Book of Practical Cats*, began to roll in.[30] Even companies that are profoundly market focused, such as Harlequin, are rarely founded for the pursuit of profit — editorial guidelines reflective of the appeal of romance to women and of the morality of the founders and editors are clear in Harlequin's history.[31] For the most part, presses are founded for no less a reason than to advance civilization or, at the very least, contribute to a world-changing discourse. True, one person's advance may be another's fall into decadence, and many publishers regard entertainment to be of equal value to information and enlightenment, but the vast majority of publishers are driven by the desire to participate in the invention and flow of ideas and outlooks in society. Publishers and authors together engage in developing and disseminating the information, concepts, and images around which societies cohere and move forward, differentiate and reintegrate, fragment and re-form. In short, the meat of book publishing is cultural activity.

Of authors, readers, and books as media

Without authors, publishers would simply not exist. Authors provide the grist for the mill of our imagination and intelligence; they set down an extended exploration of ideas, images, and information, then work with publishers to hone that exploration and place it on the public record in a form that, if not permanent, can at least be widely distributed and preserved. As creators, authors write books that, like all works of art, represent a personal-social dialectic that is original in its expressive form, yet draws from the surrounding culture. In this context, readers become intermediaries, carrying authors' ideas into the society within which the book is read and contributing to its dynamism. In turn, readers of fiction and non-fiction gain insight and contribute to the evolution of the world of ideas, and, hence, the social forms around them. As they engage with the written text, readers actively construct a reality that has generative power.[32] Words, sentences, and paragraphs create images, ideas, characters, and events that rise from the pages to enter the minds of readers and the reality they are constructing. The force of the writing gives readers intimations of alternative realities that they can extend beyond the contents of the work itself. However, it is not just their content, but the form in which it is expressed — as a sustained textual exploration of a central idea, in a preservable format — that gives books their power and influence.

John Le Carré's *Absolute Friends*, for example, so vividly captures the passing realities, interpersonal dynamics, personal circumstances, institutional actions, and deep and superficial characteristics of social movements of the 1960s that it is hard to believe the story is fiction and was written more than thirty years after the time portrayed. Wayne Johnston's *The Navigator of New York* takes readers into the character and actions of a man who might have been the first person to reach the North Pole. In doing so, Johnston queries the bizarre world of fame seekers and their financial and personal supporters. Jonathan Rose reports that most households in the nineteenth century had copies of three books: The Bible, Bunyan's *The Pilgrim's Progress*, and Defoe's *Robinson Crusoe*. Of *Robinson Crusoe*, Rose notes that there were over a hundred different versions in print, ranging from an abridgement of less than ten pages to deluxe, illustrated, leather-bound volumes.[33] Such a multifold existence suggests a profound resonance throughout British society. A similar claim might be made for the books of Dickens. The serial versions of his various works, together with the myriad editions, attest to the influence of Dickens' writing on the consciousness of the time.

The book as a medium of communication has not been studied extensively, although the modernist literary tradition of the 1920s set out in this direction.[34] Thirty years later, Marshall McLuhan pursued this line of inquiry with a vengeance. McLuhan published several books between 1951 and 1965, starting with *The Mechanical Bride* and proceeding through his two major works, *The Gutenberg Galaxy* and *Understanding Media*, in which he insisted that society consider the interaction of media form and content.[35] Books in particular, and the media in general, are symbolic forms that transmit and transform content and, in doing so, transform our consciousness, our very being in the world.

The impact of the book as a distinctive medium can be seen in the depth of involvement of readers and in the reader behaviour the book form encourages. Why is it that book readers, and far more women than men, gather to discuss books? Why do books induce socializing? Why is it that certain socially trusted people, such as Oprah Winfrey and the late Peter Gzowski, exert a tremendous influence on people's reading choices? Various studies indicate that readers often have preferred genres. Religious books, like science fiction, are a must-read for some and a never-read for others.[36] In 1984, Janice Radway brought romance reading out of the closet and gave it legitimacy as an explainable and empowering element of women's behaviour.[37] Biographies are another genre that tend to attract people of a certain mindset. Why do certain books produce specific patterns of engagement?

These questions lead to a consideration of the study of literature and the field of literary criticism. Why in our society is such attention given to literature — not information or analysis, but written narrative? And why do books, whether literary or otherwise, draw so many to intellectual challenge? (Stephen Hawking's *A Brief History of Time* comes to mind as an example.) In their marriage of form and content, books, perhaps more than any other medium, allow a multi-layered engagement with transformative and generative meaning making. But there is more to the story.

Books and literacy

In the Western world and other developed societies, including India, China, and Japan, books are a privileged creative medium, in the sense that astonishing levels of investment are made towards teaching citizens how to read and write and ensuring that books are widely accessible. In Islamic cultures in previous centuries, books and higher learning were very highly privileged until they were attacked in the period AD 1150 to 1200.[38] Consider what schools do with books. As far back as Roman times, the two main media of understanding, relentlessly rehearsed in school, were literature and mathematics, words and numbers, the latter being a specialized form of words, reading, and writing. It seems as though, through the ages, society has been shouting out Marshall McLuhan's aphorism: the medium is the message. It's letters and numbers in themselves that matter, not the content used in coming to know them. The never-ceasing hand-wringing about the literacy levels and, to a lesser extent, numeracy competence in the generation coming of age is further evidence of faith in literacy and books. At the foundation of this concern, it seems, lies a basic article of faith that to know is to write, or, at the very least, to know is to be able to express ideas coherently in words, which can be written down and examined for their veracity.

This is a powerful belief, and it has served Western societies well. Nonetheless, we know that the ability to express ideas clearly in words is *not* requisite to having an understanding of something, nor does it exhaust all possible forms of understanding. For instance, in his book *Thought and Language* (published in Russia in 1934, translated into English in 1962, and republished in 1986), Lev Vygotsky showed that there is a clear differentiation between a person having an idea and his or her ability to express it, and much psychology has explored that difference.[39] The Sapir-Whorf hypothesis, which claimed (crudely paraphrased) that one cannot have an idea without the words to express it, has been debunked. (Yet words do facilitate the extension of understanding, and new expressions provide new understandings as well as serving to communicate such understandings.) To be sure, the expression of an idea using exactly correct words is evidence of clarity of understanding. But other forms of understanding — a mother's understanding of her child's needs, abstract mathematics, a sense of right and wrong, a painting, a piece of music — lead to other forms of knowing. In developmental psychology, Jerome Bruner made a major contribution to cognitive development by pointing to three modes of understanding: the enactive mode, the iconic mode, and the symbolic mode. These are three ways of knowing, only one of which depends, to some extent, on words.[40] In short, words, writing, and books are only one form of representation of ideas, albeit a very powerful one. But they do not render all other forms of representation less adequate or redundant. Literacy does contribute to the development of an informed citizenry in a literate society, and once writing and books are established as the dominant means of meaning making, they are powerful vehicles for social and political stability, as well as for change. But literacy is not a magic bullet that causes individuals or societies to adopt democratic values, such as respect for written law and human rights.[41] Nor, on the opposite side of the issue, are non-literate societies necessarily undemocratic or prone to trampling on the individual in the name of the collectivity or the ruler.

Books and social change

Some books serve as a rallying point for social change. The economies of book production brought about by Gutenberg's development of movable type and the printing press freed the Bible from the confines of cathedrals, fundamentally changed the dissemination of religion and, later, knowledge, and arguably laid the foundation for both Protestantism and capitalism.[42] These same economics, together with Aldus Manutius's highly readable and portable volumes, freed the classics from their suppressed, scarce, and therefore elitist existence. The circulation of Chinese encyclopedias was a not-insignificant contribution to the spread of knowledge, perhaps even to the Renaissance.[43]

Another example is the oeuvre of Karl Marx, which provided a framework for understanding the position of workers in industrial society. Marx's books gave rise to social revolution in Russia and China and to disruption and instability throughout capitalist industrial society; they were also foundational to the establishment of social legislation throughout the Western world. Adam Smith and David Ricardo, political economists who were precursors to Marx, focused on the market rather than workers' interests. In his 1776 book *An Inquiry into the Nature and Causes of the Wealth of Nations*, Smith posited the existence of an invisible hand that transformed the self-interest of individuals into the common interest, while in 1817, in *On the Principles of Political Economy and Taxation*, Ricardo offered up a rationale for the advantages of international trade that remains the primary argument for free trade today. In the early 1960s, in a series of books, Jean

Piaget changed the foundations of human psychology by presenting his concept of developmental psychology. Paulo Freire's *The Pedagogy of the Oppressed* encouraged teachers and learners to make common cause and explore the world together, opening up its structures for understanding, not only through word literacy, but through the social and political literacy that can be learned concomitantly.

Not all books that are socially and politically influential are non-fiction. The dystopias described by Aldous Huxley's *Brave New World* and George Orwell's *1984* gave fair warning of the excesses of totalitarianism, no matter who is in control. From two quite different perspectives, Vladimir Nabokov's *Lolita* and William Golding's *Lord of the Flies* explored elements of our nature that continuously nuance human interaction, much as we might want it otherwise.

Being such an orderly country, and existing in the shadow of the United States and United Kingdom, to say nothing of France, Canada's socially and politically important publications have come in the usual form of books, as well as, sometimes, government reports. The books of Harold Innis provided us with a framework for understanding the early development of Canada, organized as it was around commodity extraction for (the British) empire.[44] Donald Creighton provided a different version of that same idea in his books on the Laurentian hypothesis.[45] *Obasan*, by Joy Kogawa, gnawed the Canadian cultural fabric and helped Japanese Canadians obtain redress, decades after they were disenfranchised and effectively imprisoned during the Second World War. A variety of books, including Eden Robinson's *Monkey Beach* and the published script of George Ryga's play *The Ecstasy of Rita Joe*, have done much to bring Canada's aboriginal peoples into the mainstream consciousness of Canadians. Through Antonine Maillet's Prix Goncourt–winning *Pélagie-la-Charrette*, Canadians can also understand the nature of the expulsion of the Acadians, many of whom ended up in Louisiana as Cajuns. Add to these normal books such government reports as the *Report of the Royal Commission of Bilingualism and Biculturalism*; *Northern Frontier, Northern Homeland: The MacKenzie Valley Pipeline Inquiry Report*; the report of the Massey Commission; and the Davey report (*The Uncertain Mirror: Special Senate Committee on the Mass*

Media), and it becomes clear how much we have come to know ourselves and set plans for our future through the medium of books (and book-like reports).

In short, books in their ideal form depend both upon those literary geniuses we call authors, and upon their collaborators in creation, the publishers, who serve as mediators between authors and the public. Being composed of language and written discourse, and being sustained explorations of ideas, books possess tremendous power. In their current form as almost pure text, books represent a prolonged Stradivarius moment in cultural history, in their magnificent ability to capture and communicate a certain strain of human understanding so well that we revile their burning to a degree that approaches our feeling about the same fate being visited upon human beings. Books are also the foundation of society and powerful agents for bringing about social change. Yet books are not a simple outgrowth of human existence, or even a natural adjunct of literate societies. They depend on a nurturing social context that provides opportunities for writers to write and publishers to publish.

The necessary social context for book publishing

Given the seemingly inexhaustible supply of titles from authors all around the world, it might seem an obvious verity that almost anyone is free to take up authorship on any subject and obtain a hearing from any number of book publishers, who themselves are usually free to acquire any book and sell into any market without interference. However, even setting aside certain countries where there is political or religious interference, the dynamics of the marketplace often constrain those freedoms. Initially, market constraints in Canada took a political and administrative form, as a result of the control France and Britain exercised over their colonies. However, as Canada gained its autonomy, new constraints inherent in the organization of the Canadian economy and the makeup of Canadian society took their place. Some of those constraints include: the pervasive influence of the United States and other foreign nations on Canada's affairs; the fact that we share our official languages with more-populated nations that have strong publishing industries; the existence of ever-integrating world markets and ever-more-powerful global communications technologies; the provincial control of Canada's education system; communication and copyright law and cultural industries policy; and, specifically, book publishing business practices. As a result of these variables, the importation and distribution of books written and published outside Canada is the predominant form of book "publishing" in Canada.

In some countries, such as the United States, England, France, Germany, Italy, Spain, and Japan, a host of factors, including population size and density, language, the age of the country, and standard of living, have made the organizational infrastructure necessary for book publishing intrinsic to the countries' historical evolution. In many other countries, such as Canada, Belgium, Scandinavia, Scotland, Ireland, Australia, the many countries of South America, and New Zealand, some or all of these variables have stood in the way of the emergence of a book publishing industry. As the *UN Creative Economy Report* 2010 notes, "Policy strategies to foster the development of the creative economy must recognize its multidisciplinary nature — its economic, social, cultural, technological and environmental linkages."[46] This book contends that the emergence and sustenance of a book publishing industry requires favourable conditions in a set of six environments: culture, politics, economics, law and policy, education, and technology. In Canada, these environments have at times constrained and at other times nurtured the Canadian book publishing industry. From time to time, throughout this work, reference is made to these six environments to assess the nature of developments at various times and in various situations. They are introduced below.

The cultural environment

More often than not throughout Canada's history, homegrown culture has been viewed with ambivalence. While the populations of the United States, the United Kingdom, France, and most other European powers had no difficulty seeing themselves as inherently distinct communities in the world, Canadians were not so sure. In the context of publishing and writing, there was no need, many said, for Canadians to focus on Canadian writers published by Canadian publishers, when first-class (read "non-Canadian") writers and publishers were available both as models and for the inherent value of their work.

As will be discussed in Chapter 2, such a view arose from circumstances that have pervaded Canada's history. However, by the mid-twentieth century, even as these opinions prevailed, an alternative cultural vision was being articulated by a Royal Commission chaired by Vincent Massey, a separate Ontario Royal Commission on Book Publishing, and federal departments with cultural responsibilities. As we shall see in Chapter 3, that articulation led to a cultural animation that began in the 1970s and saw structures and programs put in place that favoured Canadian production throughout the cultural industries, in spite of opposition from the dominant interests and outlook of the time. That opposition remains, and, through to the late 1980s, it constrained the development of book publishing and other cultural industries. However, during the 1980s and 1990s, the cultural environment overtook the political environment in leading governments to an affirmation of a distinct Canadian community and thereby clearing a critical path for the establishment and maintenance of domestic book publishing.

The political environment

In the twentieth century, as societies such as Canada shook off their status as European colonies to become independent nations, a collaborative political environment emerged that facilitated the establishment of national book industries. Freedom of speech is obviously foundational to such an enterprise. But freedom of speech is an individual freedom, and it does not address the need of communities and nations for the circulation of ideas that arise from within. A domestic book industry needs political will focused on internal discourse, but not exclusively; indeed, it should be freshened by breezes from the outside world. Such a discourse establishes the foundations of a distinct identity, of social cohesion among community members, and of the assertions of both freedom and community building. In book publishing terms, such political will and resulting discourse foster the exchange of ideas and information within a community. Authorship and publishing opportunities combine with the wider populace's ready access to domestically published works to build a nation.

Surely, one might think, a political environment friendly to the articulation and circulation of ideas within society is an ever-present characteristic of any free nation. On the contrary — as we shall see in the context of Canadian history and of colonial history in general, and, latterly, in the contemporary context of the promotion of international trade — in Canada it has been, and is, a constant struggle to ensure opportunities for domestic book authorship and domestic publishing, and for Canadians to read works written by members of their own community.

In Canada's colonial historical context, the struggle derived from a tension of loyalties. The United States resolved similar tensions by means of a revolution. It took much longer to replace mother-country loyalties with national loyalties in Canada, and to reach genuine independence. The dynamics of this tardiness led to whole groups of Canadian writers decamping for New York, London, and Paris, in the dying days of the nineteenth century and during the first two decades of the twentieth, to receive recognition.[47] This was only partly due to the much larger size of the U.S. market; as the biographies and writings of these authors attest, it was equally important that revolutionary America was more interested than loyalist Canada in affirming the North American experience captured by Americans and Canadians.

In the contemporary political context, the struggle to balance the internal circulation of domestic ideas with the importation of knowledge, ideas, and products remains. In the Canadian book market, far more titles distributed, sold, and read are originated in the U.S. and other countries than in Canada. This exposes Canadians to a cosmopolitan collection of writers and ideas and keeps Canadians open-minded about the world. But as subsequent chapters document, it also represents a market challenge for books written and published in Canada, by Canadians. The current organization and state of government support for book publishing is proof of the willingness of successive governments to intervene in the marketplace in order to propel Canadian writers onto the world stage, in numbers, and with reputations, that surpass what anyone might reasonably expect. Yet government interventions to support book publishing have not been enough for publishers to create a rich, profitable, export-dominated industry addressing both Canadian and universal concerns. Canadian-owned book publishers originating Canadian titles continue to exist at a survival level.

Law, policy, and derivative support programs

The existence of a collaborative political environment translates directly into the creation of effective and supportive laws and policies. Copyright law is the most salient example, and, as Chapter 5 details, a key and distinctive provision in Canadian copyright law is the so-called distribution right.[48] The Book Importation Regulations included in Canada's Copyright Act grant sole-supplier status to Canadian importers and distributors, giving them a position that parallels that of a publisher.

Few other laws, but many policies, have been put in place in the last twenty-five years to serve the distinctive needs of Canadian book publishing and to better meet the need for supportive law and policy. The development of these policies and programs is described in Chapters 3 and 4, and the current status of all programs is reviewed in Chapter 5. In brief: Beginning in 1972 and carrying through to the present day, cultural and industrial support programs of the federal and most provincial governments, for both writers and publishers, have been a feature of Canada's book publishing environment. Paragraphs 20 and 21 of the Investment Canada Act, first announced in 1985, represent a significant discouragement to foreigners thinking of acquiring or setting up book businesses in Canada. And in keeping with policies to be found in Europe, but not in the United States, Canada has established a Public Lending Right (PLR) to compensate authors for sales lost when readers borrow their works from libraries instead of purchasing copies.

These and other policies form a policy framework within which Canadian book publishing can function. If they did not exist, the book publishing industry in Canada would be a debilitated shadow of its present form.

Economics and the market environment

For the most part, the laws, policies, and programs mentioned above address market realities. But general economic conditions are also important. People must be able to afford books, and publishers must be able to actually deliver books to people. The encouragement of what is often called "the book habit" — that is to say, reading books in leisure time, which in part depends on literacy rates — is also vital. While the Canadian economy is favourable to book purchasing and reading, Canada suffers from market distortions that interfere with its citizens' access to a full range of books, especially those dealing with Canadian subjects. The predominance of imported run-on copies of foreign-published titles distorts the price of all books, resulting in a cost/price squeeze for Canadian publishers — high production costs combine with downward pressure on the price they can charge in their home market.

A second market distortion arises from normal international trading in publishing rights. Most often, as noted in the introduction, U.S. and U.K. publishers buy and sell rights to and from each other for their "domestic" markets. For the United States, "domestic" often means the North American English-language market; it may also include English-language sales in South America. For the United Kingdom, "domestic" may mean not only the U.K. and Ireland, but also the European Union and the "traditional" markets of U.K. publishers, which include Commonwealth countries. Such rights trading leaves Canadian book publishers unable to serve their own market through rights purchases. While exceptions are made, they are few. The lost opportunity for Canadian publishers amounts to about one-third of the domestic market.[49]

Yet another market distortion challenges publishers' economies of operation. Until about 1996, as a result of sheer industrial inefficiency, Canadian readers suffered from publishers' and independent bookstores' inability to keep up with demand for books that were receiving heightened publicity. Today, Canada's single national bookstore chain, Chapters/Indigo, is better at this coordination of marketing and availability, but it has erred in the other direction. Its concentration on bestsellers and sales per square metre means that the chain does a poor job of assisting publishers of non-mainstream titles to reach Canadian readers. The vast majority of Canadian-controlled publishers do not primarily publish mainstream-type products.

The financial status of firms is also important. Normal businesses borrow from banks or use retained earnings to increase their competitive position in the market. With profitability, after grants, in the order of 3 percent, Canadian-controlled firms do not have enough funds to invest to improve products and profitability. Loans are also problematic. Because publishers often earn a lower percentage on their working capital than it costs to borrow money, loans can decrease profitability rather than increase it, even when they lead to growth.

The challenges of the current market environment for Canadian publishers are reviewed in detail in Chapter 6. As later chapters also show, policies such as tax credits improve the economic environment for title origination.

The educational environment

Efforts to establish the legitimacy of Canadian culture and the Canadian book industry were and are linked to the structure and content of the primary, secondary, and post-secondary curriculum. For each new generation, the goal of establishing and maintaining a collective identity wears an educational cloak, which is patched together from at least twelve pieces, given that the constitutional responsibility for education is vested in the provinces. With no national educational presence, Canadian provinces are free to select the foundational social ideas to be included in the curriculum. This can result, as it did in the 1980s in Manitoba, for example, in the provincial adoption of a social studies series, spanning grades one to six, that devoted one year to the study of Canadian communities and five years to an informing perspective that drew on U.S. examples.[50] That curriculum has since changed. Yet it remains commonplace for provincial educational systems outside Quebec and Ontario to adopt U.S. textbooks written for the U.S. market, with little adaptation beyond the equivalent of "putting brown faces on the covers," as happens with African textbooks. Nor is this weak Canadian presence confined to the soft side of the curriculum; science books are replete with examples of U.S. scientific history.[51]

Given the main responsibilities and concerns of provincial governments, cultural realities tend to take a distant second place to learning objectives. Historically speaking, such an educational environment resulted in the importation of educational texts or their invisibles — the organizational infrastructure and pedagogy on which the materials are based — which, in turn, stunted the growth of national culture and the emergence of a domestic (i.e., Canadian-controlled) educational publishing sector. The upshot was that the organizational and financial foundation that educational publishing normally provides for the development of a vibrant trade industry was never available.

The technological and organizational environment

As noted in the discussion of economics and the market environment, it is rare for Canadian-controlled firms to have sufficient funds to address production efficiencies. This lack of access to retained earnings and loans mutes technological innovation within Canadian book publishing firms and shifts responsibility for such initiatives to the government, although, as Chapter 5 and subsequent chapters will demonstrate, a recent government/industry technological partnership called the Supply Chain Initiative has modernized distribution and proven enormously successful.

Technology and its management can play a positive role in book publishing and this is discussed in Chapters 7 and 8. Specifically, the deployment of central servers and the requisite software to provide a foundation for title management creates considerable efficiencies, as does the rationalization of printing technologies. The use of technology to organize distribution over Canada's ample geography also affects the titles available to Canadians, increasingly so in an environmentally concerned society. And, of course, we have now entered the era of tablets, those multimedia devices that have revolutionized the media foundations of what might still be called books to include sound, visuals, moving images, and morphable text: a fundamental change, for certain.

Conclusion: Preconditions and formative roots

Books rest on a series of inventions and developments that have taken place over the past 50 centuries. Those inventions and developments were themselves built on the human capacity for language invention and learning, including our mental capacity for subsequent generativity — invention through recombining letters and words. In the Western world, the main inventions that contributed to the evolution of the physical or material elements of books consisted of the phonetic alphabet, which transformed sound into visual symbols; the development of media upon which symbols could be recorded; the visual simplification and aesthetic enhancement of the representation of letters; the invention of movable type and the printing press, along with associated ink and binding technologies to allow the manufacture of books; and the development of portable book forms that, together with computers and screens, has led us to the enhanced ebook. However, such material elements are merely the substrate for authorship — human cognitive invention in textual form, and publishing — the cultural business of working with authors to reach interested readers with a work that they will fully appreciate. Far from being a mechanical enterprise, the core elements of publishing practice — acquisition, substantive editing, marketing, and exchanging rights with other publishers — together with the personalities of editors, designers, marketers, and publishers, build a cultural presence for a firm that can, from time to time, elevate a publishing house and its authors to become the primary catalyst for a substantial change in human consciousness.

Whereas the above dynamics often evolve as a component element of large and powerful nations, without any further direct intervention from an education system that teaches literacy and values individual freedom, as the preceding section underlines, the creation of a domestic publishing industry depends on a set of necessary preconditions, and putting them in place is not a simple matter. The task requires a cultural environment that affirms the collective identity of a nation. It depends on a political will and effective policies that touch on a host of variables including support for authors, trade flows, copyright, technological innovation, and distribution agreements. It is contingent on a market organization and retailing system that helps to ensure access and that rewards those involved sufficiently to sustain constant renewal. It requires an education system that not only teaches literacy, but also engenders an interest in, and commitment to, a collective identity. And it depends, especially at the beginning of the twenty-first century, on an embrace of technology, carried through to the organizational consequences that result from implementing that technology.

As of the beginning of 2012, many of the necessary policies were in place, as was relative financial stability within the industry. However, both policies and the industry itself were relatively recent developments. As the next chapter documents, the Canadian book publishing industry of the twenty-first century's second decade was made possible by the drive and determination of a small number of outstanding Canadian publishers and authors who helped to establish both an industrial infrastructure and literary roots. These individuals worked in the context of political and cultural forces affirming a colonialism of allegiance to empire first, and reflecting the powerful geopolitical influences of the United States as it pursued its self-defined "manifest destiny." Yet limited as these initial efforts were, they sowed the seeds of the slow but tenacious emergence of book publishing as a distinguishing feature of Canadian cultural reality.

Prelude to modernity: Some historical notes on Canadian book publishing and cultural development

Canada's colonial legacy   55

   Educational publishing of the colonial era   59
   Early entrepreneurship and Egerton Ryerson   60

Canadian trade publishing to 1950   62

From colonial realities to domestic publishing   67

Conceiving of Canadian culture: The Massey Commission   70

A collapse of domestic publishing amidst a cultural rebirth   73

In the years prior to Confederation, the territory now known as Canada was, in terms of English-language publishing, an extension of the British market. Most books were imported by Canadian booksellers or sold into the market by agents of British publishers. But counteracting that British-based production, and injecting some tension into the information marketplace, were the activities of colonial printers, who had access to the means of production and whose original business was to serve government. The printers began to play a central role in the exchange of information and opinion in the communities that, in 1867, would come together to form Canada. Vying with both these sources of production were U.S. printers who, with some interruptions between 1776 and 1891, ignored British copyright laws and supplied cheap reprints of British editions to the entire North American marketplace.

At various times through the eighteenth and nineteenth centuries, Canadian book publishing rose to prominence, and it played an important role in stitching Canada and Canadians together. However, it wasn't until the 1950s, in the wake of the Second World War, that Canada embarked on an explicit mission to articulate the distinct cultural underpinnings of its nationhood. The creation of the Royal Commission on National Development in the Arts, Letters and Sciences, chaired by Vincent Massey, and that commission's final report, laid the conceptual foundations for Canada to embrace its social and cultural realities. It also helped Canada establish and, later, develop its own cultural industries, including book publishing.

Canada's colonial legacy

An understanding of contemporary Canadian book publishing requires an overview of its colonial beginnings. George Parker's *The Beginnings of the Book Trade in Canada* is the major source here, and what follows derives from his writings and a few others, as noted. This section should be read as my interpretation of what others, who have delved into the history itself, recount. As Parker notes, the English-language book trade, like other forms of commerce, was organized in the eighteenth and into the nineteenth century to import books and periodicals from Britain to Canada.[52] This importing orientation was both political and economic, in that it served the interests of British publishers and Canadian booksellers while responding to settlers' cultural desire to keep in touch with their home country. There was also a certain demand for explorers' and settlers' memoirs of their time in "Canada," which were published in Britain; some were sold there and some were brought back to the colonies, thus providing ideology and inspiration both to adventuresome Brits and to early immigrants to Canada. Also, at times, booksellers in Canada would become publishers by soliciting the cooperation of their bookselling colleagues in other settlements and, jointly with them, bringing a title to market.[53]

On the surface, the general organization of the early book market was simple: Canadian booksellers would acquire imported books from Britain and put them on sale. The confirming law behind this arrangement was copyright law. Although the British Colonial Office began in 1820 to enact liberal reforms promoting colonial independence,[54] copyright law still restricted the activities of pre-Confederation book publishers to the local market. The Imperial Copyright Law of 1842, updated from the previous Literary Copyright Act of 1709, granted copyright protection throughout the British Empire for books produced in London and Edinburgh.[55] Any book published in another city within Britain, or in any of Britain's colonies, was protected only in its local market, and hence was vulnerable to being pirated by other companies within and outside the British Commonwealth. Local copyright acts that applied within their jurisdictions were passed in Quebec (1832), Nova Scotia (1838), and the Province of Canada (1841), and the Dominion of Canada passed a copyright law one year after Confederation, in 1868. (All such laws were subject to approval by the British Parliament.)

There were inherent difficulties in this market organization. Out of the colonial experience, which differed fundamentally from the status quo in class-based Britain, arose a body of opinion that was fostered in part by life in the colonies and in part by the representation of colonial life by colonial newspapers. Many in the North American colonies chafed under the restrictive British rule that advantaged such people as the members of the Family Compact. The first presses in

British North America (BNA), in 1751 and 1764, were brought for the purpose of printing government notices and other official documents, but this newly available technology created the opportunity for a community response to social conditions, not only an official response to community needs and desires. Gradually the articulation of community wishes became a stock-in-trade of the newspaper and printing industries. The content of the newspapers also reflected a mismatch between British literary traditions, with which Canadian would-be authors grew up, and the matter-of-fact realities and outlook of colonial life.[56] In fact, newspapers served as a transition medium for an evolving domestic literature. The publication and reception of single poems, stories, serialized excerpts of longer works, and other literary forms helped this frontier literature develop. Yet printers were thwarted in their attempt to respond to community desires for books because of the unwillingness of British book publishers, protected by Imperial Copyright, to grant reprint rights.

Eventually, the general discontent created by this disjuncture between colonial life and empire control manifested itself in rebellions in both Lower Canada (southern Quebec) and Upper Canada (Toronto and environs) in 1837 and 1838. Newspapermen played a significant role in both revolts. In Upper Canada the rebellion was led by William Lyon Mackenzie, the publisher of the now seemingly ironically titled *Colonial Advocate*. In Lower Canada, Pierre Bédard used his paper, *Le Canadien*, to oppose the Chateau Clique.

The critical statute in the specific tension surrounding book publishing was Imperial Copyright Law, which flew in the face of common practice in pre-Confederation Canada. It was routine for booksellers to sell pirated editions of British titles, produced in and imported from the United States, rather than importing from Britain. Like European countries of the time, the United States did not recognize British copyright law. Nor did Britain recognize the copyright laws of other countries, the United States included. The importation of pirated works into the British Empire (i.e., Canada), where U.K. copyright law clearly did hold, was a problematic but prevailing reality. Responding to intense lobbying by booksellers, and wanting some revenue rather than none, in 1847 the British passed the Foreign

Reprints Act, which allowed booksellers to import pirated editions of U.K. books for a 12.5 percent tax. This act not only officially sidelined Canadian printers, but also hijacked a significant part of the pre-Confederation market from Canadian distributors of British works. A U.S. company could print and publish U.K. copyright books to be sold in Canada, but a Canadian company, because it was within the British Empire, could not print and publish the same titles. The act therefore meant that profits, and the organizational foundation necessary to either import or reprint titles and distribute them (called agency publishing), were unavailable as building blocks for a Canadian book publishing industry. U.S.-based printer-publishers gained a further competitive advantage by their neglect in sending royalties to the originating British publishers.

But by the time of Confederation, in 1867, in spite of the disadvantages brought about by Imperial Copyright, there were lively literary, bookselling, and printer/publishing communities in Canada. As Parker notes, "Life in the new dominion unfolded not unsatisfactorily until the end of 1873."[57] In 1868, the printer John Lovell requested that the same reprinting privileges enjoyed by U.S. printers be extended to Canadian printers as well. He saw no reason why a Canadian printer should not be able to establish a reprint business and then expand into the United States. In an attempt to counter their disadvantageous political and legal position, Lovell and his fellow Canadian publishers John Ross Robertson and the Belford Brothers began pirating U.S. publications in the 1870s and 1880s, including those of Mark Twain. They sold them in Canada, and in some cases shipped them into the United States to be sold at a price that undercut the U.S. publisher. As might be expected, the Americans were unhappy, and Samuel Langhorne Clemens (Mark Twain) was personally outraged. In an attempt to claim royalties on the Canadian printings, Clemens established a Montreal residence, but the courts of the day ruled that he required a "domicile," a permanent residence in Canada, to bring a successful case of copyright infringement. Clemens also lobbied hard, and eventually successfully, for the creation of a U.S. copyright act. It could be claimed that Canadian printers played a role in his success.

Historian H. Pearson Gundy, who contributed background

research to the Ontario Royal Commission on Book Publishing, takes particular note of Lovell as the dominant publisher of the day. Lovell set himself up in Montreal in the 1830s, and by 1851 he employed forty-one printer's assistants and apprentices.[58] His printing contracts with the government and others served as a foundation for the production of a long list of novels, books of verse, popular religious works, travelogues, biographies, histories, and songbooks in both French and English. In 1877–78, Lovell established a printing plant just over the Quebec border in the state of New York to avoid penalties imposed on Canadian printers for reprinting British books without permission. While the Lovells had some initial trouble, by the time of Lovell's death in 1893, he and his two sons had published 1,500 titles in Lovell's Library of cheap reprints, and had refocused their business to concentrate on producing such books and marketing them in Canada and the United States. To a considerable extent, Lovell was an exception. As Parker notes, following 1873 "that sense of being on the threshold of greatness dimmed considerably, for the book trade, like other areas of national life, experienced a cycle of advances and traumatic setbacks ... until the prosperous years arrived in 1896."[59]

Educational publishing of the colonial era

As these trade publishing events were taking place, the market for educational books was expanding, with continuing migration mainly from the United Kingdom and the United States, but also from other European countries. Following the Irish potato famine and a mass migration of Irish people to Canada and other British colonies as well as the United States in 1846 and 1847, the clamour for free public education and inexpensive textbooks in Canada grew. To meet the demand, the (British) Commissioners of National Education in Ireland made their textbooks available to various parts of North America as tools for the education of children of the British Empire.

Satu Repo describes the ideological undercurrents driving the provision of the Irish National School Books to British North America. She sees the books as an extension of the promotion of a non-sectarian Protestantism.[60] In Ireland, they were intended to assist in suppressing

Irish culture and Roman Catholicism, and the books even went so far as to replace Irish music with English songs. The need for textbooks that promoted a common set of values, dominated by British interests, among English, Scottish, and Irish immigrants made these books attractive to British North American educators and particularly to Egerton Ryerson, Canada's most formidable publisher and the initial principal of what was to become the University of Toronto.

Early entrepreneurship and Egerton Ryerson

Egerton Ryerson had an unsurpassed early influence as both a publisher and an education superintendent (in which he acted as a true publisher, buying rights and contracting out printing to various companies). He founded a newsletter, the *Christian Guardian*, and, in 1829, the Methodist Book and Publishing House (later Ryerson Press, and then McGraw-Hill Ryerson), which by 1833 he was transforming into a significant, socially concerned, general publishing house. By 1884, the Methodist Book and Publishing House was the most prolific press in Canada, printing and binding 211,714 copies in that year. The company's list of published authors reads like a who's who of Canadian literary and political life (e.g., Nellie McClung, Charles G.D. Roberts, Robert W. Service, and Pauline Johnson), and the list of its employees was a who's who of the following generation of Canadian publishers (e.g., John McClelland, George Stewart, S.B. Gundy, and Thomas Allen).[61]

Drawing mainly from Gundy and Parker we learn that Ryerson himself retired from publishing in 1840 and became chief superintendent for education in Canada West (later Upper Canada and then Ontario) in 1845. After passage of the School Act of 1846, Ryerson opted for the Irish National School Books, mentioned above, and obtained permission to reprint and sell the books within the Canadian market for less than the British price. Having acquired the rights, he astutely sourced the reprinting among three printers. The Mackinlay Brothers was one firm that, partly on the basis of this contract, expanded from its base in Canada West to the Maritimes and parts of Quebec. Ryerson also used his position to create a central book depository for Canada West, which served as a warehouse and distribution centre for published materials that were bought centrally for both schools and public libraries. His efforts amounted to the creation of nothing less than the most powerful publishing entity in pre-Confederation Canada. Being inside government, and hence in control of market demand, Ryerson was able to organize rights acquisition, outsource printing, and distribute schoolbooks in a manner that was politically astute, culturally integrative, and technologically and economically efficient. With so much power in Ryerson's hands, and with Canadian-authored books beginning to populate bookstore shelves, publishers regularly attacked the book depository system and its exclusion of those publishers who were successful in other jurisdictions but not in Upper Canada. Those who complained wanted access to this lucrative market. Corruption was never found.

The historical, political, and cultural context within which Ryerson was acting deserves further elaboration. Many of the schoolbooks circulating in the one-room schools of what was then British North America were used simply because they were available. With constant immigration, especially from the United States, into what is now southern Ontario, many of the available schoolbooks reflected a set of republican ideas and ideals and a "strident republican and anti-British tone."[62] This was a formative period for both U.S. and Canadian societies, and tensions erupted several times in North America: during the Revolution of 1776; in the War of 1812, when Isaac Brock and Tecumseh banded together to defeat the invading Americans; and in 1837, with the rebellions in Upper and Lower Canada. They would erupt again in 1861 with the U.S. Civil War, followed by the Red River and Northwest rebellions of 1869 and 1885.[63] In this overall context, Ryerson's prodigious efforts must be seen as aimed at suppressing both Yankee influence and indigenous anti-colonialism within the school system of Canada West. By acquiring rights to books that encouraged young children to identify with Britain, he laid an operational as well as an ideological foundation for Ontario textbook acquisition and greatly influenced Canadian educational rights acquisition and publishing.

As time passed and settlement spread westward, Canada began to draw immigrants from countries outside the United Kingdom, and colonial ties were weakened. With a different balance among ethnic groups in Canada's western provinces, different ideals held sway. In education, this produced a determination not to follow in the almost exclusively pro-British footsteps of Ontario — a determination that was as strong as their resolve not to be overrun ideologically by the United States. The British North America Act, which vested constitutional responsibility for education in the provinces, facilitated this independence.[64]

In summary, as the school system expanded quickly because of both immigration and industrialization, procurement policies were developed to discourage textbook-strapped schools from using whatever books were available, since, given U.S. proximity and immigration, this would have meant U.S. textbooks filled with anti-British sentiments. This political and cultural context informed Ryerson and other educators, just as a slightly different political and cultural reality informed Canada's emerging western provinces in subsequent years. These procurement policies fostered the utilization of existing printing and publishing technology, and an industrial organization that focused on printing and reprinting, rather than publishing and the origination of titles. These developments existed against a political and economic background in which Britain strove to protect its colonial markets from the continual challenge of U.S. printing and publishing.

Canadian trade publishing to 1950

As Parker and others tell the story, after Confederation in 1867, Canada became a potential single market for books (although not in education, because of that market's constitutional assignment to the provinces), and various attempts were made to establish it as the exclusive home territory for Canadian book publishers. In seeking this exclusivity, Canadian book publishers were merely following the trade patterns of other countries, many of which were protected by a unique language or market size. But Canada's colonial relationship with the United Kingdom stymied their efforts. The Canadian Copyright Act of 1872 was an early attempt to carve out the market for Canadians. The 1872 act shut out both foreign reprints and original British editions, restricting the market to Canadian publishers. From the perspective of principle, the act had the disadvantage of condoning the reprinting of works without their authors' permission. The British Parliament disallowed the act one year later. In the ensuing years, nearly annual attempts to control Canadian copyright from within Canada were also rebuffed by the British Parliament. Finally, in 1875, an Act Respecting Copyrights was passed (in Canada); it allowed British-copyrighted titles republished in Canada to be treated as if they were originally published in Canada, thus bestowing copyright on the Canadian edition. This act remained in force until its repeal in 1924 and gave authors who registered their books in Canada control over both the Canadian reprinting and the importation of foreign editions. If authors did not register a title, they had no control over the importation of foreign editions, but did retain control over reprinting within Canada.[65] To further the aim of reserving the Canadian market for Canadian publishers, the Canadian government attempted to discourage importation by introducing a tariff in 1879. This produced considerable public backlash, as it increased the price of imports and served to illustrate how much Canadian readers relied on unfettered importation of titles. The tariff was soon withdrawn.[66]

In 1885, Britain signed the Berne Convention, the first international copyright convention, which printers in Canada resisted over the next two decades as it outlawed reprinting without permission of the author. After passage of the 1891 U.S. Copyright Act and an accompanying agreement, the Anglo-American Reciprocal Copyright Agreement, American pirated editions of British titles gradually disappeared from both the Canadian and U.S. markets, and John Lovell consolidated various reprinting firms, thereby ridding the market of cutthroat competition. Prices rose in both the U.S. and Canadian markets but, evidently, books were still affordable.[67]

The reorganization of the U.S. and Canadian markets following the passage of copyright acts in the 1870s and 1880s appeared to make things worse for Canadian book publishers. Booksellers tended to purchase cheaper U.S. copyright editions directly from U.S. publishers, rather than from the Canadian agents of the originating British publishers, who tacked on a surcharge of 10 to 20 percent, and

Canadian publishers were prevented from publishing and exporting books by U.S. authors to the U.S. by the manufacturing clause in the U.S. Copyright Act, which required books written by U.S. citizens or domiciles to be manufactured in the United States to gain copyright protection. Other Canadian publishers besides the Methodist Book and Publishing House did exist at the time — including Drurie of Ottawa, and Copp, Clark and Company and the Belford Brothers, both of Toronto — and their publishing activities, in the main, paralleled those of the Methodist House.

In 1879, after twenty years as a preacher, William Briggs followed in the footsteps of Egerton Ryerson as a vigorously entrepreneurial head of the Methodist Book and Publishing House. In his first two decades at the press, Briggs published popular fiction by Canadian authors and provided Canadian distribution services for British publishers. However, as time went on, sales figures gradually persuaded him to diminish his title origination and concentrate on agency sales. As Gundy reports, Briggs' peak year for original publishing was 1897, when the house originated thirty-seven titles, a level that was not repeated until the late 1920s. By the 1910s, the last decade of his career, title origination slowed to almost nothing.[68]

Beginning in 1890, led by "modernist" poets and novelists such as Charles G.D. Roberts, Canadian authors entered a new phase that saw their work published in New York, Boston, London (England), and Canada. Their reception outside Canada appeared to drive interest within the country, and these authors managed to strike a responsive chord in Canadian and international audiences. Some — such as the Rev. Charles W. Gordon, who, under the name Ralph Connor, wrote about the settlement of Canada's western provinces — achieved blockbuster status. Three of his books, *Black Rock*, *The Sky Pilot*, and *The Man from Glengarry*, together sold over five million copies.[69] Between 1905 and 1914, an average of twenty-one novels from various houses appeared each year.[70] The interest generated by this fiction opened the market to other genres, such as multi-volume historical and biographical series.[71]

In the country as a whole, the end of the nineteenth century and the beginning of the twentieth was a time of substantial immigration, western settlement, and individualistic, opportunity-seizing ideology. Business records reflect a matching vigour in the book publishing and writing scene.[72] Between 1896 and 1913, at least thirteen firms were founded, many of them Canadian branches of U.S. or U.K. publishers. These firms broke two major nineteenth-century patterns — one in which printers became reprinters and then publishers, and the other in which booksellers extended their activities to printing and subsequently to publishing. The new publishers established an almost-universal financial and organizational model for the book trade that survived until the 1970s. That model combined agency sales, in which the firm acted as an importer and sales and distribution agent for foreign companies, with title origination, which was financed by the agency sales. The model reduced business risk, as it allowed Canadian firms to import titles as needed, at deep discounts, and distribute them to booksellers. In some cases, where large sales were expected, Canadian firms negotiated Canadian rights, which gave them greater margins and encouraged them to invest in more extensive marketing and promotion. Many firms founded at this time survived through to the 1960s and later, after which they were acquired by larger firms. Some still remain, such as McClelland & Stewart (sold as this book was going to press), University of Toronto Press, and Oxford University Press (Canada).[73]

Far from depressing the book market, the First World War stimulated sales as readers searched for books on Germany and European powers, followed by war adventure stories and both patriotic and tragic poetry. In all, Canadian publishers distributed 1,000 war-related titles from 1914 to 1918, with sales as high as 40,000 (ten times normal).[74]

After a brief slump, the 1920s were distinguished by an ebullient nationalist spirit helped by a vibrant economy. Yet another indomitable book publisher surfaced in the person of Lorne Pierce, a literary scholar and former minister who took command of the Methodist Book and Publishing House, now renamed the Ryerson Press. Under Pierce's direction, Ryerson Press took up the glories of a new nationalism, as the publisher set about his goal of transforming school curricula to reflect Canada's growing literary heritage. Pierce was not alone in his designs — Hugh Eayrs, of Macmillan of Canada, and John McClelland (formerly of Ryerson Press but now out on his own) were also notable figures — but Pierce was certainly the leading light. Connected to the outstanding authors of the day, he toured the country looking for promising men and women of letters. Associations of authors, booksellers, and publishers were also active in this period, and the publishers reached out to Canadian artists for design and illustration.[75]

The Great Depression of the 1930s, exacerbated by a severe prairie drought, caused book publishers to retreat into their agency business, but not entirely. The book business maintained a market during those years by providing some diversion from the stark economic realities faced by many in Canadian society. The period of the Second World War was not as favourable to Canadian book publishers as the First World War had been; Britain was able to continue supplying the Canadian market through New York, and the United States, which remained out of the war until 1941, published more or less as usual. As well, market demand had shifted between the wars with the development of radio broadcasting. While the market was strong in 1945, it was followed by a recession that lasted until 1952, which, combined with a flood of U.S.-published cheap paperbacks, caused many firms to abandon Canadian title origination.[76]

Educational publishing drove the post-war book market. This sector was highly competitive, characterized by newly aggressive U.S. companies merging to become powerful publishing corporations and establishing firm business roots in Canadian provincial marketplaces. By applying emerging research into reading and pedagogy to the development of textbooks, and combining that with the marketing muscle of a large corporation, these firms relegated Canadian companies to trade publishing, where they faced competition from imported U.S. and U.K. titles sold at low "run-on" prices based on the marginal costs of extending the U.K. or U.S. print runs plus Canadian markups. They also faced Canadians' distinct lack of interest in Canadian literature as a means of establishing a national identity. This apathy was the result of a host of factors that characterized the period, including the Cold War, industrialization, the baby boom, and the trauma of centrally controlled German and Japanese (and Italian) nationalism.

As in earlier times, the period from the 1880s through to the 1950s saw the interplay of the environments discussed in Chapter 1. The economy played a formative role, but within a profoundly political context that saw two world wars. Pinched economic circumstances constrained but did not thwart the emergence of what might in retrospect be called a nascent Canadian culture, led by authors recognized in the United States before they were in Canada. Law and international policy brought some stability, even as they constrained Canadian book publishers who wanted to claim Canada as their market.

From colonial realities to domestic publishing

This overview of Canadian publishing history illustrates how difficult it was for Canadians to establish a domestic book publishing industry. Faced with two powerful countries, each desirous of supplying the Canadian market, Canadian book publishers have had difficulty carving out an ongoing market segment, let alone a dominant presence. The same difficulty exists in former colonies around the world, exemplified by the situation of English-language book publishing in colonial and post-colonial India.[77] The Indian Charter Act of 1813 established a British curriculum and a new educational system in which the printing of educational textbooks played a major part. In 1835, following the recommendation of Thomas Macaulay, one of the advisors to the Governor General of India, English became the official language of education in India. Books were imported from Britain, and British publishing houses gained positions of dominance and outright monopolies in different school subjects. In post-Independence India, English textbooks remained the standard for education, in part because they avoided the politics of privileging one indigenous language over another. Similarly, colonialism on the African continent was stabilized, in part, through the establishment of education systems and colonizer-published educational materials, with the result that European books and their publishers were firmly ensconced in both schools and universities, in spite of Africa's centuries-old artistic and literary histories.[78]

The Indian, African, and Canadian experiences reflect the difficulties former colonies face as they attempt to make the transition from importing and distributing the books of the colonizer (and other larger, developed economies) to establishing a stable and thriving domestic book industry designed to meet the emerging country's needs and ambitions. It is a market structure problem. When, as happens among European countries with different languages, rights are sold from one domestic publisher in one market to another domestic publisher in another market, publishers on both sides of the trade become representatives of both domestic and international authors. When, on the other hand, the market is structured in such a way that foreign-owned publishers represent "international" titles, and domestically owned publishers are confined to publishing domestic authors and books on domestic issues, invidious comparisons come to the fore, and competitive forces tilt towards the marginalization of domestic authors and their publishers. In such a situation, domestic authors are encouraged to write about domestic affairs, because it is unusual for the domestic publisher to have the necessary connections to publishers in other countries to properly exploit an international title. As a result, domestic authors who wish to address world issues tend to seek international companies to represent their work.

This domestic-international dynamic extends into educational publishing. In response to perceived opportunity, foreign-owned educational publishers establish branch plant operations, instead of selling rights to a domestic company that could then translate or adapt learning materials as it saw fit. These foreign-owned publishers hold on to the educational sector tenaciously, making it unavailable to domestic trade companies as a financial, technological, professional, and organizational building block. Trade publishing (and, often, professional and reference publishing) thus becomes divorced from educational publishing. As time passes, the differences between the sectors become more apparent, and it is increasingly difficult to carry on both trade and educational publishing within a single firm. Without the opportunity to purchase rights and add localisms in examples and styles of thinking, with an audience grown used to the wares of the colonizer and other dominant societies, and in an open society where firms from anywhere are free to set up and do business, to say nothing of economies of scale that severely disadvantage domestic production for domestic markets, Canadian publishers are reduced to importing and distributing foreign works while originating only a limited number of titles that speak specifically to the Canadian condition.

To illustrate this, when Leonard and Virginia Woolf founded Hogarth Press and drew on the ideas and writings of the Bloomsbury Group and what its members saw as exciting and dynamic, they produced a literature of the transition from the Victorian era into modernism that was meant for the whole world to read. When Canadian book publishers of the same era developed a list, they were more often applying and adapting such ideas to the Canadian context, in order to keep this country abreast of the changing world. When Canadian authors and stories dealing with Canadian realities gained international audiences, as happened with Ralph Connor's *The Man from Glengarry* or L.M. Montgomery's *Anne of Green Gables*, they achieved that status in the hands of a U.S. or U.K. publisher — Grosset & Dunlop in the case of Connor, L.C. Page of Boston in the case of Montgomery. In the main, the result was then as it still is today: a local industry focused on the domestic market.

This interpretation complements the central argument of this book. From time to time, authors and other talented creators and entrepreneurs coalesce around the writing and publishing of books. That can lead to a rich and generative interaction and an emergent identity at a time in history when opportunities for change exist. The 1840s through the 1860s saw a struggle to establish a Canadian publishing industry. Canadian authorship and Canadian book publishing flourished at the turn of the century and during the 1920s, only to be restrained by financial realities in the subsequent years until the 1970s, mainly because Canadian society was looking outward rather than inward. With both flourishings, national vigour appears to have been concomitant with market vigour. Indeed, one might say that in such times, a confluence of political, cultural, legal, economic, technological, and educational variables creates opportunities for a national book publishing industry to surge ahead, and that surge is maintained in the years following, even though it may weaken somewhat with the passage of time and the passing of the conditions that encouraged its emergence.

This confluence of supportive environments was as influential, if not more influential, in Canada in the late 1960s through the 1980s as it was in the previous notable periods of Canadian book publishing. In the late 1960s and the 1970s, a new wave of publishers came into being, led largely by the forerunners of the baby boom. They re-established national priorities in domestic book publishing in Canada by insisting, against a background of the worldwide assertion of civil rights,[79] on the importance of Canadians being able to write for and read one another.

Conceiving of Canadian culture: The Massey Commission

In 1951, even if few saw it this way at the time, Canada began to wend its way towards a cultural awakening. The document that initiated Canada's first steps down this path was the report of the Royal Commission on National Development in the Arts, Letters and Sciences.[80] Chaired by Vincent Massey, the Commission produced an encyclopedic report on an Arnoldian concept of culture, in which the arts were seen as the pinnacle of human achievement. It also sang the praises of research and knowledge before the idea of a knowledge society had been articulated. The commission took a broad perspective and delved into language, the press, writing and publishing, the humanities, philosophy, psychology, the social sciences, national history, archives, historical societies and museums, the natural and physical sciences, medical research, applied sciences, music, theatre, painting, and architecture — usually in both languages. The five commissioners saw a need to create national cultural and research institutions to nurture the development of artistic, literary, and scientific activities that Canada could call its own and to serve as the public face of Canada on the international stage.

The commission was enormously successful at framing its narrative in a national context, and its report changed the nature of Canada and the role of government in modern Canadian society. The commission recommended the creation of the National Library, primarily to collect all Canadiana and to serve as a lending library of last resort. The Canada Council and its academic cousin, the Social Sciences and Humanities Research Council, which the commission recommended be established as one body, are still in existence today, as is the National Library, which has been joined with the National Archives to form Library and Archives Canada.

Basing much of its framework on Vincent Massey's understanding of British exemplars, the commission set the stage for the twentieth-century development of national cultural institutions and processes that were unknown within Canada to that time.[81] With respect to writing and publishing (as well as other artistic endeavours and social scientific research), the commission advised:

> That a body be created to be known as the Canada Council for the Encouragement of the Arts, Letters, Humanities and Social Sciences to stimulate and to help voluntary organizations within these fields, to foster Canada's cultural relations abroad, to perform the functions of a national commission for UNESCO, and to devise and administer a system of scholarships as recommended in Chapter XXII.[82]

The Chapter 22 recommendations were "that there be created a system of grants for persons engaged in the arts and letters (including broadcasting, films and the press) for work and study either in Canada or abroad; that arrangements be made for grants to artists, musicians and men of letters from abroad for study in Canada; that these grants be administered by the Council for the Arts, Letters, Humanities and Social Sciences mentioned above; and that funds be made available for these purposes."[83]

The commission heard from writers and publishers about the state of book publishing, and the report included a table that noted that the output of fiction titles in Canada for 1947 was approximately 2.6 percent that of the United States and 2.0 percent that of the United Kingdom (not 10 percent that of the United States, as the comparative populations of the two countries would suggest it might be). In 1948, the commission noted, the output of fiction diminished to 1.3 percent that of the United States and 0.8 percent that of the United Kingdom. Yet, in part because the testimony from authors, publishers, academics, librarians, and the general public was so various, the commission offered no recommendation dealing with book publishing. Perhaps the commissioners believed that somewhere amidst the evidence was a solution for these seeming shortcomings — of timid publishers not wanting to take a chance on Canadian writing; of Canadian writers' lack of success at creating bestselling works; and of the possible establishment of Canadian-book-reading clubs in English Canada. Perhaps, alternatively, they saw no problem, although such blindness would be inconsistent with the orientation of the commission. Or maybe the Arnoldian sense of culture the commissioners brought to their task, and with it the tendency to venerate British and French writing, conflicted subconsciously with the idea of providing opportunities for Canadians. Such a conflict would have affected the commission's ability to see how government might reasonably assist the private enterprise of book publishing.

Insight into the commissioners' states of mind and the realities they faced can be gained not only from the commission's report, but also from a subsequent book written by one of the commissioners. Hilda Neatby's *So Little for the Mind* was a call to Canadian educators and parents to help them understand how little their children were being taught of disciplines such as mathematics, geography, history, grammar, and the sciences, and how little about Canada they were learning.[84] While it is clearly the work of a person intent on preserving the perceived, British-derived, rigour of the educational past, including rote memorization of the foundations of the disciplines, the book amply documents the lack of an informational base in the school curriculum of the time, which was influenced by American pragmatist philosopher John Dewey, whose writings depicted children as bio-social beings without any cultural roots or identity.[85] This approach neglected both culture and the recognition that children existed in a particular time and location, tied to a particular religion, community, or tradition. Neatby's book indicates that she saw progressive, experience- and curiosity-based education as completely inconsistent with the national cultural project Vincent Massey was promoting through his chairing of the commission.

Neatby's national and discipline-based concerns were not seen as salient by Canadian publishers in this post-war era, as wartime shortages disappeared and trade in all manner of goods resumed. In educational publishing, U.S. branch plants began moving into the Canadian markets created by educational administrators, many of whom had received their advanced training in the United States at Stanford, Columbia, and the University of Chicago.[86] Trade publishers were busy representing foreign publishers, acting as exclusive, under-licence importers and distributors. Bookstores were engaged in providing British, U.S., French, and Canadian titles to Canadian readers. Authors of literary fiction, including Morley Callaghan, Mavis Gallant, and Mordecai Richler, were migrating in and out of world meccas, such as London and Paris, to join in the rejection of materialism, while their more materialistic colleagues set their sights on New York, where book contracts could be had and subsidiary rights sales had some chance of emerging. Non-fiction authors were writing for magazines such as *Saturday Night* and *Maclean's*. The trade books that were appearing tended to derive from those sources (in the 1960s, books by Pierre Berton, for example) or from a social class or institutional connection of some sort, such as the memoirs of community leaders. Informal reviews of bestseller lists of the time indicate that in 1958, about half of bestselling fiction titles were set in Canada. In 1962, in fiction, only three Canadian authors appear, two writing about Canada and one writing about an American in Rome. In 1970 one Canadian — Hugh MacLennan — appears, with a historical novel. Only McClelland & Stewart (M&S) and Macmillan of Canada dot the lists as publishers. In non-fiction, 1958 brought Churchill's *The Great Democracies* and William Whyte's *The Organization Man*, while also bringing three Canadian writers forward. In 1962, six Canadian authors made an amalgamated bestseller list, with history and memoirs being predominant. Here we see much more frequent participation by M&S, Macmillan, Copp Clark, Ryerson, and Thomas Allen, all Canadian firms. In a list from 1970, at the beginning of the new wave of publishers, nearly half are Canadian titles published by M&S and the new publishers.

A collapse of domestic publishing amidst a cultural rebirth

If the above titles reflect the reading habits of the time — and they seem to, to some extent — then it is hardly surprising that few publishers chose to focus on Canadian culture. McClelland & Stewart, the University of Toronto Press, Clarke Irwin, and Ryerson Press were notable exceptions, but none of them was entirely financially successful. By 1970, Ryerson was costing the United Church of Canada half a million dollars a year.[87] To much general consternation, the church sold the company to McGraw-Hill, a U.S.-based publisher, which named its new acquisition McGraw-Hill Ryerson and positioned it closer to market demand, abandoning its Canadian trade program to concentrate on educational publishing. In the same year, the owners of Toronto-based educational publisher W.J. Gage sold the company to U.S. educational publisher Scott Foresman (Gage had published Scott Foresman's *Dick and Jane* series in Canada). In response to public consternation of the same kind that had greeted the sale of Ryerson, Scott Foresman put Canadian Ron Besse in charge of the company, and Besse later emerged as the on-paper owner of Gage — with financing provided by Scott Foresman. In contrast, McClelland & Stewart presented itself as an aggressive proponent of Canadian culture, albeit one that needed financial assistance. Clarke Irwin branched out from educational to trade publishing and managed to maintain itself for over a decade. The University of Toronto Press carried on with its scholarly titles and journals, but was constrained considerably by a limited market.

Outside the publishing milieu, the post–World War II years saw the beginnings of a cultural and political resurgence in the context of a vibrant and growing economy that was leaving a decade of rather frightening politics behind. Amidst this market vibrancy, the Massey Commission defined a cultural project to bring Canada into full nationhood. It laid foundations for development in Canadian arts, letters, and sciences that served as a blueprint for action and further policy development for the latter half of the twentieth century. The perspective that the Massey Commission articulated was not unique in Canadian history. It echoed the nationalist sentiments that existed in book publishing in the 1840s and at the turn of the century, when Canadians became preoccupied with affirmations of nationhood. The light of the Massey Commission shone rather faintly onto book publishing, but it was sufficient to be magnified by the Ontario

Royal Commission on Book Publishing, twenty years later, into a cultural reading of book publishing that has persisted. Like the Massey Commission, the Ontario Royal Commission laid out principles and a concrete agenda for government actions that were ever more powerful and effective. The support thus made available was seized upon by individual firms to develop the publishing industry we have today.

CHAPTER 3

Establishing a book publishing industry: from the 1960s to the 1990s

Introduction 79

Landmarks of Book Publishing Policy 82

Guaranteeing Canadians a voice: The Ontario Royal Commission on Book Publishing 84

    The reports of the commission 86
    General recommendations 90
    Marketing 91
    Educational publishing 92
    Government publishing 96
    Copyright 96
    Summary 97

From inquiry to direct action 98

The industry responds 101

Insistent industry demands I: A 1980s stocktaking and strategy 101

Insistent industry demands II: An assessment 104

Baie-Comeau in a political context 107

***Vital Links*: Tilting against globalization in a time of free trade** 109

Insistent industry demands and government studies at the turn of the decade 110

Government and book publishing to the early 1990s 115

A penultimate ACP policy paper 116

A policy paper to close the century 118

A discontinuous policy history 1952-98 119

Introduction

Beginning in the late 1960s and continuing into the 1990s, authors, book publishers, some booksellers, the federal and some provincial governments, and some members of the book community mounted a concerted effort to establish a domestically owned and controlled book publishing industry. The purpose of such an industry was to serve Canadian readers, and society in general, by providing publishing opportunities for Canadian authors and room in the marketplace for books that addressed Canadian concerns. In the late '60s and early '70s, a host of new firms were founded, and in 1971 the Ontario Royal Commission on Book Publishing laid essential groundwork for the development of Canadian book publishing and government support programs.

The Royal Commission established the primacy of the cultural (rather than economic) contribution of the industry. It espoused the need for cultural support, which targeted genres (e.g., poetry, fiction, biography), as well as business support. And the commission's report made public the vast array of variables that contribute to the flourishing of a book publishing industry in any country. The report of the commission had a dual focus on Ontario, as the primary site of publishing in English, and on the national scene. Arising from the Royal Commission were a succession of federal and provincial government support programs that began as extensions to arts funding, evolved into industrial support, and were complemented by structural interventions such as changes to the Copyright Act. On a foundation of industry determination and public demand, government support has nurtured a stable, if marginally profitable, industry that has encouraged a national discourse and allowed many Canadian authors to flourish. Most importantly, both the federal and provincial governments have provided this support in a manner that has not interfered with the creativity and freedom of expression of both authors and publishers.

The importance of the establishment of a Canadian-owned and controlled industry comes from its desire and ability to provide talented Canadian writers with public exposure and an abundance of opportunity to write about what they know — in their minds, in their bodies and emotions, and in their very being. That opportunity has imbued Canadian authors with the self-confidence to develop their talent and achieve literary success. Just like the Wayne Gretzkys and Sidney Crosbys of hockey, and the "own the podium" investment in Canada's athletes in preparation for the 2010 Winter Olympics in Vancouver, the publishing industry has incubated a generation of writers that has been given sufficient opportunity to achieve international recognition, and a second generation is coming up. As indicated in the introduction, the matter is not just one of printing, binding, and distribution. Among other roles book publishing plays, even after it has brought a title to market, is the provision of exposure to the targeted publics, to critics, to the media, to libraries, and to expert readers on awards juries.

The key decision that the Ontario Royal Commission, helped by the Massey Commission, carved into Canadian policy was this. Compared to many larger countries with developed economies around the world, Canada has a national domestic market that is insufficient to warrant publication of new literary voices. But rather than foreclosing on literary opportunity, Canadian governments have put in place cultural subsidies, in the form of financial support for both writing and publishing. Through these subsidies, the Canada Council for the Arts and provincial arts councils help cover the deficits publishers incur by producing literary titles. However, cultural subsidies provide publishers with sufficient welfare only to live for another day. They do not allow publishers to retain earnings and reinvest those earnings in the development and expansion of their firms. To complement cultural subsidies, Canadian governments put in place a second category of assistance: industrial support programs. The original intent of these programs was to help firms establish themselves in the market and gradually become independent. What industry and government both discovered was that because of Canada's open market, the price brackets for books are established by the print runs of U.S. publishers with the result that Canadian book publishers are at a continuing competitive disadvantage for their main stock in trade: original titles directed primarily at the Canadian market. The need to amortize origination costs over a Canadian-only print run, combined with a dominant price range established by imported books, precludes the possibility of normal profits; hence there is a need for continuing industrial support.

The type of policy that could have changed this inherent competitive disadvantage is powerful structural intervention. Structural intervention is effective for three reasons: it changes the rules within which industries operate; after a short period it becomes invisible; and it does not draw on financial subventions from government. This chapter introduces, and Chapter 5 extends, the discussion of a structural intervention that assists Canada's book publishing industry. Mainly, that structural intervention has been the inclusion of the distribution right included in Canada's Copyright Act, an intervention that helped address, but did not neutralize, the competitive disadvantage of Canadian book publishers. The federal government's more recent considerable investment in technology to serve the industry can be seen in the short term as industrial support, but in the long term, it is also a structural intervention.

Three additional points may help readers keep their bearings in this chapter: From the 1970s to the present day, Canadian-owned book publishers, recognizing the cultural value of their industry, have been committed to keeping it heterogeneous in terms of firm size, location, and genre orientation. Such a commitment works admirably in the pursuit of cultural goals. At the same time, it works against concentration of ownership. An emphasis on profitability, which inevitably leads to ownership concentration, exerts pressure away from cultural expression, favouring firm growth that results in economies of scale sufficient to generate profitability and further financial growth. As we will see, these two approaches to doing business — cultural development versus financial profit — have been in continual tension. As we will also see in Chapter 4, the emergence of the notions of social capital and the creative economy have decreased this tension. A second point: it is useful to view the various government policies and programs detailed in this chapter as inventive ways of supplying needed financial support to book publishers. With this bottom line in mind, the reader may gain a clearer sense of the intent of these government efforts. Third: A major identifiable trend in government support has been a gradual acceptance that book publishers cannot attain self-sufficiency without abandoning cultural value, and that the whole *raison d'être* for the industry was, and is, to create cultural value. Over time — that is, from the 1970s to 2000 — successive governments came to understand that while publishers accepted financial realities, they did not embrace profit and market share as the primary goal of book publishing. In turn, governments gradually changed their policies so that publishers were increasingly able to seek support for their own publishing priorities, rather than for business priorities or those of the funding agency.

This chapter deals primarily with the development of cultural subsidies and industrial support from the 1970s to the late 1990s, while Chapter 4 focuses more on structural intervention and the emergence of the notion of cultural industries as part of the creative economy. To capture the nature of the development of industry and policy over thirty years, this chapter identifies the major issues and actions of government and industry by reviewing major government inquiries, policies, and support programs; industry policy papers; and the sociopolitical context.

There is considerable detail in both chapters; the following list of landmarks may assist the reader in tracking significant developments.

Landmarks of Book Publishing Policy

Chapter 3

1. 1971: Ontario Royal Commission articulates the importance of book publishing as cultural.
2. 1971: Ontario government provides the first business support (loans) for cultural production (without evaluation of titles published).
3. 1972: Funds are provided to the Canada Council for cultural subsidies, i.e., support for titles, and groups of titles, through an arts agency operating at arm's length from government.
4. 1972: Association for the Export of Canadian Books is established to assist publishers in developing exports markets.

5. 1974: Foreign Investment Review Act restricts foreign ownership (reaffirmed in the Baie-Comeau policy [1985]).
6. 1975: Cooperative promotion and distribution program is established.
7. 1977: First public support for celebration/awards program is established — National Book Week.
8. 1979: Book Publishing Development Program, a full industrial support program, is created. In 1986, it is redesigned and renamed, becoming the Book Publishing Industry Development Program. BPIDP established a workable, stable regime for financial assistance to Canadian publishers, primarily for title origination. It remains in force (as of 2012), but has been renamed the Canada Book Fund.
9. 1987: the Free Trade Agreement is passed with a cultural industries exemption.

Chapter 4

10. 1994: Ontario government releases *The Business of Culture*, emphasizing a market participation model.
11. 1997: Canadian Copyright Act is rewritten and includes the distribution right.
12. 1999: Federal government begins efforts to create a separate international agreement and, hence, trading regime for cultural products without final economic arbiters (as in FTA and NAFTA).
13. 2001: Universal Declaration on Cultural Diversity is created and signed.
14. 2002: BookNet Canada is established to provide title-based sales information on a weekly basis. The aim is to provide market information to all participants, to increase distribution efficiencies and reduce returns.
15. 2005: The Convention on Cultural Diversity establishes that signatory nations must take into account cultural diversity in the signing of any other international trading agreement.

Guaranteeing Canadians a voice: The Ontario Royal Commission on Book Publishing

In the absence of specific recommendations for book publishing from the Massey Commission, culture building in writing and book publishing remained minimal between 1952 and 1967. Existing Canadian publishers such as Copp Clark; Clarke Irwin; and McClelland & Stewart (M&S), which had grown on the foundations of their agency businesses, plied their trade largely by selling imported books spanning education, scholarship, and leisure reading. Within this business was a ticking time bomb. By monitoring the sales of such Canadian agents, some U.S. publishers recognized a growing market for their books and, in response, cancelled agency agreements and set up branch plants to serve the Canadian market directly. This trend was especially strong among educational publishers.

Yet the cultural project outlined by Massey and his fellow commissioners slowly took hold. It did so as the biggest population bubble in Canadian history, the postwar baby boom, proceeded through its childhood and adolescence. It also did so amidst a stark absence of opportunities for Canadian authors to write and be published in Canada, or to write on Canadian subjects. During that era, Jack McClelland was one of the few publishers offering significant numbers of publishing opportunities to Canadian writers. Indeed, on the foundation of M&S's agency business, McClelland became a significant force in bringing enticing new authors into print, who demonstrated creativity every bit the equal of their foreign counterparts.

Beginning in the mid-1960s, new firms were established that were not interested in building their business by importing and distributing foreign titles. Publishers such as Coach House, House of Anansi, New Press, and Peter Martin were founded and staffed mainly by young authors determined to have their voices heard without compromising their ideas or the form of the businesses they were setting up. A changing consciousness, expressed in the writing of such authors and reminiscent of the dynamics of the Bloomsbury Group, began to emerge. For example, Coach House's 1967 motto was "Printed in Canada by mindless acid freaks," and its books combined the unexpected designs of Stan Bevington with visionary poetics, such as *LSD Leacock* by Joe Rosenblatt.[88] House of Anansi embraced a political as well as creative agenda, focusing on new poets and prose writers such as Dennis Lee, Margaret Atwood, and Michael Ondaatje, and topics that included a manual for draft dodgers from the United States. One of Anansi's founders, Dave Godfrey, went on to found New Press with James Bacque and Roy MacSkimming as partners. Their initial title, *The Struggle for Canadian Universities* by Robin Mathews and James Steele, which documented the inundation of Canadian campuses by foreign — mostly American — professors and the lack of Canadian content throughout the curriculum, was a huge market and political success that successfully pressured the Association of Universities and Colleges of Canada to set up a commission of inquiry into the Canadian content in post-secondary curricula. Subsequent titles dealt with aboriginal rights and environmentalism, giving David Suzuki his first shot at fame. Peter and Carol Martin initiated a Canadian book club that gave Canadian titles wider circulation to Canadian readers and competed with U.S. and British book clubs that offered few, if any, Canadian writers to their Canadian members. The Martins also worked tirelessly at collaborative marketing projects, and delved into Canadian publishing history for some of their titles. Underlying all these efforts was a vigorous and assertive nationalism.

The demonstrable creativity of the publications produced by these authors and publishers, and their constant insistence on their right to publish, disrupted the gentleman's gatekeeping business of Canadian book publishing. Publishing became a site of social struggle that pitted idealistic young Canadians against a cultural *status quo* that was under attack in this specific industry and in the general national milieu. Equally disruptive, for much the same reason, was the sale, introduced in Chapter 2, of two venerable book publishers, Ryerson and Gage, to U.S. firms. The young publishers weighed in, with imagination and without respite, on the side of nationalism, banding together to form the Emergency Committee of Canadian Publishers (later the Independent Publishers Association and eventually the Association of Canadian Publishers), as the sale of Gage and Ryerson tipped the scales of tolerance of foreign investment in the industry and led to the creation of the Ontario Royal Commission on Book Publishing.

On December 23, 1970, the Ontario provincial government named Richard Rohmer (Canada's most militarily decorated citizen, a lawyer and, subsequently, author of over twenty books), Dalton Camp (Conservative party strategist and political commentator), and Marsh Jeanneret (University of Toronto Press publisher from 1953 to 1977) to the Royal Commission and charged them with conducting an examination of

(a) the publishing industry in Ontario and throughout Canada with respect to its position within the business community;

(b) the functions of the publishing industry in terms of its contribution to the cultural life and education of the people of the Province of Ontario and Canada;

(c) the economic, cultural or other consequences for the people of Ontario and of Canada of the substantial ownership or control of publishing firms by foreign or foreign-owned or foreign-controlled corporations or by non-Canadians.[89]

Both Ontario and Canada were included in points (a), (b), and (c), for it was clear in the minds of those who commissioned the report and of the commissioners themselves that they were to address national as well as provincial realities. Many of the recommendations made in the report were eventually followed, some by the federal government and others by the Ontario government. A summary of the main points made by the Royal Commission and the other main policy documents discussed in this and following chapters is presented in the Appendix.

The reports of the commission

No sooner had the commission been created than it was faced with another falling domino, alongside Gage and Ryerson. In February 1971, Jack McClelland announced his intention to sell his company, McClelland & Stewart (M&S) — known, by then, as the foremost Canadian publisher of fiction written by Canadians — to the highest bidder. On March 12, 1971, the commission received the firm's financial statements and set them before Clarkson Gordon and Company, a reputable firm of chartered accountants with known nationalist sympathies. Having done their due diligence in three days, on March 15 the accountants reported to the commission that the company was vulnerable to bankruptcy and needed immediate access to approximately $1 million in working capital. The commission recommended that the government give M&S access to that amount, representing one-third of the company's assets, through the Ontario Development Corporation. The government came forward with the loan, which was eventually forgiven.[90]

The commission's final report recommended extending the special treatment given M&S to other Canadian-owned, Ontario-based book publishers. Recognizing that banks were loathe to extend loans to publishers (Section 88 of the Canadian Bank Act at the time gave printed books, most publishers' sole collateral, no more value than dirty paper), the commission recommended the creation of a program of government-guaranteed loans to provide working capital for Canadian-owned, Ontario-based book publishers. (Generally speaking, book publishers need to use approximately 37 percent of their annual sales to finance a continuing program of new titles.[91]) The commission recognized that such actions could reduce the value of book publishers to foreign buyers, who would lose access to such loans. But the point was to build a Canadian-owned industry, and so government assistance came at the price of a financial disadvantage were a publisher tempted to sell to a foreign buyer. To emphasize its intent, the commission advised that sales of book publishers to foreign buyers be expressly discouraged, if not forbidden.

The cultural core of the commission's perspective can be found throughout the seventy recommendations in its final report, but particularly in Chapter 6, "Nurturing a Canadian Identity." The text of the report indicates that the commissioners started from the belief that the importance of book publishing was cultural, with the corollary that book publishing contributed to the articulation of culture by providing Canadian authors with access to Canadian audiences. The commissioners heard and accepted evidence that the existing state of the book market strangled Canadian authors and books, imposing a form of censorship by impeding both production by

Canadian authors and consumption by Canadian audiences. Without Canadian book publishers bringing books on Canadian matters to market, any discourse in book form on Canadian culture was subject to the investment priorities of foreign businessmen. In other words, the commissioners as well as the young publishers saw book publishing not just as a site of struggle, but also as a foundation of Canadian culture and Canada itself.

More specifically, the commission was alarmed at the possibility that "a country's national consciousness [might sag] until its sense of cultural identity vanishes," with "the resulting vacuum . . . sooner or later . . . filled by a cultural idealism imported from somewhere else."[92] The commissioners saw the need to provide opportunities for the creation of books that focused on key ideas and elements of Canada. They also saw the need to stimulate Canadian unity and diversity, an orientation that was both a departure from, and an allusion to, the social goals found in Section C of Canada's Broadcasting Act of 1968, where the words "unity and identity" are used.[93] Books could contribute to this end by helping all Canadians to understand Canadian culture and its mosaic of values and people.

The commissioners saw a need for assistance, not only for publishing in general, but also for specific enterprises aimed at developing Canadian school textbooks, scholarly publishing, review publications, and literary criticism, with the latter two serving as both marketing devices and means for authors to understand the reception of their books among expert readers. The commissioners acknowledged the distinction between books and broadcasting, but argued that, in parallel with the aims of the Broadcasting Act, there was a need to intervene in the marketplace to ensure that books would be created that would appeal to Canadians in all parts of the country, of all ages and backgrounds, and of all tastes.

Commissioners Rohmer, Camp, and Jeanneret also took note of practices used by other countries to promote arts and culture, specifically mentioning the United States, Japan, the United Kingdom, France, the Soviet Union, and Sweden. They saw the position of books as similar to that of magazines, noting that the Royal Commission on Publications, which had reported eleven years previously, had recommended against allowing split runs of magazines — that is, the foreign magazine publishers' practice of interrupting a press run to insert ads directly targeted at Canadians, but changing little of the editorial content. On the matter of Canadian or foreign manufacture (i.e., printing), the commissioners argued that each sector of the industry should not be indifferent to the health of the others and noted that each was critical to the health of the overall enterprise; they therefore recommended that support to publishers be contingent on printing in Canada.

In assessing the impact of the Royal Commission, it is easy to overlook a point of considerable significance. No one from the commission examined McClelland & Stewart's catalogues to determine the salesworthiness of its titles or to judge their cultural contribution. The firm's public reputation was taken as an indication that the list was worthy. The matter was simply that a *bona fide* trade book publishing company, led by a champion of Canadian authors and literature, was threatened. Given that it was a private business, and perhaps because the government did not want to be seen as subsidizing the books themselves, a loan was issued with no mention of the long-term viability of the company, nor of its publishing program. Subsequent actions made by the government in light of the commission's report extended this *prima facie* acceptance of the legitimacy of existing book publishers. For instance, the 1972 loan guarantee program, called the Ontario Book Publishers' Assistance Program, which was a direct extension of financial support given to M&S, was extended without regard to content. Neither the commission nor the program administrators ventured into an examination of titles. They defined the issue as one of assisting publishers to carry on a stable business.

The commission made a number of overarching general recommendations, followed by a more focused set covering marketing, educational publishing, government publishing, and copyright. A discussion of all seventy recommendations would court tedium and take up far too much space, but many are worth reviewing for several reasons: a very high percentage of them came into effect, either soon after the report's release or eventually, and they became a foundation for policy. They also set the stage for future reports and policy papers.

And they also illustrate not only the commission's micro-managerial perspective, but also its determination to wrest control of the education market from the foreign-owned sector.

To provide readers with an appreciation of the encyclopedic nature of the report, the following subsections explain the nature of the recommendations and describe their implementation or impact. This allows a review of both the nature and the importance of the commission. A few specific recommendations that are not germane to the argument and topic of this book are excluded. The exact formulation of the recommendations, and, indeed, the problems that the commission saw and addressed, are in the report of the commission. A summary of all recommendations is to be found in Chapter 7 of the report.[94]

General recommendations

So extensive were the needs of the industry, and so particular to the industry were these needs, that the commissioners began their general recommendations with a call for the Ontario government to establish a Book Publishing Board that would oversee the development and maintenance of the industry, in the interest of all Canadians. (The report was silent on the inherent contradiction of a provincial body acting on behalf of all Canadians.) The proposed board would address Canadian authorship and the financial foundations and competitiveness of the industry. With the notion of an overseeing board in place, the commission proposed the following: establishing title grants (subsidies based on the number of titles published, also tied to Canadian manufacture) and a royalties insurance scheme for authors; providing export sales assistance; restricting ownership of existing and new publishing firms to Canadians; and encouraging the development of scholarly publishing. To aid market development, the commission suggested publishers bring out-of-print titles back into circulation to increase audience awareness of Canadian writing and publishing. It also called for governments to fund literary awards, display Canadian books in London (U.K.), and fund a review journal of Canadian books. Believing that people interested in literature were untrained for and not focused on business, the commission recommended training programs for publishers, along with professional business consulting services. The three key areas — finance, marketing, and training — along with technology and distribution would prove to be key categories of government attention over the subsequent three decades.

Over time, the thrust of virtually all the general recommendations was implemented. The royalties insurance scheme was implemented indirectly. Publishers receiving grants were required to be up to date in their royalty payments.

Marketing

Marketing was addressed further in fifteen more specific recommendations. Six suggested ways to increase awareness and acceptance of Canadian titles among librarians, educators, and the media. Their reception? The recalcitrant librarians, whose received notions of literary quality carried overtones of British imperialism and class consciousness, were finally won over by Francess Halpenny, a leading librarianship educator, who articulated the notion of a "Canadian collection" — a phrase that librarians could understand.[95] While educators continued to resist "pandering to Canadian titles," a modicum of acceptance developed. For instance, from the mid-1970s to the early 1990s, libraries in British Columbia could apply to a school library book purchase plan, instituted by the provincial government, for a set of recently published Canadian trade titles. In the early 1980s, a Canadian Learning Materials Centre — at first with offices in several different cities and eventually based in Halifax — increased awareness of Canadian-authored and -published titles among teachers and other educators, marketed such titles, and lobbied to expand the number and types of firms eligible to submit materials that would be considered for course adoption. As of 2011, there was certainly substantially greater Canadian content in most provincial curricula, even though, as noted in Chapter 4, the production of Canadian school textbooks remained mostly in the hands of large Canadian-based, foreign-owned textbook publishers. Also, in general, as Chapter 5 reveals, the Canadian media have responded well in helping to promote books, fostering an environment open to publicizing books in newspapers, on television and radio, and through awards and festivals. Educational television has also played a role, as have the CBC and private talk-radio programs.

In two recommendations, the commissioners called for industry data monitoring to measure the characteristics of the industry and the nature of change. This activity has emerged in three forms. From 1973 forward, Statistics Canada carried out an annual statistical survey of book publishing. However, in recent years it dropped exclusive distributors from its survey, thereby introducing some major difficulties in understanding the size and nature of the industry (discussed in the Introduction). Since 2004, the Department of Canadian Heritage (DCH) has been monitoring the performance of its clients — the vast majority of Canadian-owned publishers — in a document called *Publishing Measures*.[96] As well, BookNet Canada (see Chapter 5) and Book Manager, a software system sold to independent booksellers, monitor sales of titles through bookstores. The recommended market research studies have been commissioned from time to time, and some have served the industry well.[97]

The commission's proposal that government provide assistance to build exports led to the federal government establishing the Association for the Export of Canadian Books, which assists with the development and maintenance of international sales. The commission's recommendations on distribution and fulfillment resulted in a number of ventures, including the Publications Distribution Assistance Program (PDAP) and, perhaps most successfully, BookNet and the Supply Chain Initiative. The system in place in 2010 was effective and stable, although the major irritant of far too many books being returned by retailers remained (see Chapter 6 for a fuller exploration of returns).

Educational publishing

The eighteen recommendations dealing with educational publishing addressed the following: the competitive position of Canadian-owned publishers; the need for training at all levels to sensitize educators to the need for the inclusion of Canadian social realities in Canadian education materials; the use of appropriate non-fiction titles; the provincial balkanization of the education market (which diminished the influence of educators on learning materials); and the need for more interaction among educators, Canadian-owned publishers, and Canadian authors. This last recommendation encouraged communication, so that authors would understand the market forces acting on publishers, and publishers and educators might shift the study of literature to focus on its creation, rather than on the products of notable foreign writers.

The response to the various and sundry educational publishing proposals was that, in due course, various provinces followed the commission's advice and allowed officials from their ministries of education to participate in the development of school materials, with the intent of increasing the learning materials' relevancy to students. Involving education ministry bureaucrats was not entirely successful in producing learning materials that teachers favoured, although it certainly helped some Canadian-owned trade publishers gain a temporary foothold in the educational market.[98] As the commission suggested, the Ontario Institute for Studies in Education became more involved in the development and evaluation of educational materials, but this did not have the intended impact of increasing the desired cultural or Canadian content of learning materials. This appears to be because educational theory, which develops in an international context, severely and fundamentally undervalues group or cultural identity.[99]

Periodic reviews of the curricula of teacher training programs have shown that there are few courses in materials evaluation that focus on the cultural identity, and hence socialization needs, of students, in spite of the commission's advocating such training. For a time, Ontario, British Columbia, Newfoundland, and the Maritime provinces extended research and development assistance to Canadian-owned publishers for textbook projects, with federal support from the Canadian Studies program of the Department of Canadian Heritage (and its predecessors). The Canadian Studies program still exists,[100] but the latest funding disbursed was for 2008–9 and the level was relatively small. The products of the multinational educational book publishers remain dominant.

Various surveys have examined the development of school textbooks and the degree to which national and cultural realities were prioritized in teacher training programs.[101] In general, these surveys, undertaken to determine why cultural context was so lacking in Canadian learning materials, found that Canadian social, cultural, and literary realities were simply not considered when textbooks and training programs were developed.

Since the 1970s and even the 1980s, partly as a result of the commission's report, there has been increasing support for Canadian writing and book publishing in government, in the media, and in society in general, as well as widespread support for other cultural industries. Canadian content has increased across the curriculum, and the racial, ethnic, and gender plurality of Canada is now better reflected in school materials published by multinational firms. All educational publishers have made efforts to encourage the use of Canadian-written and -published materials in schools, although one must be careful when reviewing statistics: educational publishers (the vast majority of which are multinational companies) report a 90-plus percentage of Canadian authorship of learning materials to Statistics Canada, but these authors are often contract employees of the publishing companies; they play a relatively minor role, selecting content that is then massaged into the form that the company deems appropriate.[102]

The strength of the preference for Canadian-developed materials on the Ontario Ministry of Education's Circular 14 (a Ministry-approved list of learning materials, now called the Trillium List, that still today defines what can be purchased for each course) has been strengthened. And programs to encourage development of materials for Native peoples and Franco-Ontarians have also been put in place.

To summarize the impact as well as the intent of the commission's effort: A rich literary resource remains untapped in Canadian school education. The need for educators to introduce Canada to immigrant students, who arrive from a wide spectrum of countries and often do not have English or French as a first language, has become more apparent to Canadian educators in schools, faculties of education, and government education ministries. However, in general, Canadian contemporary fiction and non-fiction reflective of Canadian realities and imaginings do not find their way into schools or, for that matter, college and university classrooms.

In retrospect, as later developments in book publishing indicate, the commission underestimated the degree to which a separation between trade and educational publishing is endemic to book publishing in Canada and, for that matter, around the world. The thrust of the commission's thought was to transfer market dominance from the hands of multinational companies, who appeared to be unconcerned with schools as community- and nation-building institutions, and place it in the hands of Canadian-owned publishers, for whom nation and culture were a priority. As explained in Chapter 2, the separation between the two sectors stems from the historic dominance of the import-and-distribution structure and, later, the branch-plant structure, over a rights-trading structure in Canada. Since 1972, the differences and distance between trade and educational publishing have increased. Rather than bringing human creativity to market, as trade publishers do, educational publishers produce learning materials to specifications. While trade publishers receive unsolicited submissions, or commission books by known authors and rely upon the author to bring his or her voice and vision to the work, the development of learning materials demands much more extensive involvement. Educational publishers are continuously reading the market, which reflects social values, disciplinary content, and educational theory. They import the development and management systems surrounding textbooks from their parent companies. Even with those systems in place, they can spend several million dollars developing and marketing their materials, whether for a multi-grade language arts or social studies series, or for peripheral subjects such as French as an additional language,[103] all the while keeping the preferences of various provinces in mind. They also provide ancillary teaching materials and post-publication marketing and assistance to ensure sales, and there is still a chance they will fail. In this framework, authors (and even teacher editor/compilers) play a relatively small role, which is why they are contracted rather than given a percentage royalty. The capital required for such an enterprise is, obviously, considerable, and even the largest educational publishers have consolidated in order to survive. As of 2011, three dominated the Canadian educational market — Pearson, Cengage (Nelson), and McGraw-Hill Ryerson — while others, such as Wiley-Blackwell, Oxford University Press (Canada), Simon & Schuster, and Scholastic, specialized in particular subject areas. Educational publishing is a rich company's game, a game that is very much market driven. Originality and creativity exist, but they exist within an educational context, not, as in trade publishing, with authorial vision.[104]

Government publishing

The commission made three recommendations aimed at dissuading the Ontario and federal governments from self-publishing and thus competing with trade publishers in the marketplace. By and large, governments pulled back from publishing for public consumption, but there are still situations where governments and other public agencies publish informational works for niche markets that they can better serve in this fashion than by collaboration with book publishers. For example, Canada's International Development Research Agency has an extensive publishing program that has embraced online publishing in an effort to get its valuable content into the hands of those who need it most.[105]

Copyright

In contrast to the brevity of its coverage of government publishing, the commission's report contains nineteen recommendations with respect to copyright. They were designed to strengthen the position of Canadian-owned publishers and improve the financial rewards to Canadian writers. The recommendations are now dated, but their significance is that the commission set a direction that has guided actions and development in the decades since.

At the commission's behest, the government bolstered the ability of Canadian publishers to purchase territorial rights by setting down the principle that all changes to the Copyright Act should reinforce this ability and, specifically, by calling for the strengthening of laws protective of works for which there were Canadian editions. In the

1990s, to bolster import control, the federal government included distribution rights in the Copyright Act to help ensure that books for which Canadian publishers/distributors had distribution or publication rights were not imported by others into the country — a move that also enhanced the protection of Canadian editions. Importation of single copies was exempted, to guard against government interference in the flow of ideas.

A public lending right (PLR) was established to compensate authors for sales they might otherwise have expected if copies of their books were not available for borrowing from public libraries, and a publishers' and authors' rights collective (Access Copyright) was established to collect payment for the photocopying of copyrighted works beyond what is allowed under fair dealing provisions. Consistent with the commission's opening principle on copyright reform, to protect Canadian publishers and distributors, Canada has refrained from signing international agreements that would allow an unimpeded flow of books across our borders. Finally, as recommended, Canada is a full participant in international agreements on copyright, and all elements of copyright practice are defined within the Copyright Act, rather than in a range of Canadian statutes.

Summary

Sweeping in scope, the many recommendations of the Ontario Royal Commission addressed elements within the political, cultural, economic, legal, educational, and technological environments that affect the establishment and operation of a book publishing industry (outlined in Chapter 1). That Canada embraced the task of creating the necessary infrastructure within which a book industry could establish itself, rather than dismissing the task as unachievable, is quite remarkable.

In general, with the exception of educational publishing, where its recommendations to increase Canadian context and cultural content were fulfilled in spirit rather than substance, the Ontario Royal Commission on Book Publishing succeeded in setting an agenda for policy and program development in English-speaking Canada, and complemented developments in Quebec. The thrust of that policy and program development, as well as the general development of the book publishing industry in Canada, can be best understood by considering, as the remainder of this chapter does, several industry reviews and policy proposal documents generated by the federal government and by the Association of Canadian Publishers, the national association that represents Canadian-owned firms.

From inquiry to direct action

While official government inquiries are powerful, and royal commissions are the most powerful form of government inquiry, they do not always lead to immediate action, especially when they are provincial inquiries that call for federal as well as provincial involvement. Yet in the case of the Ontario Royal Commission on Book Publishing, action, perhaps triggered by the immediate bailout of M&S, was not long in coming. Shortly after the commission's report was published, the federal government began to take steps supportive of book publishing. In 1972, Secretary of State Gérard Pelletier recommended that cabinet spend $300,000 to establish a Canadian Publishing Development Corporation — the overseeing "Canadian Book Publishing Board" the commission had proposed. No doubt the funds were spent, but the initiative did not seem to result in anything of substance. Later that year, $1.7 million was allocated to the Canada Council to provide assistance over three years to "encourage the expansion of Canada's own publishing industry."[106] These funds allowed the Council to increase financial assistance to publishers through a program of block grants, which provide assistance for a group of titles (i.e., the publisher's program for the year) rather than for single titles. A book purchase program was also put in place, along with export marketing assistance and a co-publishing policy that encouraged government ministries to work with publishers rather than publish on their own. In 1975, a co-operative promotion and distribution program was added, and in 1977 the Canada Council received additional "national unity" funds for the book purchase program, a special fund for children's literature, translation assistance, and a National Book Week.[107] The Foreign Investment Review Act of 1974, which discouraged foreign ownership of book publishing firms, helped at least one firm expand its distribution of foreign books.

It is easy to miss the significance of these policies and the pattern of support that grew from them. They indicate, for the first time, a willingness of the governments of the day to extend arts and culture funding to assist for-profit business firms through an arts organization. While such cultural support has its drawbacks for building businesses, it allows government to keep its distance from the content of the publishers' programs by means of two arm's-length relationships. The first is the arm's-length relationship the Canadian government has with the Canada Council for the Arts. The second is the arm's-length relationship arts funding agencies, including the Canada Council, have with grant recipients through the use of juries of peers.

Extending arts funding to book publishers without linking it to judgement of content was an auspicious way to start the flow of funding aimed at facilitating ideas, creativity, and the cultural industries. Value judgements regarding the content, expression, and ideas contained in the books published were left to the marketplace, and evaluation of professionalism (in editing, layout, cover design, publicity, marketing, distribution) was left to publishers' peers. To determine half of each publisher's grant, the Canada Council used a formula based on the average deficit incurred by all publishers active in the genre. The other half of the grant was based on comparative professionalism, judged by the publishers' peers. In confirmation of the wisdom of this arm's-length approach for book publishers, in 1973–74 the Ontario government increased the Ontario Arts Council's budget from $50,000 to $340,000, with the understanding that some of the funds were destined for book publishers.[108] (These funds were in addition to the loan program, based on the M&S bailout, that was extended to other Ontario book publishers.) As time passed, the principle of supporting publishing activity for its ability to make the creative writing of Canadian authors available to Canadian readers became entrenched, as other provinces adopted the Canada Council model for assessing applications and followed Ontario's lead.

While arm's-length, culturally oriented funding to reduce incurred deficits had its advantages in privileging cultural production over sales, both the publishers and governments knew that such funding could never be enough if the industry was to grow and earn normal profits. Only if a book publisher outperformed its competitors year after year would there be a chance that such subsidies would result in a profit.[109] If Canada expected to have a viable, financially independent book publishing industry, the possibility of profit had to exist.

In 1979, the federal government took its first significant step beyond arts subsidies and created the Canadian Book Publishing Development Program (CBPDP; also abbreviated BPDP in many documents) to provide industrial assistance for the financial and economic base of the Canadian-controlled sector. The CBPDP targeted marketing, project research and implementation, company analysis and implementation, foreign rights marketing, and professional development. In addition, it funded associations that carried out data analysis for the industry. These initiatives appeared to be motivated by the need to reduce the peripheral costs of doing business, so that book publishers could turn a profit on their core business of originating new titles. Unfortunately for the industry as a whole, the CBPDP succeeded only in averting a major financial hemorrhage in the industry during a time of high inflation — at one point, interest on loans soared to over 20 percent.

As it turned out, in the eyes of an industry composed of many small publishers, the program was not well designed. Five large publishers with total sales of $6 million received 61 percent of the support, approximately $1.5 million of the CBPDP funds. In addition, when cultural-industries support programs moved from the Secretary of State to the Department of Communication, the fast-paced, highly competitive, market-capturing technological environment in telecommunications seemed to seize the imagination of Francis Fox, the Communications minister of the day. The result? Fox mused about creating centres of excellence in publishing — that is to say, large, profitable, and stable firms. For an industry that saw its strength to be in its heterogeneity (in terms of size, location, and genre orientation), this was not a welcome message.

The industry responds

The early governmental responses to the needs of book publishers are evidence of the governments' willingness to assist what they, the Canadian public, and the book publishers saw as a strategic industry engaged in the project of advancing Canada as an independent centre of ideas and literary expression in the world of nations. In the wake of the Ontario Royal Commission, a series of government initiatives in Ottawa and Toronto, and also in Quebec, affirmed the efforts of the book publishers and kept them afloat. In hindsight, the programs announced and implemented by governments appear impressive, even if the publishers themselves struggled mightily to make ends meet. The long list of initiatives also should not camouflage the fact that the industry expended prodigious effort to ensure that these much-needed government policies and programs were in place.

Both before and after the Ontario Royal Commission, but much assisted by the supportive action defined as necessary by the commission's report, the locus of industry development and of the design of government policy and support programs appeared to shift from government to industry. Every five years from 1980 to 1995, the Association of Canadian Publishers (ACP) convened a policy forum for its members and invited government policy and program personnel to take stock of the industry and define priorities for its development. The policy forums were instrumental in setting priorities for lobbying and for pressuring government to create a benign rather than hostile business environment for book publishing. The industry-government interaction began with the most impressive of four policy papers generated by the Association of Canadian Publishers in 1980, 1985, 1990, and 1995, which were followed by an additional paper in 1997.

Insistent industry demands I:
A 1980s stocktaking and strategy

In 1980, Canadian-owned English-language book publishers and their organization, the ACP, detailed what they saw to be the industry's successes and its problems.[110] *Canadian Publishing: An Industrial Strategy for Its Preservation and Development in the Eighties* presented its subject as a cultural success story, noting that this was a Canadian-owned entrepreneurial industry run by numbers of small business owner-managers who were taking extraordinary risks for almost no financial reward. Moreover, the ACP claimed, these individuals possessed a set of financial, editorial, production, and marketing skills that allowed them to reach their audiences, and, increasingly, they were co-publishing and exporting titles and books. The priorities and behaviour of such an industry, the paper noted, were in distinct contrast to those of industries dominated by corporate concentration and foreign ownership, with their one-form-fits-all products.

In spite of the cultural achievements of the industry, the ACP emphasized that its member publishers were facing a substantial net loss before grants, essentially, it noted, because their costs as a percentage of revenue were higher than those of the foreign-owned sector, in both trade and educational publishing. Echoing elements of the royal commission's report, the paper identified what it saw as the key problems, beginning with limited subsidiary markets — that is, the ability to acquire and sell such rights as movie rights, translation rights, excerpt rights, and foreign territorial rights. The problems also included foreign competition and its associated competitive advantages, such as lower-risk importing and distribution of titles owing to the greater economies of scale associated with larger markets. The paper highlighted the fact that book production costs were rising, yet there was little opportunity, given foreign competition, to raise prices. It also noted increasing interest rates (at the time around 14 percent and rising), and mentioned the problem inherent in publishers' commitment to culturally valuable publishing that was also usually unprofitable. According to the ACP, these problems were exacerbated by the lack of a distribution right within the Copyright Act, inadequate distribution mechanisms, weakness in all traditional book markets, and an unfriendly banking system that limited access to working capital.

The ACP recommended a five-year strategy to address these structural weaknesses in the market using a shopping list of measures reflective of many of the same concerns identified in the Royal Commission's recommendations, notably financial underpinnings, competitiveness with imports and foreign-owned publishers, marketing and market development, distribution, and audience awareness. The one area not addressed was training and professionalism. The recommended steps included increasing grants and augmenting them with structural measures; establishing a government-funded, arm's-length mechanism to develop industrial support measures such as bank loans, subsidies, market development, and regulatory measures; making market development a priority; establishing an electronic information-based distribution system; encouraging gradual repatriation of the industry through foreign investment restrictions; and organizing related groups, such as writers, booksellers, librarians, teachers, school trustees, students, parents, Quebec publishers, and other cultural industries, in support of book publishers' major aims. It was a veritable call to arms.

Reviewing this strategy document in full provides insight into the industry of the early 1980s. The ACP asked for a great deal of government intervention without putting forward a plan for its members to address their market realities. It portrayed its members' projects and publishing programs, in principle, as praiseworthy cultural commitments and attributed their lack of viability to an ill-advised market structure resulting from insufficient or inadequate legislation.

As was the case with the Royal Commission recommendations, reading the full list of proposals advanced by the ACP reveals the extent to which many small factors stood in the way of Canadians having a financially independent, domestically owned and operated book publishing industry and the many different points in the publishing process at which there are constraints on the emergence of such an industry. The shortcomings of the document lay in the fact that while it lobbied in the interests of its members, it saw no need to craft or cast its proposals within the context of the interests of all Canadians. Like Charles Wilson, president of General Motors in the 1950s, who told the U.S. Senate that he believed "what was good for the country was good for General Motors and vice versa,"[III] the publishers were fully convinced that their needs were Canada's needs.

Overall, the ACP document reinforced and updated the Ontario Royal Commission and provided a foundation for understanding exactly how much needed to be done, in what areas, to establish a culturally focused, financially stable industry and create conditions that would allow Canadian publishers to compete with foreign publishers. The document's major achievement was to attract enhanced levels of support from the federal government through the 1980s. Increasingly, that support was directed at placing publishers on a firm financial footing. That is to say, it took the form of proposing industrial development.

Insistent industry demands II: An assessment

In 1984, the ACP met to evaluate how the industry had progressed since the CBPDP, the main support program it had negotiated, had been put in place in 1979.[112] In the report that followed the meeting, entitled *A Mid-Decade Assessment*, the ACP noted with approval the redesign of the CBPDP's sales-based grants, which had been contributing to corporate concentration led by profit-driven rather than culturally oriented publishers. Both the ownership concentration and the business orientation of the firms the grants had originally benefitted, notably Macmillan Canada, had been counteracted by changing the CBPDP's formulas and implementing size- and location-based corrective factors. The ACP believed the CBPDP now had the potential to address its industry's needs for growth. The report made little mention of the cultural grants of the Canada Council and the Aid to Scholarly Publishing Program, partly because they were much smaller in size. Underlying these statements of support for the redesigned CBPDP was the desire to advance the interests of the majority of the ACP members, as opposed to the interests of only a few larger, more powerful firms within the organization.

The report accepted the inevitability of an imports-based market rather than a rights-based market, and it suggested that Canadian-owned firms could capture a larger share of the income from such activities if the government established policies that would set targets of 10 percent increases in importing and distribution revenues for the Canadian-owned sector every five years, based on a starting point of 20 percent of the market. The report reiterated the call for a distribution right and for Canadian editions of foreign titles to have unchallenged rights in Canada. It also urged strengthened participation by Canadian-owned firms in all sectors of book publishing, and it requested that financial resources be made available to help Canadians — including the Canadian management of foreign-controlled companies and, perhaps, ambitious ACP members — to acquire the Canadian divisions of these comparatively gargantuan foreign-owned multinational companies. The report supported the export marketing assistance being provided by the government-funded but industry-controlled Association for the Export of Canadian Books (AECB), noting that attendance at foreign rights fairs was a top priority, followed by assistance for marketing and promotion, trade missions, and international liaison.

The report recognized the need for robust electronic data interchange (EDI) to provide an electronic-based ordering and distribution system in Canada. It also recommended government procurement of books through public sources, specifically qualified retail outlets carrying a minimum of Canadian-published titles as a means to enhance wide availability, if not sales. (Such a model was, and is, in place in Quebec.) Independent bookstores were seen as important to literary houses, and an unrealistic tax rebate was proposed for retailers based on their sales of Canadian books. The report also expressed alarm over centralization in buying by bookstore chains, a pattern that has only increased in the years since the report.

Other recommendations included the commissioning of a purchasing study to reveal the profile of readers of Canadian-authored titles. Such a study was carried out in 1996 for the ACP[113] and a second study, commissioned by the Canadian Publishers' Council,[114] whose current roster of eighteen members includes eleven foreign-owned companies in addition to seven Canadian-owned companies, examined purchases of all books.[115] As well, the ACP recommended that researchers look into the competitive challenges facing regional publishers, the barriers to participation in various genre markets, and the effectiveness of financial and fiscal support measures.

Such recommendations, along with the report's criticism of the limitations on access to both components of the CBPDP — financial assistance based on sales, and innovative projects — suggested some tensions among members.[116] The recommendations also included the expansion of eligibility criteria to include tour and promotion money for children's book illustrators and non-fiction authors; funding for a travelling book fair; and the creation of financial incentives for aggressive marketing initiatives. All these recommendations showed an expanding concern with marketing and promotion. The recommendations for assistance for development and marketing in the educational sector reflected the strong set of recommendations of the Ontario Royal Commission, then eight years old. Finally, again reflecting the commission's recommendations, the report called for a more accessible national loan guarantee program and further suggested a program be established that would attract private equity, venture capital, and takeover financing.

While many of the recommendations were sound and were implemented, the industry was also lobbying for measures that, while culturally justifiable, were financially and organizationally unfeasible. The call for all firms to have access to all categories of funding, with no mention of financial and other business-based eligibility criteria, demonstrated a lack of appreciation of the need for a stable infrastructure at the firm level for innovation and firm development. Some of the recommendations in *A Mid-Decade Assessment* were also weakened by the ACP's excessive self-confidence — unrealistic, given the relative financial strength of its member firms — and the setting of unachievable goals. For example, the report called for Canadian-owned educational publishers to hold 90 percent of market share of the el-hi and post-secondary market (with 25 percent local participation) by 1994.

Unrealistic though some of the recommendations were, industry lobbying just before and at the time of the report's release helped convince the federal government to draw up the Baie-Comeau Agreement in 1985.[117] This agreement, which eventually found its way (in a somewhat weakened form) into paragraphs 20 and 21 of the Investment Canada Act, placed Canadian book publishing firms and other businesses, such as bookstores, off-limits to foreign buyers. The agreement (dubbed Baie-Comeau by the industry) also called for the patriation of foreign-owned subsidiaries operating in Canada when they changed ownership. Such forced divestiture ran so counter to the normal policies of capitalist economies that the measure was weakened under continuing threats, including that of a "scorched earth policy" by Gulf + Western in 1985.[118] (A "scorched earth policy" is a financial term derived from the wartime practice of evacuating land and leaving nothing of value behind for the invading forces. It refers to a company making itself an undesirable takeover target by selling off desirable assets or saddling itself with excessive debt. The rhetorical value of the term should not be ignored. A more common modern synonym is "poison pill.")

Baie-Comeau in a political context

In 1984, a Conservative government under Brian Mulroney replaced the Liberals, and Marcel Masse was appointed minister responsible for culture. Masse was also a member of the inner cabinet, the circle of ministers that dealt with policy and planning. Masse took the policy framework for supporting book publishing a bold step forward. Inspired by the French government's 1982 pronouncement that "economic progress must be subordinated to cultural goals, and culture must be recognized as a source of development and of progress," Masse announced in September 1985 that culture must be part of the mainstream of government, even as economics must be the foundation of an argument in support of culture. Emboldened also by the Ontario Royal Commission, which justified industry support to attain cultural ends, and lobbied intensively by the cultural sectors in both French and English Canada, Masse declared that he wished to see 50 percent Canadian control of the book publishing industry within two years. To bring this about, in 1986 he announced a new program, the Book Publishing Industry Development Program (BPIDP), and promised the unheard-of sum of $75 million for cultural funding for book publishers, to be spent over several years. By 1989, he had found another $110 million for direct grants to publishers; some of this money became available as a result of the winding down of the very rich but somewhat ineffective postal assistance program.

The BPIDP program was designed to provide industrial incentives balanced with cultural support; it was also an attempt to open up the provincial textbook markets.[119] This long-lasting program provided much of the support publishers needed, delivered in a form that was acceptable to the industry.[120] The significant innovation in the design of its financial assistance was that it was based on a set of accounting measures that took into consideration industry norms as well as sound business practice. Eligibility criteria included Canadian authorship and a minimum level of title output and sales. Support provided was sales based, but was weighted in favour of small, specialized firms and, hence, author and small community development.

If the primary framework for government support of cultural industries had not previously shifted from an arts-subsidies approach to industrial support — and it had certainly been delayed by the ravages of runaway inflation in the mid-1970s and early 1980s — then certainly the money that Masse brought to the table produced that shift in 1986. The viewpoint that emerged, which placed economics at the centre but culture at the core, set aside the marketplace perspective reflected in Francis Fox's musings on centres of excellence.

During the latter part of the 1980s and the early to mid-1990s, Canadian federal publishing policy and the industry itself struggled to survive expanding free trade policies that were gaining an ever-more-secure foothold in government, especially during the prime ministership of Brian Mulroney (1984–93). Support for the cultural industries, and specifically the development of a domestic book publishing industry, existed in direct contradiction of such policies. Nevertheless, government allocations to assist book publishing flourished and were consolidated from 1985 forward.

The 1985 Baie-Comeau Agreement was not insignificant. It sent an apparently strong signal that Canada was determined to protect its cultural industries to the point of defying received wisdom on what was acceptable in developed capitalist economies. Baie-Comeau's disallowance of further foreign investment in book publishing also benefitted at least one Canadian firm: It allowed Key Porter to gain temporary putative control of Doubleday in Canada.[121] However, other successful implementations of the policy did not follow, perhaps because the Canadian government was unprepared for the vehement reaction to the policy in the United States, or perhaps because it was more posturing than policy commitment. However, if the policy was posturing, it was determined posturing, because it was complemented by another government paper that was released in the months just prior to the October 1987 signing of the Canada–United States Free Trade Agreement (FTA).

Vital Links: Tilting against globalization in a time of free trade

It is interesting, if not a trifle bizarre, that in an aggressively pro-free-trade federal government, *Vital Links*, an explicitly anti-globalization document, was published in 1987 under the signature of the Minister of Communications, Flora MacDonald, just prior to the finalization of the FTA. Admittedly, the FTA kept culture off the table, and *Vital Links* could have been intended to contribute to that effort. But in contrast to the federal government's promotion of global free trade as the order of the day, *Vital Links* portrayed globalization, for culture, as the evil of the day.[122] Its discussion of how the federal government had protected, and planned further to protect, Canada's cultural industries could have stood on its own with little reference to globalization, but the report portrayed globalization as a threatening presence on Canada's doorstep. Perhaps its rhetorical force was a response to the strength of the U.S. reaction to the Baie-Comeau agreement. The publication of *Vital Links*, and the exclusionary clauses in the FTA, may also have been designed to play to the self-interests of the cultural sector and thereby ensure the sector's relative quiescence in a time of vigorous debate and social turmoil. Whatever the exact explanation, cultural protectionism (inherent in *Vital Links*) and the promotion of free trade are contradictory, at least within an economic framework.

The paper opened with a condemnation of "corporate gigantism in the cultural industries," noting that a limited number of risk-averse entertainment conglomerates dominated international markets with formulaic, often violent or titillating, culturally non-specific entertainment products that scooped up disposable income. "The result," according to the report, "is the potential marginalization of Canadian culture within Canada itself."[123] *Vital Links* recounted the major problems which to that date had received government attention — the limited market share for Canadian titles; the dominance of foreign-owned subsidiaries and imported titles distributed in Canada by the foreign-owned sector; the dominance of large firms; the suppression, by importation and distribution, of a rights-based market; parallel importation by institutions and libraries; Canadian-owned companies' lack of access to capital; and Canadian-owned publishers' inability to establish themselves in educational publishing. It also reviewed government measures that had been introduced to address these issues, including title-based production subsidies, industrial development measures, and provincial publication grants.

The paper's "positive approach for the future" consisted of two elements. The first was a provision for increased self-financing by Canadian-controlled firms, especially larger ones, that would theoretically lead to diversification, including participation in the more lucrative sectors of the market, such as educational publishing. The second was support for publishing "of cultural importance but low commercial viability,"[124] which would ensure the continued development of authorship by smaller firms across the country and, hence, the creation of an ongoing literary heritage. There was nothing new in principle here. But *Vital Links* played a strong card in favour of fostering the growth of large Canadian firms by means of financial support, while limiting the support of small firms to publishing of "cultural importance." Such a policy direction would have substantially diminished support for heterogeneity and thus would not have found favour among the many small, Canadian-owned firms.

Insistent industry demands and government studies at the turn of the decade

In 1989, following an ACP conference in Saskatoon and the presentation of at least twenty papers, mainly by ACP members, the organization developed a strategy paper, which was not released until 1991, entitled *Book Publishing and Canadian Culture: A National Strategy for the 1990s*.[125] Like its predecessors, this paper described the strengths of the Canadian-owned sector of the book publishing industry, characterizing the industry as culturally vibrant; diverse in terms of size, location, and genres represented; owned and operated by Canadians; entrepreneurially driven; professionally skilled; and active in developing export markets. The paper noted that this structure ran against the homogenizing trends of the increasingly large multinational firms and concentration of ownership and control, and it called for a government strategy containing three specific elements: the creation of an environment in which Canadian books could be published profitably by Canadian firms for Canadian readers; an increase in the market share of Canadian firms; and the maintenance of cultural and regional diversity.

The 1990 paper acknowledged the structural weaknesses of the industry, which were recognized by both industry and government: the industry's lack of competitiveness with foreign firms; the overwhelming presence of imports; the cost/price squeeze in which origination costs needed to be amortized over a small print run, and where prices were set by run-ons of U.S. and U.K. titles; the industry's participation predominantly in the high-risk trade market; its limited access to capital; and its limited access to more lucrative market sectors such as education and reference publishing. The paper acknowledged that existing government support programs reduced deficits, but it argued that the programs had not built profits, and, as a consequence, industry development had not been forthcoming. The 1990 report claimed that federal policy failed to embrace publishers' use of business means to pursue cultural ends. Reflecting statements made by Marcel Masse, it recommended the primacy of cultural ends and, hence, the profitable publication of Canadian titles designed for and of specific interest to Canadian readers by Canadian-owned publishers; the enhanced distribution and availability of Canadian-authored titles to the widest spectrum of Canadian readers; the support of publishers' chosen areas of expertise; and support of the development of export markets. Such recommendations might be summarized by saying that the report called on government to put the necessary money and structures in place for firms to earn a profit by publishing as they saw fit. Such an orientation illustrates the difficulty of giving primacy to cultural values. Should bills arise, someone must pay them. If the bill payer has no right to question those who incur the expenses, the system is inherently unstable.

The report called for the federal government to enact measures in three areas: fiscal support, industry structure, and legislative change. (In the terminology being used in this book, the first two measures are industrial support, while legislative change is structural intervention.) The fiscal measures called for included the creation of an investment tax credit to assist with title development costs (equivalent to research and development assistance). The report recommended that 50 percent of investments be refunded, regardless of the tax status of the firm, with bonuses for companies located outside Ontario and for small companies. It also called for the enhancement of Canada Council and Social Science and Humanities Research Council (SSHRC) funding for culturally significant and scholarly books respectively; the enhancement and increased flexibility of the loan program of the Cultural Industries Development Fund; and the removal of the GST on books. Non-fiscal industrial measures called for were the stimulation of exports and increased access for Canadian-owned publishers to education and public library markets through a federal-provincial learning materials development program. Legislative measures (structural intervention) called for were the strengthening of the Canadian-owned distribution system by the establishment of distribution rights for Canadian publishers, and the development of a foreign investment policy to arrest and reverse globalization in the book industry.

At more or less the same time, the government called in the consulting arm of a firm of accountants, Peat Marwick, to undertake an analysis of the strengths and vulnerabilities of the English-language industry.[126] Noting the shift in orientation from the financial assistance of the CBPDP to the industrial orientation of BPIDP, and thus the potential for self-financing, Peat Marwick noted: "It is hard to imagine significant permanent benefit for Canadian-controlled publishers who remain confined in the high-risk, low-profit sector of the total publishing spectrum." It continued: "Canadian publishers can hardly survive on such a small domestic market, unless foreign publishers let them distribute more foreign books, or bid for the rights to publish them, or unless they can expand into the lucrative education market."[127]

Immediately following the Peat Marwick study, the federal

government commissioned former ACP executive director Paul Audley to undertake a thorough review of the performance of the industry and present the government with various policy options.[128] Audley reviewed Statistics Canada data for the period, noting in particular the dominance of imports in the market, the relative inaccessibility of imports to Canadian-controlled English-language publishers, the barriers to Canadian firms' entry into educational markets, and the increased costs to Canadian-controlled firms that resulted from their having a higher percentage of title origination than their foreign-owned counterparts. He also reviewed federal, Quebec, and Ontario policies prior to presenting analysis of major policy considerations.

The first consideration was market share held by Canadian-owned companies and the policy being used to increase that share, namely, Baie-Comeau. The central issue Audley considered was whether Baie-Comeau should be strengthened or weakened in the face of lack of government action consistent with its original principles. His preference (among four alternatives he discussed) was to strengthen Baie-Comeau by means of facilitative tax credits, distribution support, and access to loans that would help Canadian-controlled publishers finance title origination. He argued that strong support in these areas would increase the possibility of Canadian firms taking an ownership position in companies that came up for sale as a result of the application of Baie-Comeau.

The second consideration was how to strengthen the opportunities for Canadian-owned firms to participate in book importation and distribution. Here, Audley emphasized improving the financial position of Canadian-controlled firms and revising the Copyright Act to deal with "buying around" — now termed "parallel importation." This refers to booksellers' practice of purchasing books directly from U.S. sources, usually U.S. wholesalers, instead of ordering them from Canadian distributors who hold the rights to those titles. At the time, Audley reported, bought-around purchases constituted about one-third of the Canadian market — an astonishing amount, given that a policy was in place designed to prevent such importation.

In his conclusions, Audley predicted pressure towards continental integration of markets as a result of the 1988 Free Trade Agreement;

uncertainty in the industry "arising from the failure to consistently and effectively implement the 1985 Baie-Comeau policy"[129]; and "continuing financial vulnerability of Canadian-controlled publishers, who account for 80 percent of the new Canadian-authored titles published,"[130] exacerbated by high interest rates, lack of access to imports, and the implementation of the GST.

Following the Peat Marwick study, but prior to the release of Audley's report, Minister Masse took action in three areas discussed by Audley. In July 1990, Masse announced the creation of a Cultural Industries Development Fund to provide publishers with greater access to loan capital (a fund that, the publishers noted in their 1990s strategy document, was lacking sufficient flexibility); moves were made to discourage "parallel importation" in the Copyright Act, which came to fruition in 1996; and Masse initiated consultations with the provinces to open educational markets to Canadian-controlled publishers. His efforts with the provinces bore no fruit.

The turn of the decade certainly saw copious study of the book industry and its possible future. Looking back at those industry and government reports, the obvious question is: what happened? The boldest recommendation of the ACP, for "the creation of an environment in which Canadian books can be published by Canadian firms for Canadian readers profitably," was nothing less than a declaration that the government had a societal obligation to respond to the demands of authors and publishers. To some degree, the Peat Marwick and Audley studies were an attempt to determine how that might be possible. Perhaps the most surprising element of book publishing policy is that, federally and in some provinces, governments did indeed put into place sufficient resources to create an environment in which Canadian books *could* be published by Canadian firms for Canadian readers profitably. However, there were, and are, unspoken caveats. The first was that "profitably" is understood to mean "not at a sizable direct financial loss to publishers." The second was that authors were willing to accept whatever royalties happened along, based on sales and standard royalty rates. A third was that industry workers would accept low wages. When the distribution right was included in the 1996 revisions of the Copyright Act, it strengthened the protection of Canadian editions and, hence, access to imports by Canadian firms. But three other major policies were not pursued — trade publishers did not gain greater access to education markets, globalization was not reversed, and the GST was not removed from books.[131]

Government and book publishing to the early 1990s

Prior to 1992, and largely as a result of efforts during Marcel Masse's ministership, it appeared that a stable, long-term relationship had evolved between the book publishing industry and the federal government. But as events would prove, it remained vulnerable, as it still is today, to sudden political and economic changes of priorities. Two shocks hit the industry during the 1990s. Towards the end of the Mulroney government's time in office, the already-weakened cultural commitments of the government were dealt a blow by a 1992 evaluation of BPIDP submitted by Fox Jones, a consulting firm.[132] In the opinion of the consultants, BPIDP had failed fundamentally, because the financial health of the industry and individual firms had not changed substantially despite the considerable funding brought forward by Marcel Masse. Indeed, the consultants argued, by placing culture first, the policy had undermined the industry's ability to become viable and self-sustaining. The authors cast the report as a direct repudiation of Masse's informing framework of supporting business operations (economics at the centre) to facilitate cultural expression (culture at the core).

The shortcoming of the evaluation was its circularity. It discarded the key idea that the Ontario Royal Commission had established and that had been rephrased by Masse: the cultural core. Instead, Fox Jones regarded the financial performance of publishing firms as the sole legitimate measure of the value of the industry, and did not even take industry output into account. While the report seemed to have been disregarded by government officials in the weeks and months following its release, it nonetheless presaged the suppression of a Paris-inspired attempt to place culture on a par with the economy in the design of a post-industrial state.

Perhaps in response to the Fox Jones study, when the BPIDP program was renewed in 1993, government officials increased pressure on the industry to become profitable and lose its dependence on grants by establishing financial criteria for industrial support eligibility. The criteria were as follows: minimum increasing eligible sales; minimum sales-to-inventory ratios; minimum sales of own titles per employee; and minimum debt-to-equity ratios. This support existed alongside export marketing assistance, a postal subsidy program, and continued Canada Council funding for cultural titles. Subsequently, the financial rules were relaxed somewhat and, with provincial government assistance in many provinces, the industry settled into an equilibrium that stabilized the resulting earnings-plus-grants culture.

The second all-but-knockout blow came in 1995, followed by a threatened, but never delivered, further skewering in 1996. Both were aimed directly at book publishing by the finance ministry under Paul Martin and officials in the Department of Canadian Heritage.[133] Cuts in government spending to tame Canada's debt were the order of the day, as the Liberals fought to stay ahead of apparently surging popular opinion, represented by Reform Party leader Preston Manning.[134] In 1995, the publishers lost 56 percent of their overall federal funding, which, in the end, amounted to $23 million. In 1996, they were slated to lose a further 6 percent. Other cultural industries were also cut, but by roughly half as much.

A penultimate ACP policy paper

The 1990 ACP position paper represented a major effort of animation, with many publishers presenting papers on their areas of emphasis, and it therefore involved a significant number of ACP members. In April 1994, against a background of sound industry performance, a smaller but still significant number of industry members met again in Vancouver, and a draft summary statement was prepared, followed, after some time, by a policy paper that was not released until November 1995. By that time, the industry was much subdued, as a result of the 55 percent cut to publishers' federal grants.

New Directions: Rethinking Public Policy for Canadian Books opened with the following question: "Given the outstanding success of Canadian writers with Canadian readers, what is the best way to meet the public's strong and continued demand for Canadian books?"[135] The emphasis on the reading public, rather than the industry itself, was far from accidental.

The paper reviewed the industry's contribution to the increase in the numbers of Canadians reading books — from 29 million in 1987, to 59 million in 1991 — and in the number of hours the average Canadian spent reading books — from 3.3 hours per week in 1978, to 4.1 in 1991. It noted the growth of book publishing over twenty-five years — from annual sales of $222 million, to nearly $2 billion — and the current employment it then generated: nearly 7,000 full-time employees, complemented by many freelancers and much author opportunity. It drew attention to recent increases in title production (nearly 10,000 new published titles were originated in Canada in 1993–94), to the 85 percent contribution to those titles by Canadian-owned firms (in 1992–93), and to the 71 percent share of titles written by Canadian authors, over 80 percent of which were published by Canadian-controlled firms. It pointed to healthy export sales — then near 20 percent — and experiments with new media, such as encyclopedias on CD-ROMs. Alluding to small market size, the paper identified the need for both cultural and infrastructure (i.e., industrial) support, highlighting the increases in production that had resulted from increased assistance and noting the participation of book publishers in the information highway initiative.

The paper suggested four guiding principles for policy: First, to respect the value of an open market to the pluralistic, democratic society that is Canada. Second, to acknowledge the irrefutable economics of the Canadian book market, where, for example, the cost of sales for Canadian-controlled publishers came in at 57.7 percent of all costs compared to 48.5 percent for foreign-controlled companies. Third, to recognize the undeniable gap between market revenues and the amount of investment capital required to develop and publish Canadian authors, a gap that could only be closed with government funding. (In this context, the paper mentioned that in 1993–94, $38 million in grants generated over seven times that amount in industry sales.) Fourth, to proceed with the understanding that the object of book publishing, and hence public (i.e., government) support, was Canadian cultural development, not the growth or profitability of the publishing industry.

The paper then laid out five policy options, paying by far the greatest attention to the first: financing. It prefaced its suggestions by arguing that the GST on Canadian-authored books alone amply covered the funds needed to support the Canadian-owned sector of the industry. To finance the production of enough titles to meet market demand for Canadian books, the report said, there were two main alternatives: a production investment fund that would subsidize most titles but see funds from successful titles returned to government, or a refundable tax credit where a percentage of origination costs would be returned by government to publishers. In addition to either of these, the paper argued for further supplements through loans and increased funding of deficits of cultural titles. Beyond financing, the paper pointed to the need for foreign investment regulation benefitting the Canadian-controlled sector, the protection of creators' rights in copyright, and the inclusion of a distribution right in the Copyright Act. It lauded the announcement of international cultural relations as a new pillar of foreign policy. The paper closed with a call for a national action plan to strengthen book publishing.

A policy paper to close the century

In the context of the Manning-inspired, Martin-delivered 51 percent cut, the industry sat down with the Department of Canadian Heritage under Sheila Copps to craft a recovery plan consisting of three points — enhanced direct funding, a loan guarantee program, and a new structural measure — summarized in a 1997 ACP paper called *Setting Priorities for Federal Book Publishing Policy*.[136] The paper noted that the government had already promised to restore $5 million to BPIDP and had pledged an additional $15 million to follow. It had also agreed to establish a loan guarantee program and to work with the industry to create long-term structural solutions to its financial problems.[137] The paper recommended to the government that the additional $15 million be used mainly for marketing and distribution support and further funding of BPIDP's aid to publishers. While in the end the money was delivered slightly differently, effectively, as of 1998, "publishers were ... receiving a similar level of departmental funding."[138] The new structural measure they requested was the refundable investment tax credit, which the publishers refined into a request for an equity investment tax credit. In the end, the tax credit failed to materialize at the federal level.

A discontinuous policy history 1952–98

Looking over these many different initiatives, we can see that in the wake of the 1952 Massey Commission, the 1971 Ontario Royal Commission laid essential groundwork for the development of government support programs for Canadian book publishing. The commission established the primacy of the cultural (rather than economic) contribution of the industry, embraced the need for cultural support (of titles or genres) as well as business support, laid out the need for structural intervention, and made public the vast array of variables that contribute to the flourishing of a book publishing industry. With the commission's report in place, and in the context of a sympathetic socio-political milieu, Canada's book publishing industry lobbied for support by means of its strategy and policy documents, as well as its negotiations with various governments.

The ACP was largely successful in its lobbying from the late 1960s through to the mid-1990s. It persuaded governments to put programs in place that allowed firms within the industry to maintain themselves. The calls for policy and support of the ACP papers were bold. With the announcement of the Baie-Comeau policy in 1985, publishers threatened the legitimate existence of established foreign firms, just as it was becoming clear that they were overestimating their capacities to take a significant position in educational publishing. Baie-Comeau incurred the wrath of major U.S. corporations and the U.S. government because it challenged the basic assumptions of open economies.

The governments of the day were not handmaidens to industry. For the most part, they listened to the sometimes strident voices of publishers and designed programs of support that were a compromise between industry demands and evolving government principles and priorities. Over time, and through to 1994, the support provided by the federal and provincial governments became a robust infrastructure that served Canadian authors, Canadian-owned book publishers, the people of Canada, and the Canadian nation well, by fostering national self-awareness, a sense of belonging and opportunity, and the establishment of a distinct identity. After a hiccup in 1995, the support programs returned to assist the publishers in building a stable industry.

In retrospect, the nationalism that drove the development of modern trade-book publishing clouded the very real differences between, on the one hand, educational publishing, and on the other, trade-book publishing oriented to a modern articulation of a national spirit. In the same way that Canadian trade publishers sought to give Canada its own distinctive voice, they also sought to compensate for the weak embrace by educators of their cultural role and their contribution to social cohesion. Had Canadian-owned trade publishers attained a footing in educational publishing, no doubt cultural content in today's curriculum would be stronger; however, responsibility for the cultural orientation of education lies with educators, not with publishers. Nothing has ever prevented Canadian provinces from following Quebec's lead in overseeing the development of educational materials that are appropriate to Canadian classrooms and to the important task of connecting students to their country and culture. The Canadian-content-in-broadcasting model has simply not been embraced by educators.

The genius of Canadian support for book publishing is that even though culture has played a leading role, policy and support have steered clear of making judgements on what was published. The industry brought an unswerving faith that Canadians were well served by its activities; bolstered by this, it argued, and the government came to accept, that substantial funding was needed to maintain its viability. The institution of the GST on book sales made it possible for the Mulroney Conservatives, led by Heritage Minister Marcel Masse, to provide the funds needed for industry consolidation.

Luckily for the industry, in the second year of the 1995–96 cuts, the minister in charge of Heritage, Sheila Copps, began rebuilding funding to the level it had achieved prior to the cuts. Clearly the Chrétien cabinet of the day was not of one mind.

CHAPTER 4

Reconceiving book publishing from the middle 1990s forward

A socio-historical introduction 123

The foundations of a reconception of book publishing 127

    An Ontario assessment 129

The beginnings of a new vision: *The Business of Culture* 131

    The (presumably) last gasp of traditional economics 136

A new federal policy direction: The Cultural Industries SAGIT report 138

A new international instrument for trade in cultural products 140

From NIICD to CCD: The wash of international diplomacy 145

Moving forward on emerging foundations 147

Responding technologically to a retail crisis 149

Conclusion 153

A socio-historical introduction

As evidenced by the Massey Commission, the post-war changes that took place in Canadian society in those years were not limited to the gradual replacement of agriculture and resource extraction with manufacture. The Massey Commission represented the dawning of a cultural awareness among Canadians that Canada was connected to other nations through immigration and institutional structure, but was also distinct in itself.

In the context of this new vision and self-confidence among Canadians, Canada's nineteenth prime minister, Pierre Elliott Trudeau, came to power. As biographies of Trudeau begin to appear,[139] his character and actions are coming to be more understandable, as are the times in which he led the country. Trudeau was a prime minister like no other. He held the post from 1968 to June 1979 and then from March 1980 to June 1984. He was a charismatic man of ideas and passions, an accomplished student of law and philosophy, and an ambitious politician and political actor who gave English Canadians, if not all Canadians, the impression that he had interests larger than the partisan politics of the day. He was driven by principled beliefs, whether these involved private morality (as when he noted, as Justice Minister, that "there's no place for the state in the bedrooms of the nation") or the national good (as when he patriated Canada's Constitution in 1982 and included with it a Canadian Charter of Rights). The latter action placed Canada, for the first time in history, completely in charge of making appropriate amendments to its Constitution to meet the country's needs.

Trudeau emerged as a political figure just as the post–World War II baby boom advanced towards adulthood. This was a generation for which social institutions were adapted or created to deal with its needs and numbers.[140] It was a period of massive change. Those interested in advanced education found, as they reached maturity, that Canada lacked the institutions to provide it. Thus, in pursuit of graduate degrees, many travelled to Britain, France, and the United States, where they became aware of differences between Canada's muted national self-image and the cultural confidence of former colonial powers like France and Britain, or the cultural imperialism of the United States. Some embraced the values of their educators; others returned determined that Canada should carve a place for itself, both among its own citizens and in the world at large.

Trudeau also came to power at a moment of general social change, as the leading edge of baby boomers were completing their high school education. Great crowds of hitchhikers were on the move, exploring their country, smoking marijuana, and swallowing other drugs such as mescaline, LSD, and the psychotropic compounds contained in magic mushrooms. The economy was manifestly unable to absorb such numbers, even with temporary summer jobs. Management of this disruptive social movement was a genuine but unacknowledged political and social priority, and the *cri de coeur* of the young publishers and authors demanding to have their voices heard seemed little distant from the more general demands of youth for opportunities in a society they could value. In the summer of 1971, the Ontario Royal Commission on Publishing released its report; that year and the next, Trudeau's federal government put in place programs to support self-directed job creation. The 1971 program was called Opportunities for Youth; the 1972, Local Initiatives Projects. As well as managing youth overall, such funding gave starts to at least two publishers, Harbour Publishing and Kids Can Press.[141]

The impact of the United States' imperial bullying in its war in Vietnam must also be taken into account. Although it claimed the moral upper hand, America's assertion that it was protecting the free world from Communism in Vietnam was highly tenuous, and the televised carnage resulting from the clearly desperate and vicious actions of the U.S. Army led many Canadians to thank fate, or their preferred deity, that they had been born in or had immigrated to Canada. (Many who had not been so fortunate dodged the U.S. army draft by immigrating to Canada.) The determination throughout Canada to carve out a separate national identity was palpable. It was also in evidence in the titles being brought out by such publishers as House of Anansi and New Press. And while the baby boomers were especially vociferous, the desire for a separate Canadian identity spanned all ages.

The period from the late 1960s to the end of the 1980s — like the

1840s, the turn of the last century, and the 1920s — was disruptive of the status quo. To some degree, Pierre Trudeau, in personifying the ideals of many, was a pied piper. In book publishing, with ideas at the forefront, authors, professors, investigative journalists, and those who struggled to make their ideas public — mainly, Canada's new wave of young book publishers — were the heroes of the day.

Canadians were setting out to define a national or bi-national imaginative universe,[142] a project that carried echoes of Bloomsbury's rejection of Victorianism. The discourse on book publishing paralleled this national discourse, the core of which was its demand for self-definition and, derived from that, independence. Both required an infrastructure that allowed Canadian publishers to partner with Canadian authors to address questions of priority to Canadians, and to get works of fiction and non-fiction into the hands of Canadian audiences.

Creating policy and support programs for Canadian-owned book publishing was an uphill struggle. The first task was to unravel laws, policies, and ways of operating that had been put in place to support importation. The second task was to replace them with structures that encouraged domestic production, without overly impeding importation. The third task — persuading allied agencies such as bookstores, libraries, and educators that Canadian writing and Canadian-produced materials were not just equal in quality, but were more appropriate options for students, borrowers, and the reading public — was not an easy one. Phrases such as "world-class authors and literature," "customer preferences," "second-rate production," "curricular demands," "world citizenship," "the dangers of nationalism," and "needless provincialism," among others, were used to resist change. Some, but far from all, members of government and industry began to understand more about the nature of the "free market." They saw that laws, policies, and practices were put in place to secure the interests and markets of producing nations, and other nations were then exhorted to abide by those rules — while the sponsoring countries worked around them.

One of the major issues the book industry had to address, as did other cultural industries such as film, music recording, magazine and newspaper publishing, and broadcasting, was that the *sine qua non* of its activity was cultural rather than economic. Although the Ontario Royal Commission, and the various policies that arose from its recommendations, embraced that reality (some more strongly than others), the ever-aggressive U.S. entertainment industries sought to protect an important export market. In their protests, they virulently insisted that the discourse remain economic; they were being denied access to markets.

For a long time there was a disconnect between, on one side, the Canadian business community and, on the other, book publishers and other cultural industries that were looking for government support. The business community, on the whole, was simply unable to accept that the cultural industries were seeking assistance parallel to what traditional industries received in the form of transportation infrastructure, tax credits for research and development, regulatory and standards-setting agencies, competition law, and a host of other normal "free market" interventions to keep their businesses viable.

Yet another struggle with which the publishers had to contend was the design of financial assistance. Though they consistently asked for laws to allow them to fairly compete with foreign companies, for a long time they were provided mainly with direct financial aid. That aid began with title subsidies, designed to compensate publishers for the deficits they incurred in publishing titles that were widely recognized as making valuable contributions to national discourse and providing exposure to promising new authors, but that lacked sufficient sales to make them a paying proposition. The industry made its dissatisfaction with title assistance known by calling such grants "welfare-style grants." The ACP papers discussed in Chapter 3 demanded financial assistance that would allow publishers to operate as profitable businesses. Such support would provide funds for marketing, distribution infrastructure, and access to government procurements; it would reward good business practice, assist in the opening of export markets, and serve as a catalyst in publishers' needs for working capital.

Government discourse and supportive action was dominated, almost to the end of the century, by an emphasis on the temporary nature of assistance. Governments continuously sought formulae that would obviate financial assistance; grants were temporary measures designed to get the industry established. The publishers consistently responded that they, too, wished to be financially independent and stable — knowing full well, and often arguing vociferously, that without radical change in Canada's open marketplace, such independence was a distant hope. They made this point most clearly in the 1995 policy paper.[143] Too, it was a long time before governments began to accept that, like authors, artists, musicians, filmmakers, dancers, and actors, publishers were publishers because they wished to participate in the celebration of creativity, the production of meaning, the enhancement of social justice — or, more generally, in the pursuit of ideals that were the very pinnacle of human existence. They were in business to facilitate such goals.

In the end, government and industry came to an accommodation, but it was an accommodation subject to sudden reversals, such as those imposed by government cuts or the pursuit of free trade, which threatened all cultural policies. This chapter describes a re-visioning that began in the 1990s. This new vision is still being refined and established as we look forward to the second decade of the twenty-first century. But it is a vision that gains strength as each year passes.

The foundations of a reconception of book publishing

A transition in thinking and policy with respect to book publishing began to emerge in the mid-1990s. Several factors contributed to this change. The market environment of the 1990s was reliable enough to allow book publishers to combine their sales with grants in order to run low-wage-paying, slightly profitable businesses. Publishers had become accustomed to marrying two motives: addressing market demands in proven categories, while also making a cultural contribution. Various technological initiatives were being considered while the publishers gradually sorted out distribution. And the unrealistic dream that Canadian-owned trade book publishers might obtain a significant share of educational markets was gradually being set aside.

In 1994, a policy paper, entitled *The Business of Culture*, which was the report of Ontario's Advisory Committee on a Cultural Industries Sectoral Strategy (ACCISS), was released in Toronto.[144] In its examination of the cultural sector as a whole, it focused on how Canadian cultural industries could position themselves for participation, growth, and increased market share amidst technological change. By emphasizing the nature of markets and market participation, *The Business of Culture* introduced a change of focus that entailed an acceptance of consumer patterns and preferences, rather than an implicit belief that if only mass consumer products were less available, Canadians would turn to books written about Canada on Canadian subjects or, at the very least, including Canadian realities.

As we will see, *The Business of Culture* was followed, three years later, by a federal international trade initiative that aimed to prevent globalizing free trade laws from further marginalizing the participation of domestic cultural industries in domestic markets. Underlying both these initiatives was a nascent understanding in social theory, and within governments and agencies such as UNESCO, of the role played by cultural production (for all ages and tastes) in facilitating social cohesion and generating social capital, both contributors to quality of life and a well-functioning economy. This conception of cultural production was developing in academic circles in the 1990s, and it has begun to influence policy and the manner in which book publishers see themselves in this new century.

These very positive conceptual and governmental emergents did not represent a sudden and complete turn of events. Ironically, one year after *The Business of Culture* argued that the needed supports for market participation were in place or would be put in place, the federal government slashed book publishing support by a purported 56 percent (which MacSkimming calculated to be actually about 36 percent). Then, in 1999, amidst developments on the international trade front, a crisis brought about by giant bookstore chain Chapters' monopoly in the retail sector triggered a review by the parliamentary Standing Committee on Canadian Heritage. The committee's response, which in the short run seemed like inaction, in the long run, surprisingly, turned into a technological fix: the Supply Chain Initiative (SCI), was inaugurated and continues to serve the book publishing industry well (partly as BookNet Canada). And even six years after *The Business of Culture*, in 2000, the federal government — knowing full well the profit-driven perspective that economists were likely to bring to the table — commissioned two economists to examine the state of the industry and recommend interventions for future design of support. The details of the key events follow.

An Ontario assessment

In the late 1980s, a misalignment developed between federal and Ontario policies and attitudes regarding the cultural industries. Whereas the 1987 *Vital Links* paper discussed in the previous chapter was meant to trumpet to the cultural industries and other interested parties the success of Conservative federal government support and plans for a glorious future, a May 1989 paper prepared for the Ontario Ministry of Culture and Communications presented a much more pessimistic analysis of the impact of book publishing support programs.[145] This paper began with a review of several positive developments, including the growth in the number of Ontario book publishing companies, from 53 in 1971 to 266 in 1987, and the five-fold growth in income (uncorrected for the changing value of the dollar). However, the paper noted, from 1984 to 1986, while the total Canadian domestic book market grew by 28.5 percent, the domestic sales of English-language trade books grew by only 7.8 percent, and the sales of English-language books published in Canada in all commercial categories actually declined by 14.4 percent. The paper noted further that while net domestic sales of all trade books from Canadian-controlled, English-language publishers increased by 6.4 percent between 1986 and 1987, sales of their own titles declined by 13.1 percent. It concluded that while the value of the trade book sector and of Canadian-owned firms remained high in terms of cultural participation, from a commercial perspective the Canadian-owned sector was losing ground to its foreign competitors.

The key problem for English-language, Canadian-owned publishers, the report noted, was competition from foreign imports. For every dollar increase in sales, year over year, of Canadian-originated own titles (written mainly by Canadian authors), imported books gained a dollar as well. Moreover, between 1975 and 1989, Canadian-owned firms lost market share for the imports they distributed under agency agreements. Overall, between 1975 and 1987, Canadian-owned firms lost 4 percent of the total sales of own-title production and import distribution, slipping from 48 percent to 44 percent. Foreign firms gained that 4 percent, moving from 52 percent to 56 percent. By 1989, the share of all imports for Canadian firms had decreased to 13 percent. Overall, net domestic sales of own and exclusive agency titles of Canadian-owned firms occupied only 20 percent of the total book market in Canada at the time. (The traditional claim to market share, repeated in *Vital Links*, was just under 25 percent.)

The paper concluded:

> In general, Canadian-owned publishers devoted to culturally significant trade publishing depend on government grants to sustain their operations; they experience difficulty raising both debt and equity capital and are unable to expand their distribution, publishing programs or marketing and sales efforts because of their inability to generate working capital from their own operations.
>
> ... [T]he major and obvious need in the publishing industry is for equity capital. In an industry with such low profit margins, there is seldom money to reinvest in the company for development, and investment partners are difficult to attract.
>
> **Because there is no question as to the quality of Canadian books, the problem now is how to continue financing production.** [Emphasis added][146]

The realities presented in the Ontario paper were stark. Its diagnosis focused on the financial state of the industry, without falling back on textbook-standard economic remedies. It noted the competitive disadvantage of Canadian producers in a market open to imports and, hence, disadvantageous economies of scale. But in accepting "the quality of Canadian books" and, implicitly, the expertise of Canadian publishers, it defined the problem as a need to finance production in pursuit of sustainability. Its solution was to redefine the target market at which Canadian publishers should aim — the consumption patterns of not only Canadians, but also consumers in international markets.

Considering this paper alongside the positive vision of *Vital Links*, and, nearly a decade later, *The Business of Culture* (commissioned by the NDP government of Ontario), book publishers might have been forgiven for thinking that the traditional notion that their industry needed to be restructured in the name of the rationalization of the costs of production economics had been laid to rest. But economists' constant readiness to ignore the economic minutiae that play such a crucial role in book publishing, the cultural success of publishing in providing opportunities for writers, as well as the cultural value of titles — because, until recently, they have had no framework that measures cultural value — seems to grant the notion of restructuring to achieve economies of scale everlasting life. Thus, in 2000 (a little over ten years after the 1989 Ontario paper), a Department of Canadian Heritage report once again documented the industry's economic weaknesses, but, instead of recommending the preservation of cultural value by ensuring access to capital, revived the tired idea of a restructuring of production within the industry to increase production economies.

The beginnings of a new vision: *The Business of Culture*

From 1972, when the Ontario Royal Commission released its report, to 2000, two governmental approaches to book publishing vied for supremacy. The first emphasized the cultural contribution of the industry and made the case that the very survival of Canada depended on supporting cultural production without judgement or interference. A frequent corollary to this idea was a rejection of global products, usually from the United States, that spoke to Canadians as generic members of a rich, western, democratic society, rather than appealing to any distinguishing Canadian characteristics, attitudes, or tastes. The second approach emphasized the industry side of the cultural industries equation. It argued that Canadian producers of cultural goods and services should compete in the marketplace for consumer dollars. Proponents of this perspective often purported to believe that Canadian cultural producers were fully capable of competing with such megacorporations as Bertelsmann, News Corp, and Pearson, if only they pursued market opportunity rather than their own sense of cultural worthiness. Although existing at opposite poles, these two approaches shared the assumption that placing culture first meant rejecting the judgement of the marketplace.

The Advisory Committee on a Cultural Industries Sectoral Strategy (ACCISS) appointed by Bob Rae's NDP government of Ontario was different from previous inquiries into book publishing, in that it was charged with proposing a general strategy for all cultural industries in Ontario. The new vision this committee put forward emerged from the interplay of understandings brought to the table by the assembled experts and representatives of various industries.

The results of the advisory committee's deliberations was the comprehensive policy report *The Business of Culture* (released in 1994), which set the culture/industry dichotomy aside, gathered data on the nature and extent of cultural markets, set specific financial targets, forecast employment and economic outcomes, and proposed ways to achieve those objectives.[147] The report outlined the committee's vision and strategic goals; recommended strategies to make Ontario's cultural industries more effective in the marketplace and to position them for the future; and suggested changes to the operating environment that would support and strengthen the cultural industries. The report was all-encompassing and, reminiscent of the Ontario Royal Commission, addressed all relevant variables, including training, financing, market information and audience analysis, technology, exports, copyright, taxation, trade, distribution, foreign investment policies, government procurement, Canadian content requirements, industry/government relationships, building Ontario as a production centre, and linking the various cultural industries to each other and to tourism. And rather than focusing on cultural production as a means to achieve the social goal of providing Canadians with a sense of themselves, it focused on the cultural marketplace, and the many and varied elements that are part and parcel of cultural production and cultural markets.

The thrust of the argument in *The Business of Culture* was that, as the centre of national business activity and finance, Toronto had been leading English Canada's efforts to establish its own cultural industries, including book publishing. Government programs to support this effort had aimed to establish substantial and profitable industries, but they had not been entirely successful. A few programs, such as the Capital Cost Allowance for film investments, were fiascos.[148] In spite of government efforts, Canada's cultural industries had remained economically marginal, both nationally and globally. On the other hand, there had been enormous cultural gains. Canadian writers and recording artists were now well known to Canadians, and their sales were good. Cultural producers were gaining access to world audiences.[149]

The Business of Culture called for the Ontario government to strengthen Toronto's position as a cultural industries centre so it would be in the same league as Hollywood, New York, London, or Paris, where deals are done and products are created and launched globally and domestically. The report suggested that the resulting increased visibility would contribute directly to the regional economy, drawing from and contributing to its cultural diversity, and creating productivity multipliers that would stimulate substantial secondary economic growth in allied services.[150]

The report identified three strategic goals. The first was to increase Canadian-owned, Ontario-based producers' share of the international and domestic markets for entertainment and information products. The second was to develop an industry/government framework that would let Ontario's cultural industries take a lead role in developing new multimedia products and in using new systems to distribute traditional cultural products. The third was to create a stable, equitable operating environment for Canadian cultural industries in Ontario.[151]

It is easy to see these general cultural industry goals as an updated articulation of policy goals outlined in previous book publishing and other cultural industry policy papers. For instance, increasing market share had been a long-standing aim of both government policy and cultural industries. However, there were two key differences in this new formulation. The first was found in the words "international and domestic." Placing "international" first signalled a focus on world market participation, in contrast to the previous emphasis on domestic market share for distinctly Canadian products, with attempts to market such products internationally seen as secondary. A second key phrase, "entertainment and information products," further emphasized participation in the whole range of markets, both mainstream and niche. These phrases signalled the prioritization of market participation over the goals of cultural expression or of rationalizing the structure of production to create large, profitable firms. Neither of these were precluded; rather, it was left up to producers and market forces to pursue them, in much the same way that Toronto filmmakers, for example, competed with Hollywood and Bollywood in mainstream entertainment markets, or musicians tried to build a worldwide fan base. This new vision was clearly influenced by sectors such as film and video, music recording, and even magazine publishing, where the objective is audience building rather than cultural expression for its own sake, regardless of market response.

In the context of the first market-share goal, the second goal's emphasis on Canadians taking a lead role in multimedia and in electronic distributions systems reinforced the difference between this and previous policy documents. It aimed to bolster market participation by embracing technological change, both in product creation and in the infrastructure necessary to compete in world markets. Also in the context of the first goal, the third goal's emphasis on "stability" and "equitable treatment" suggested that policy should not be subject to swings in political priorities and consultants' opinions, nor should certain sectors be favoured for non-cultural reasons. Instead, it called for serious treatment of cultural enterprises and a general strengthening of all sectors, with their inherent interdependencies.

To achieve these three goals, the report listed thirty-one recommendations and forty "tactics." Many were familiar, particularly the call for an arm's-length agency to administer policies and support programs. The report also called for the government to establish a Centre for Culture Industries and Technology that would link the cultural industries and other sectors (the Ontario Media Development Corporation has taken on that role), and a Cultural Industries Advisory Council (currently the Minister's Advisory Council on Arts and Culture) to oversee the implementation of a to-be-developed strategy.

By concentrating on the characteristics of cultural markets rather than railing against globalization and large international entertainment conglomerates, and by portraying the participation of Canadian-owned firms as crucial to the establishment of Toronto as a world-renowned cultural production centre, the report cast government and the cultural industries as partners in the development of goods and services that have undeniable economic benefits and substantial productivity multiplier effects. These multipliers do not just take the form of restaurant meals for audience members on a night out on the town. They are also evinced in tourism; in the pride Canadians take in their musicians, filmmakers, and authors; and in the attention Canadians have achieved on the world stage as creative, insightful people — all of which have economic consequences, as we will see. The report also portrayed market interventions, such as Canadian content regulations in broadcasting, as powerful mechanisms capable of building cultural participation.

Of greatest importance, and, again, echoing the Ontario Royal Commission's *prima facie* assumption of the value of McClelland & Stewart and subsequently other firms, was the tacit acceptance in *The Business of Culture* of the various industries as functioning operations appropriately adapted to their various business environments, and, hence, as worthy actors in a proposed government-industry partnership. The report was a significant departure from the parade of consulting economists whose textbook theories told them how wrongheaded the thousands of cultural industry participants were in acting as they did. This is not to say that it did not offer incentives and opportunities for greater market participation and profits; it did. But these offers were not camouflaged attempts to restructure the industry; rather, they were genuine opportunities to strengthen those members of the industry who wished to take advantage of them.

It took some time for the perspective expressed in *The Business of Culture* to emerge in the cultural programs of Ontario and to be reflected in federal policies and programs. There was a change of provincial government shortly after the report was released, and culture was no longer a priority at that level.[152] The federal lag appears to have happened for two reasons. First was the desire of the Chrétien Liberal government to undermine the growing influence of Preston Manning and the Reform Party, which was calling for smaller government and the reduction of government debt and deficits. Finance Minister Paul Martin's cuts to book publishers occurred in 1995, the year following the publication of *The Business of Culture*. The second probable reason for the initial neglect of the report at the federal level was the Department of Canadian Heritage's preoccupation, from 1993 to at least the end of 1999, with a spirited defence of Canadian magazines, which were facing aggressive competition from Canadian editions of U.S. magazines, whose editorial costs could be recovered in their home market. Indeed, this largely unsuccessful defence — Canada was defeated in a World Trade Organization decision and appeal (discussed below) — appears to have convinced the cultural arm of federal government to accept the need for a long-term protective partnership with the cultural industries, so that Canada would ultimately benefit. But a further stumbling block, this time from the finance and industry sectors that shared responsibility with the federal government for cultural industries, was yet to come.

The (presumably) last gasp of traditional economics

As discussion continued about the possibility of tax credits,[153] in 2000 the Department of Canadian Heritage commissioned two consultants with backgrounds in economics, Arthur Donner and Fred Lazar, to study the competitive challenges facing book publishers.[154] Donner and Lazar strung their consultations over two reports. Their first report highlighted "some of the limitations of public policies designed both to support Canadian ownership and to promote Canadian authors, and some of the structural problems which have emerged in the industry — particularly the long-run picture of poor profits and inadequate capitalization."[155]

The report reviewed the competitive position of Canadian publishers in comparison with their foreign-owned counterparts, noting that prices were externally determined, firms were small and hence had no competitive economies of scale, there was lack of access to needed investment capital, and the dominance of the giant bookstore chain Chapters weakened publishers' bargaining position with the retail sector. The report concluded that while federal policies had achieved certain gains, they did not address the disadvantageous competitive position of the Canadian-owned sector. Additionally, the consultants pointed out that consolidations to achieve economies of scale were not occurring, and the succession issue was emerging as the owner-operators approached retirement. To a considerable extent, the 1989 Ontario paper had already tilled this soil.

In a second report, Donner and Lazar suggested that Canada's book publishing industry required internal structural adjustment.[156] To address long-term financing needs, the consultants called for industry consolidation, by which they meant concentration of ownership, accompanied by incentives to attract new players to the industry. Small firms would be encouraged to sell out to the larger and more profitable companies, resulting in fewer, larger firms. This, they argued, could be accomplished by means of a refundable investment tax credit to cover 30 percent of pre-production costs (rather than the 50 percent tax credit that the ACP had called for in its *National Strategy for the 1990s* paper).[157]

Not surprisingly, the proposal went nowhere. It would have incentive-marched many small Canadian-owned publishers into penurious retirement, or the opportunity to work under the direction of former competitors more interested in the bottom line than they were.[158] It also failed to take into account the net benefits incentives inherent in Investment Canada regulations that resulted in foreign-owned publishers taking on Canadian authors. As well, it paid no serious attention to start-up costs for new firms. Non-recoverable start-up costs are in the neighbourhood of $200,000. Why Donner and Lazar[159] imagined that a retreaded consolidation proposal would be favourably received, and how they thought the larger firms could parlay themselves into a position to expand, even with a 30 percent refundable tax credit, is a mystery. (Tax credit programs established in Ontario in 1997 and in BC in 2002 have not led to cross-sectoral expansion.) With the vast majority of Canadian-owned book publishers strongly committed to cultural goals (i.e., publishing new original titles), the report's bleak future was sealed.

The industry's opposing view of consolidation was captured in the ACP 1991 strategy document that stated: "The indigenous sector ... now comprises a highly creative, culturally and regionally diverse, and entrepreneurially driven industry from coast to coast. This is a stunning achievement in an industry increasingly dominated throughout the world by multinational media conglomerates."[160] The fact that of the four firms that could have led the consolidation — McClelland & Stewart, General Publishing, Key Porter, and Douglas & McIntyre — only one (D&M) still survives independently is an indication of the wisdom of the recommendations.

A significant reality unexamined by Donner and Lazar was the size and nature of the foreign-owned companies operating in Canada. The top three active book publishers are Random House Canada, part of Bertelsmann, a conglomerate with annual revenues in the order of US$10 billion; Penguin, part of Pearson, another international conglomerate with annual sales of US$8.4 billion; and HarperCollins, part of NewsCorp, with annual sales of US$33 billion. Other large firms also have a significant presence, including Simon & Schuster (part of US$14 billion CBS), Wiley-Blackwell ($1.7 billion), and McGraw-Hill (US$6.8 billion). Even with a willingness to jettison cultural value, the belief that internal industry consolidation of Canadian-owned firms, which account for about one-half of a $2 billion market, could have transformed any Canadian trade book publisher into a firm competitor with these giants suggests that the consulting economists were arithmetically challenged.

A new federal policy direction: The Cultural Industries SAGIT report

In 1999, five years after *The Business of Culture* was published, in the midst of the losing trade battle on the magazine front, mentioned above, and the ongoing sandcastle renderings of Messieurs Donner and Lazar, the federal government created an advisory group to consider the position of Canada's cultural industries in the context of international trade and report to the federal Minister of Heritage. The report prepared by this Cultural Industries Sectoral Advisory Group on International Trade (Cultural Industries SAGIT) was titled *New Strategies for Culture and Trade: Canadian Culture in a Global World*, and it proposed to fortify the space carved out for the exemption of culture in the free trade rules negotiated for the Canada-U.S. Free Trade Agreement

(FTA) and the North American Free Trade Agreement (NAFTA).[161] The main targets were the most all-encompassing and most powerful rules of international trade, which were administered by the World Trade Organization (WTO), to which Canada was and is a signatory.

The SAGIT report, *New Strategies for Culture and Trade*, identified two main strategic instruments that were needed to allow Canadian domestic cultural production to survive and grow. The first was a cultural exemption, like the one used in the FTA and NAFTA, that took culture "off the table" in international trade negotiations; the second was a new international instrument that specifically addressed cultural diversity and acknowledged "the legitimate role of domestic cultural policies in ensuring cultural diversity."[162]

The second strategic instrument, the call to legitimize the "role of domestic cultural policies in ensuring cultural diversity," amounted to taking culture off the table in any negotiation of international trade in goods and services, and legitimizing government support for cultural production. The acceptance of such a principle in international trade negotiations would give federal cultural policy a new life. Instead of leaving cultural industries to play valiant Davids in battle with the global mainstream entertainment Goliaths, the principle of ensuring cultural diversity legitimized the Canadian government's acknowledgement and protection of the considerable social and economic benefits provided by domestic cultural industries. Representatives of the cultural industries had been promoting these benefits for years, and this new focus on the part of government put the numerous micro-measures the book industry required into a new framework. Legitimizing domestic cultural policies by positioning them as a means of promoting international cultural diversity transformed the fostering of Canadian content and perspectives, changing it from a tactic for protecting products that underperformed in the domestic market in the face of international competition into a fundamental tenet of interaction among the communities of the world. *New Strategies for Culture and Trade* called for nothing less than respect for the plurality of human communities and their mores, the diversity of human societies, and the richness and overall social gain inherent in cultural distinctiveness and the interaction of such plural perspectives.

Following the release of *New Strategies for Culture and Trade*, senior Department of Canadian Heritage officials, along with the minister, Sheila Copps, poured considerable energies into bringing the proposed policy instrument — which came to be known as the New International Instrument for Cultural Diversity (NIICD) — onto the world stage. The conceptual and policy foundations for this endeavour deserve some space in this book, because they represent new thinking that may bring about a fundamental turn in international cultural relations.

A new international instrument for trade in cultural products

Those involved in framing rules to govern international trade in cultural products have been at loggerheads since the 1960s, when a variety of countries, including France and Canada, put in place various measures to restrict the market share of foreign productions and to ensure the survival of domestic production in the face of aggressive exporting. In the 1980s, UNESCO's MacBride Commission depicted the ongoing struggle as a quarrel over "fair flows versus free flows."[163] Exporting nations, led by the United States, have claimed that normal market theory (i.e., the encouragement of free trade) should apply to what they call "entertainment products." France, Canada, certain other importing nations, and UNESCO have asked that non-marketplace rules prevail, to preserve the obvious social contribution made by "cultural products." The Canadians and their allies base their arguments on the connections of cultural products both to their creators and to the cultural context in which the product is created, a dynamic recognized in such defining documents of cultural production as copyright law, where expressions of ideas belong to creators and the ideas behind them belong to society.

Based on his first-hand experience in providing legal counsel to Canada's cultural producers and governments, Peter Grant, with the help of journalist Chris Wood, has summarized the development of the new international instrument called for by the *New Strategies for Culture and Trade*. In *Blockbusters and Trade Wars*, published in 2004, Grant and Wood give detailed insight into the thinking of governments and industry leaders, beginning with an economic analysis of the cultural marketplace.[164]

Citing economists such as Richard Caves,[165] Grant and Wood note that the cultural marketplace has a "curious economics" with a number of unusual features. They note that most cultural products (in the neighbourhood of 80 percent) fail to achieve commercial success, and it is virtually impossible to predict ahead of time which products those will be. As well, successful cultural products can produce a much higher reward than ordinary commodities, because while first-copy costs are enormous, the costs of subsequent copies are minuscule. In addition, because cultural goods are "public goods" — they are not destroyed when they are consumed — they can be reused or consumed simultaneously by many without degradation.[166] Further curious economics come from the fact that cultural products that are produced for large, rich, and heterogeneous domestic markets have a lower risk and a much greater potential reward than those produced for smaller, more homogeneous, markets. Too, in large-sized markets that generate high consumption levels, first-copy costs can be completely amortized, and there is little "cultural discount" to diminish consumption in export markets.[167] ("Cultural discount" is the degree to which a product lacks appeal outside the producing culture because of its foreignness — its language, for instance, or its portrayal of personal interactional dynamics peculiar to a specific culture.) These product and market characteristics set up a risk-reward dynamic that causes cultural industries to cluster in certain areas, such as Hollywood, where all services are readily available — for publishing, this would mean editing, design, printing, and marketing.[168]

Contributing to the unusual economic status of cultural products in the international marketplace is the fact that David Ricardo's justifying notion of comparative advantage in international trade does not hold up well for this sector. This is for two reasons. First, although domestic and foreign cultural products compete for leisure time, domestic cultural products are not interchangeable with foreign cultural products from a social or cultural perspective. One builds community; the other invites identification with outsiders. Economists have traditionally not measured such social impact. The opportunity to produce and make available one's own imaginings to one's own community is a basic criterion for human social cohesion[169] and self-determination. These are elements for the generation of Bourdieu's social capital. The second reason derives from the curious economics of cultural products. Given that first copy costs are immense and those of subsequent copies extremely low, and that consumption of a cultural product rarely destroys the item and others can use it afterwards (for instance, a DVD can be viewed repeatedly), run-on exports from large foreign countries are destined always to have the economic advantage over domestic products in the international trade of cultural products. The application of Ricardo's law leads to the undermining of domestic cultural production.

Insisting that Ricardian comparative advantage does apply, representatives of the U.S. entertainment industries argue that while no two cultural works are the same, in the marketplace they compete with one another for audience share. Audiences are then sold to advertisers, and it is the economics of price-based, time-swallowing consumption and sales of audience attention that are fundamental to trading rules. This was the argument U.S. negotiators used at the WTO when the United States challenged Canadian legislation supporting domestic magazines. The United States argued that even though *Maclean's* was a Canadian-owned newsmagazine, containing mainly articles about Canada, or looking out at the world from Canada, written by Canadians, it was also a newsmagazine competing for audiences interested in the news and was therefore interchangeable with the (now-defunct) Canadian edition of *Time*, a U.S. newsmagazine with a few extra pages inserted that dealt with Canadian affairs.[170] Consistent with the U.S. way of thinking, economists and international trade tribunals (mostly involving economists, trade lawyers, and judges) tend to focus solely on the sale of audiences to advertisers.[171]

This is too narrow a focus for addressing trade disputes in cultural goods and services, because there are four markets involved in the production and sale of cultural goods. In the first market, creators sell their content to cultural middlemen such as publishers.[172] Here, the cultural character of the product matters greatly, since a publisher is committed to a certain type of content in a certain area, and the distinctiveness of the product forms the basis for the author/publisher partnership. In the second market, producers sell to retailers — in this example, publishers sell to booksellers. Although all books must compete for bookstore space, here, again, the content of the product matters.[173] For example, mainstream bookstores are unlikely to stock scholarly monographs, because they are selling leisure reading to the general public. While independent bookstores may focus on the content itself in making purchasing choices, national chain stores such as Chapters/Indigo or Barnes & Noble use surrogate measures, such as sell-through ratios (i.e., how many copies will sell, as a percentage of how many the bookseller orders) to make their choices.[174]

The third market involves the sale of the content to the audience: the book to the purchaser, the movie to the patron. This is indeed a competitive marketplace — the consumer has only so much to spend and so much leisure time. That spending is highly influenced by the nature of the product. A customer interested in music-related biographies who plans to purchase *George Gershwin: His Life and Work* is unlikely to buy *Here, There and Everywhere: My Life Recording the Music of the Beatles* as a substitute, although the customer may choose to buy the second, on an impulse, in addition to the first.[175] Generally speaking, the evidence (presented in more detail in Chapter 6) is that book purchasers choose by author, subject, and genre, not by price point.[176] In the fourth market, an audience is sold to advertisers, whether the advertising takes the form of a conventional advertisement, a product placement, or a sponsorship.[177] While advertisers are interested primarily in access to a target audience, the nature of the product also matters in this case. The frame of mind audiences bring to products is what makes *Vogue* magazine more desirable to marketers than *The Walrus*. A case in point: while the Print Measurement Bureau's reliable surveys indicate that over 300,000 males read each issue of *Chatelaine* magazine, it would be unusual for an advertiser to attempt to reach a male audience through *Chatelaine*. More recently, research has shown that television viewers have less recollection of products advertised on shows such as *Sex in the City* than they do of products advertised on shows that are "less sexy."[178]

Grant and Wood argue that the international trading rules should respect all of these markets that, together, form the cultural marketplace. Doing so, they assert, would simultaneously allow trade in entertainment products and leave room in the marketplace for cultural goods that are purchased for their distinctiveness. They go on to describe a set of policies, which they call a cultural tool kit that promotes cultural diversity and takes into account the global market for entertainment products. This tool kit is derived from Canadian cultural policies that have developed over the past four decades. A quick review of all six policies provides an instructive context, although only the last two listed here apply directly to book publishing, as well as to other cultural production.[179]

> **Public ownership** (of broadcasting) aims to serve an informed, entertained, and enlightened citizenry, of all ages and tastes, in all regions, and, in Canada's case, in both official languages and to some degree in other languages.[180]
> **Content and scheduling quotas** — such as the famous Canadian content broadcast regulations — ensure that domestic artists have a chance to gain access to domestic audiences; screen quotas in movie theatres are another example.
> **Rules governing spending by retailers** of cultural content (i.e., broadcasters) redirect the profits of distribution to domestic productions and allow them access to the national market.
> **Competition policy** can be set up to avoid excessive corporate concentration and thereby strengthen the receptiveness of the market to heterogeneous content from creators.
> **Ownership restrictions** can limit ownership of privately held culture-producing institutions to citizens of a country. They don't forbid foreign ownership altogether, but prevent foreign owners from overrunning the market.[181] Investment Canada restricts new foreign participation in the book market and, when it is allowed, requires any foreign takeover of a book publishing business to show net benefit to Canada.[182]
> **Granting subsidies** to domestic firms for specific cultural purposes is a mechanism that has been fought over and accepted by the United States in the case of magazine and book publishing.

Together, these six policies allow nations to create a market space for national culture and populate it with successful content. Grant and Wood note that all six have been generally accepted in an international marketplace governed by free trade agreements and WTO policies, and show that this grounds the Canadian government's case for a seventh policy, the New International Instrument for Cultural Diversity (the NIICD), derived from the SAGIT report, which would govern international trade in cultural goods and satisfy both global entertainment producers and those who wished to ensure access for domestic production.

The proposed NIICD was initiated by Canada through the Department of Canadian Heritage (with encouragement from France and UNESCO), developed in international discussions, and finally placed in the hands of UNESCO, which shepherded it into a full-blown international agreement: the Convention on the Protection and Promotion of the Diversity of Cultural Expressions — in short form, the Convention on Cultural Diversity (CCD). The purpose of the NIICD was to embrace the way in which the curious economics of cultural production work, and to balance the economics of the business with its cultural dimension. Such a balance, supporters argued, would create a much healthier foundation for international peace and understanding. Underpinned by a belief in the desirability of human and cultural diversity, the NIICD was meant to encourage governments to sustain the diversity of thought and expression essential to societies' "resilience, adaptability and capacity for regeneration and growth."[183] It was not intended to stop societies from changing, or even to stop foreign products from being heard or seen in any country. Nor was it intended to restrictively guarantee the circulation of locally created contributions, circumscribed by a government's sense of a country's cultural identity. Rather, its purpose was to invigorate culture by creating an interaction between domestic and international creativity and productions.

From NIICD to CCD: The wash of international diplomacy

In the years immediately following the 1999 report of the Cultural Industries SAGIT, *New Strategies for Culture and Trade*, there was not a great deal of progress apparent in Canada other than consultations and trips to UNESCO's home in Paris by senior Department of Canadian Heritage officials, along with the minister, Sheila Copps. And it is interesting that, given its considerable role in policy development, the Association of Book Publishers was not at all active in advancing the development of the NIICD. In Paris, however, UNESCO officials were at work. In 2001, the organization adopted the Universal Declaration on Cultural Diversity,[184] which laid the groundwork for the UNESCO Convention on Cultural Diversity (the CCD). This convention was passed in October 2005 and subsequently ratified by approximately ninety nations and the European Community, more than enough for it to attain legal status in 2005.[185]

Encouraging and precedent setting as this was, the formulation of the CCD, and the actions following its ratification, have been somewhat disappointing. The preamble to the declaration is inspiring — affirming twenty different statements on cultural diversity, beginning with "cultural diversity is a defining characteristic of humanity," and noting that cultural diversity should be cherished and preserved because it is a mainstay of sustainable development for communities, peoples, and nations and is important for the full realization of the rights and fundamental freedoms proclaimed in the Universal Declaration of Human Rights. However, in the body of the Convention is the statement: "Nothing in this Convention shall be interpreted as modifying rights and obligations of the Parties under any other treaties to which they are parties." Without naming them, this refers to such trade agreements as the WTO and NAFTA.

As Garry Neil, executive director of the International Network on Cultural Diversity, notes, the result is weak support for cultural diversity.[186] Although the CCD obliges states to "foster mutual supportiveness between this Convention and the other treaties to which they are parties" and to "take into account the relevant provisions of this Convention" when interpreting "other treaties to which they are parties or when entering into other international obligations," Neil notes that this "falls well short of [the WTO] standard," which requires states to make "specific and concrete commitments," and which contains an obligatory dispute resolution mechanism with enforceable decisions. Further, there is no room for third parties to intervene. Rather, control is left almost totally in the hands of states themselves.

Neil's analysis is pessimistic, at least for the short-term consequences of the convention. But the CCD does set in place mechanisms that, with determination, Canada and other countries can use to prevent their cultural products from being overwhelmed, especially in their home markets, by global products. And it does lay the groundwork for countering appeals to commodity competitiveness that take no account of the difference between normal commodities and services and cultural ones. Progress in the international arena will be gradual and will depend on the interplay of world events. The near-collapse of the financial system in 2008–9 revealed the greed of business and its inability to self-regulate. Ultimately, this may have weakened the influence of the business sector in society, creating room for other sectors, including arts and culture. Whether a UNESCO-based convention will have any long-term positive impact on the evolution of an international marketplace for cultural products that accommodates cultural diversity is yet to be seen.

Moving forward on emerging foundations

Notwithstanding the 1995 Manning-inspired, Martin-delivered cuts, and Donner and Lazar's musings of 2000, government officials in Ottawa and Toronto (and Quebec) appear to have come round to at least considering the view that the cultural industries, and book publishing in particular, create a form of capital, first identified by Bourdieu, called social capital.[187] This can be defined as the value that arises from "networks together with shared norms, values and understandings that facilitate cooperation within or among groups."[188]

The notion of social capital has been taken up by social scientists spanning a variety of disciplines, including Harvard political scientist Robert Putnam and, more recently, Canadian economist John Helliwell. Putnam's understanding of the nature and value of social capital derived from his study of the efficiency and effectiveness of regional governments in Italy.[189] In his famous interview-essay on the subject, Putnam claimed that America was experiencing a dramatic collapse in participation in civic groups. As a result, he said, social capital was diminishing substantially, a loss for the country as a whole that might even be leading towards its potential disintegration. In his research in the United States, following his work in Italy, Putnam found that communities that scored highly in terms of social capital — e.g., those that were home to networks and cooperation based on shared norms, values, and understandings — also scored well on many indicators of economic, physical, and community welfare.[190]

Canadian economist John Helliwell used the idea of social capital to discuss the operation of the Canadian economy, Canada's place in the world, and the role of government in a globalizing world.[191] His research showed that Canadian interprovincial trade was far more extensive than economists believed. The corollary to this finding was that the Canada/U.S. border represented a far more significant barrier than economists supposed. He explained his findings in terms of trust, which, following Putnam, he suggested is a form of social capital. Helliwell's conclusion was that a major job of governments is to foster cooperative networks and norms within the country, leading to interpersonal trust and community and national well-being.

As the tangible and sizable social and economic contribution made by the cultural industries became more apparent to scholars, governments, and industry members forward from the later 1990s, the federal and most provincial governments seemed to accept their long-term betrothal, if not permanent marriage, to culture and to cultural industries. It was becoming apparent to them that the support they had put in place was providing needed infrastructure for the creative sector to contribute to both social and economic development. In light of the work of Putnam and others, the federal government began to recognize that cultural production is really of greater value to a community than anyone, especially economists, had thought. And in terms of fiscal responsibility, as previously noted, the imposition of the GST on books in 1991 gave the government far more revenue than it has ever thought of giving back to the industry to generate the sense of recognition authors feel in being published, the sense of satisfaction the publishing industry gains from bringing ideas and authors to market, the social cohesion that evolves from these various activities, the jobs created, the income tax earned, the articulation of existence inherent in cultural expression, and the networks of trust and cooperation that lead to the generation of social capital for Canada.

Responding technologically to a retail crisis

It would be fitting to end this chapter on such a dramatic policy achievement: the international acceptance of the principle of cultural diversity that has been implicit in Canadian cultural policy dating back to the Massey commission. Even if that acceptance must be tempered by an acknowledgement of the constraints on the CCD's power, it is, nonetheless, a significant achievement. But there remains a final policy development that arose from an inconvenient particularity in retailing. Surprisingly, it led, seemingly indirectly, to government action equal in effectiveness to the long-lived BPIDP and to the inclusion of a distribution right in Canada's copyright act.

The retailing of books has always been a problem in Canada. The ability of publishers and bookstores to ensure that both newly published and backlist titles were available to consumers was nowhere near perfect. Delivery delays of six weeks for a requested in-print title were common, publicity and retail availability were often uncoordinated, the stocking decisions of small independent bookstores were difficult to predict and highly conservative, and the tendency of the chains to stick with mainstream product depressed both consumers and the market. Efforts to increase the efficiency and effectiveness of the distribution system were continual. In 1994, for instance, the federal government spent just short of $21 million on the Publications Distribution Assistance Program (PDAP). Yet still, the system fell short of being satisfactory.

The issue at the forefront of publishers' minds as they entered the twenty-first century was the impact of a single national chain on the market for books. In its aggressive attempts to capture market share, the national bookstore chain, Chapters, was demanding higher discounts and returning far more unsold books than were normally allowed under the industry's terms of trade. Chapters was also wielding its market power absolutely in its own short-term interests. Such was the level of concern within and outside the industry that the parliamentary Standing Committee on Canadian Heritage undertook a series of hearings in response to entreaties from the Canadian Booksellers Association (CBA). The committee released a report on the Canadian book industry in June 2000.[192]

The report noted that both the CBA and the Department of Canadian Heritage were calling for a re-examination of legislation and regulation in the book industry with an eye to public policy objectives,[193] given that Chapters, through its power in the marketplace and its mode of behaviour, could not only undermine independent bookselling but also destroy Canadian-owned book publishing and negate thirty years of public policy in support of the domestic industry. The report also noted that the committee's mandate was "to monitor the link between the Government of Canada's support to the book industry and the provision of increased choice of Canadian-authored materials to Canadian readers."[194] In taking up its task, the standing committee pointed out that not all forces were lined up against Chapters. Authors and publishers with an interest in the general or mainstream market — including author Peter C. Newman, HarperCollins CEO Claude Primeau, executive director of the Canadian Publishers' Council (which represents foreign-controlled publishers) Jackie Hushion, and even Roch Carrier, the bestselling children's author who was serving as National Librarian at the time — supported Chapters. Opponents pointed to the fragile ecology of literary creativity and experimentation, on one hand, and the barely profitable status of book publishing and independent retailing on the other. For its part, Chapters' main tactic was to deny that it was doing anything unusual. It also called into question any data presented to the committee, particularly with regard to market share.

Canadian-owned publishers and independent retailers hoped that the Standing Committee would find a way to limit the growth of Chapters and its ability to wield its market power. The difficulty the Standing Committee had in taking such direct action was that any move to constrain the company would have been subject to court challenge, and the possibility of forcing the none-too-financially-stable Chapters into bankruptcy was quite real. As a result, while the committee called for better information and set up a limited-term monitoring of the chain's behaviour, no direct action was taken to limit Chapters' market power.

In receipt of the report of the Standing Committee, the Liberal government, in its speech from the throne at the end of January 2001, announced its commitment to help the book publishing and sound recording industries prepare digital content and make the transition to what was then called the new economy. In April 2001, citing the recommendations in the final chapter of the standing committee's report, the federal government offered the book publishing industry the opportunity to receive partial funding to capture point-of-sale purchase data, institute electronic payment and invoicing (otherwise known as EDI), and benefit from a web-based survey of book readers and buyers. Effectively, this proposal set aside the Chapters issue and became the foundation for the Supply Chain Initiative (SCI) that, in turn, established BookNet Canada.[195] This proposal was also undertaken in the context of a long-term concern with book distribution and, as late as 1994, the $21 million investment in PDAP to assist with effective distribution of books throughout Canada.

According to a report prepared by Heather MacLean while working in government, the SCI was an attempt to modernize business communication and product data collection and dissemination in the publishing industry in order to achieve multimillion-dollar efficiencies.[196] Of immediate concern was modernization of product data exchange to provide timely information on title availability, both internally — to publishers, wholesalers, or retailers — and externally between business partners. Just as pressing was the need to monitor sales data, so publishers could see what was selling and be able to order timely reprints. The long-term goal was to collect sales data for every title, ultimately producing a database of past sales, with rich metadata markup, that would be accessible to all members of the industry. The metadata markup would make it possible to analyze the data based on price, genre, format, author, subject, release date, time period, and a host of other variables, singly and in combination.

Such a system was already in place in the United States, the United Kingdom, Ireland, Australia, New Zealand, and South Africa, so the

Canadian industry contacted U.K. experts and formed two committees to improve on what the British and others had created. Along the way, the publishers began to realize that the database they were creating for business-to-business communication could also be used in-house for tracking all necessary information on existing titles and even for title development management. In December 2002, a not-for-profit agency, BookNet Canada, was formed specifically to oversee supply chain management for the industry, assist with transition and training, create and administer standards, liaise with parallel organizations in other countries, administer subject categories appropriate to Canada, and help achieve supply chain efficiencies.[197]

As the supply chain initiative took hold, book publishers and retailers soon recognized the strategic value of the technologically based information system that resulted. In the absence of that system, the information collected and made available to all publishers and retailers — that is, the number of copies sold each week of every title, with extensive analyses of trends over time — could only have been assembled and analyzed by Chapters.[198]

Like the NIICD and the CCD, the creation of the Supply Chain Initiative and, subsequently, BookNet Canada further confirmed and elaborated the evolving policy perspective articulated by *The Business of Culture*. Responding to a retail crisis that threatened to undermine government policy supporting a rich cultural dynamic connecting readers, authors, and publishers, the architects of the Supply Chain Initiative established a powerful information infrastructure that achieved five ends. It precluded Chapters' monopolization of industry-wide sales information; it prevented the monetization of that information, which would have given Chapters an even greater competitive advantage; it gave industry-wide access to all industry members, large and small, both bookstores and publishers, allowing the publishers to take advantage of new printing technologies; it preserved the trans-Canadian heterogeneity of book publishing; and it tied the health and growth of a traditional cultural industry to technological innovation.

Conclusion

The socio-political context within which book publishing policy developed in the 1970s and 1980s was, to say the least, distinctive in the history of Canada. A population bulge known as the baby boom was reaching adulthood in an atmosphere of social discontent with the status quo. Social disruption in the late sixties alerted governments to the immense challenge of dealing with a vast inflow of birth-controlled, libidinous, prepared-to-be-violent, job-and-meaning-seeking adolescents. Canadian governments did more than a passable job of creating meaningful engagement opportunities, and they were assisted in their endeavours by a prime minister who reflected adolescent values in his demeanour, his apparent love of reason and principle, his (denied) passion, and his rights-based and individualist policies on, for example, sexual practices.

In this context, the emphasis on self-expression and national identity in resistance to global corporate products played well and helped foster the emergence of a national consciousness and confidence. Government policy and programs of the 1970s and 1980s made it possible for Canada to develop a writing community and a literary heritage. That community is now established, and the sustained origination of Canadian titles by emerging and established Canadian authors is a credit, not only to those authors, but also to the cultural achievement of the Canadian-owned sector that continues to develop new talent. Supportive government policy instigated a variety of mechanisms that allowed cultural production to claim a marketplace presence, in the face of primarily U.S. determination to maintain open access to Canada's mainstream entertainment markets.

The blooming of cultural confidence and the maintaining of a presence in domestic markets, alongside international success in such areas as music, television programs, and film, laid the foundations for innovation in cultural thought and policies, and for industrial expansion, beginning in the mid-1990s and, since the turn of the century, proceeding in the context of both internationalization and technological change. In its time, *The Business of Culture* (1994) represented a major achievement in governments' view of cultural industries. It set aside reservations about whether governments should continue bolstering such industries, as well as the claim that without cultural industries Canada would be swamped by global products, lose sense of itself, and begin a long slow march to integration with the United States. Instead, the report focused on the conditions necessary for Canadian producers to participate in international and domestic markets.

The new vision — articulated in *The Business of Culture*, the cultural SAGIT and the notion of social capital — built on the policy foundations of the 1970s and 1980s. For instance, the government had implemented the Public Lending Right (1986), giving authors a payment for the presence of their book in libraries; facilitated legislation to allow the formation of Access Copyright, a creators' and publishers' rights collective (1988), to capture revenue from user-initiated reproduction; and legislated changes to the Copyright Act, including the distribution right (1996). These initiatives, which will be expanded upon in Chapter 5 in the context of current policies, represented structural interventions in the market that strengthened the position of the industry and authors, and allowed book publishing to finally achieve a state of equilibrium that was tolerable to both government and industry. By the end of the twentieth century, the various support programs described in Chapter 3 — BPIDP, AECB (now Livres Canada Books [LCB]), and a much better funded Canada Council — were ticking along nicely, providing a comfortable amount of money to publishers to maintain their operations.

Hiccups notwithstanding, this stability seemed to provide a platform for new thinking. Canadian book publishing, like other cultural production, is continuously under threat from trade-oriented aggressiveness and is bolstered only to some extent by Canada's second major achievement, the ninety-nation-plus acceptance of the Convention on Cultural Diversity. Within Canada, the Supply Chain Initiative represents a third major innovation, having modernized distribution control based on almost-real-time data. New investments and new resolve are still required to adapt further to technological change. At the same time, that change will open doors to opportunities that can barely be imagined. With Canada's long-term leadership in thinking about and developing communications technology, Canada's book publishing and other cultural industries are in a good position to develop and thrive. The translation of that good position into an operating reality will require an understanding of the nature and operational dynamics of the book publishing industry in the 2010s — the basics of which are to be found in Chapters 5 and 6 respectively.

CHAPTER 5

The current state of Canada's book industry, government policies, and cultural partnerships

Introduction   159

A quantitative portrait of the state of Canadian book publishing   160

    Table 5.1 Book publishing revenue 2008   163
    Table 5.2 Canadian book publishing summary, 2008   163
    Table 5.3 Canadian- and foreign-controlled firms compared (French- and English-language), 2004   165

Structural support   166

    Copyright   166
    The Public Lending Right   169
    The Convention on Cultural Diversity   170
    BookNet Canada   171
    Summary   172

Industrial support   174

    Book Publishing Industry Development Program/ Canada Book Fund   174
    Support for organizations and associations   176
    Table 5.4 Aid to publishers sales coefficients   177
    Association for the Export of Canadian Books/ Livres Canada Books   178
    Financial totals   179
    Table 5.5 Aid to publishers: Statistical overview   180

Cultural support   181

    Table 5.6 Canada Council for the Arts book publishing support programs in 2005-6   183

International marketing programs   183

Technology-inspired programs 185

Writing programs 185

    Federal industrial grants summary   186

Provincial support 186

    British Columbia   186
    Ontario   187
    Other provinces   188

Support-derived drawbacks and constraints 189

Cultural partners of the book industry 193

    Newspapers   195
    Libraries   197
    Education   198
    Magazines and scholarly journals   199
    Publishing and writing programs   200

Awards: Cultural partners and government acting together — or at least in parallel 201

Conclusion 202

Introduction

The evolution of the book publishing industry, policy frameworks, and support programs has been substantial in the past forty years. Beginning with extensions of arts funding, policies emerged that recognized book publishers as business operations that required both subsidies and access to capital. The vision of the 1970s and 1980s was that, with government assistance, Canadian book publishers would gradually assume a dominant market share and become vibrant and profitable, cross-subsidizing their trade operations with profits from educational publishing. Some publishers had dreams of changing the nature of education — and, as a result, Canadian society — by infusing Canadian cultural realities into school materials, a dream that has not entirely disappeared.[199] However, a dramatic shift in the market position of domestic publishers vis-à-vis foreign competitors was not to be. Foreign-owned educational publishers, alongside the Canadian-owned Nelson Education (now part of the foreign-owned Cengage), have maintained their market position, as have foreign-owned trade publishers, albeit assisted by concentration of ownership. Working with government support, the Canadian-owned trade book publishing sector has expanded in terms of number of firms and title output, but not in its market share or profitability. The main contribution of Canadian-owned publishers has been to bring forward new authors and open up noteworthy areas of discussion in Canadian society.

Beginning in the 1990s, the gradually emerging "creative economy" perspective recast cultural industries as net contributors to the overall economy through their generation of social capital. This was complemented by the view, articulated in the Convention on Cultural Diversity, that the world's diverse cultures are a global public good that should be promoted and protected. In 2012, this creative economy/social diversity duality is still in the process of being established; however, substantial foundations have been laid. A domestic book publishing industry is no longer the unrealistic dream of a few idealists. With a presence in bookstores and a significant profile on world literary stages, Canadian publishers and authors have attracted a following of public and cultural partners who expect nothing less than a continuing literary community in Canada. An examination of the current state of book publishing, of structural, industrial, and cultural support, and of the activities of book publishing's cultural partners bears this out.

A quantitative portrait of the state of Canadian book publishing

The Introduction laid out a brief quantitative overview of book publishing. The following description elaborates on it, beginning with the Statistics Canada data and adding further context as well as additional data. In summary, book publishing is a $2 billion industry, with three-quarters of its revenues coming from Canadian sales. Canadian-owned and foreign-owned firms both participate, with Canadian-owned firms, publishing in both French and English, having the (slight) majority market share. In lucrative sectors such as education, in lucrative areas such as bestsellers, and in importing and distribution, foreign firms predominate to a substantial degree. Canadian firms specialize in own trade-title production and originate nearly fifteen times as many titles as their foreign counterparts. Many of these are written by authors who have yet to establish themselves, and, on average, per-title sales for Canadian-owned firms operating in French and English are in the neighbourhood of $15,000–$20,000. Foreign-controlled firms operating in English are based mainly in Ontario, while Canadian-owned firms are spread more widely, if unevenly, across the country.

To begin from a broad view and with the most recent data available: There are a substantial number of Canadian-owned and -controlled book publishers operating across the country, in all major cities, and in every province. There were 282 such firms in 2006, the vast majority of which were engaged in the publishing of non-bestseller trade books. The most notable exception was McClelland & Stewart, but other firms, such as D&M Publishers, occasionally land bestseller titles. There are also eleven foreign-owned book publishers including, on the English-language side, Pearson, with its major trade imprint Penguin; Random House, with such imprints as Doubleday and Knopf; HarperCollins; Simon & Schuster; Elsevier; LexisNexis; McGraw-Hill

Ryerson; Scholastic; and Wiley. Most Canadian firms are small, doing from $50,000 to $15 million in business, but there are exceptions, such as Harlequin and Nelson Education (the latter being a Canadian-controlled company with considerable foreign investment).

In total in 2008, the industry generated just over $2 billion in sales, $1.5 billion of which was within Canada. This $1.5 billion translated into $3 billion in consumer spending. The net receipts to all firms operating in Canada in French and English for the sale of trade books — that is, titles sold to consumers through such retail outlets as bookstores — was $761 million. While just over half (53 percent) of sales in Canada of all categories of books were made by Canadian-controlled firms operating in French and English, exactly half of trade book sales, $381 million, went to Canadian-controlled firms operating in both official languages. Having said this, it is important to recall that all comparisons between foreign- and Canadian-owned firms are subject to an 8 percent reduction,[200] since as of 2004, as many as 20 percent of books sold in Canada are not taken into account. This is because Statistics Canada does not treat distributors as book publishers unless they originate or co-publish titles. This restriction means that data from firms such as Canadian-owned Raincoast and foreign-owned Simon & Schuster, for example, are not included in the statistics. Because English-language firms accounted for $578 million in trade book sales (which is 76 percent of all trade book sales), a reasonable estimate is that Canadian-owned English-language firms also had the same market share of 76 percent of all Canadian-owned firms, which is $293 million in trade book sales. In order to maintain the confidentiality of sales figures for individual firms, because there are so few firms involved, Statistics Canada does not report any figure for English-language Canadian-owned firms.

Of the $381 million in trade book sales by Canadian-owned firms in Canada overall, $293 million (77 percent) came from sales of own titles — that is to say, titles that either originated in Canada, or for which publishing rights had been acquired by a Canadian-owned firm. Of these own titles, approximately 80 percent were written by Canadian authors and, hence, were probably originated in Canada. In comparison, foreign-owned firms sold $89 million (23 percent) of their own trade book titles, clearly demonstrating that their main stock in trade is imports.[201]

With respect to imports (i.e., agency sales), Canadian-controlled firms accounted for $88 million in trade book sales, while foreign-controlled firms accounted for $255 million, nearly three times as much. This ratio would be higher if distributors were included in Statistics Canada figures.

Of the $233 million in revenue earned by foreign sales of books, $220 million was earned by Canadian-owned firms — probably over half by Harlequin alone through "other foreign sales," a designation that includes books manufactured and sold outside the country by a Canadian firm. International rights sales, which, like grants, are included in "Other revenue," were reported to be $154 million for Canadian-controlled firms. This leaves the eleven foreign-controlled firms with earnings of $151 million. If we set grant and donation income at $60 million, rights sales by Canadian-owned firms can be estimated to have been $93 million — approximately one-third of all rights sales, presuming that the Statistics Canada data includes no other significant sources in this category.

Table 5.1 provides the foundational data for the above analysis.[202] Comparative data for 2004 and 2006 can be found in the Appendix, Table 5.1 (extended). (One cannot go back further for comparative purposes because of Statistics Canada's changed definitions. As well, the portion of the industry that was sampled rose from 94 percent in 2006 to 98 percent in 2008, so percentages are the best foundation of comparisons.) All cells mentioned are bolded in the table to assist identification. Domestic sales of educational and trade books by Canadian-controlled firms are not provided by Statistics Canada and are my guesstimates. All percentages are calculated using the surveyed data. Ontario figures on the right-hand side of the table provide an indication of the dominance of foreign-controlled and Canadian-controlled Ontario publishers in the Canadian market. In quick summary, of all domestic sales earned by firms operating in English in 2008, Ontario book publishers (Canadian- and foreign-owned) accounted for 88 percent of all sales, 80 percent of own-title sales and 68 percent of own trade title sales.[203]

Table 5.1
Book publishing revenue 2008 (in thousands of dollars)

	Survey portion	Eng-lang surveyed	Share of sales French & English	Cdn-controlled firms		Share of total foreign- and Cdn-controlled	Ontario	% Ont
Total Operating Revenue 2008	**$2,118,182**							
Survey portion	$2,067,642	$1,622,190	78%	$1,177,757	57%		$1,310,577	81%
Sales in Canada	$1,528,412	$1,180,388	77%	$803,626	53%		$1,038,441	88%
Own titles	$939,880	$684,319	73%	$617,975	66%		$549,665	80%
Educational	$432,732	$306,446	71%	$244,829	57%		$299,894	98%
Trade incl. children's	$417,551	$303,922	73%	$292,997	70%		$206,194	68%
Schol., ref, prof.,tech	$89,597	$73,950	83%	$80,148	89%		$43,576	59%
As exclusive agents	$588,533	$496,069	84%	$185,652	32%		$488,776	99%
Educational	$219,301	$207,742	95%	$86,000	39%		$207,400	100%
Trade incl. children's	$343,432	$274,242	80%	$87,623	26%		$267,648	98%
Schol., ref, prof.,tech	$25,800	$14,084	55%	$12,029	47%		$13,727	97%
Combined Own and Agency titles	$1,528,413	$1,180,388	77%	$803,627	53%		$1,038,441	88%
Educational	$652,033	$514,188	79%	$330,829	51%		$507,294	99%
Trade incl. children's	$760,983	$578,164	76%	$380,620	50%		$473,842	82%
Schol., ref, prof.,tech	$115,397	$88,034	76%	$92,177	80%		$57,303	65%
Exports & other foreign sales	$233,292	$210,742	90%	$220,335	94%		$154,963	74%
Other revenue (grants, rights)	$305,938	$231,060	76%	$153,796	50%		$117,173	51%

Source: Statistics Canada Catalogue no. 87F0004X

Table 5.2 summarizes the 2008 data in a slightly more user-friendly form.

Table 5.2
Canadian book publishing summary, 2008

All book sales by firms operating in Canada	
$2.1 billion	Total revenue of Canadian book publishing
$2.0 billion	Total revenue included in Statistics Canada survey
	(NB: All figures that follow are survey figures)

CHAPTER 5 The current state of Canada's book industry

$1.18 billion	Revenue of Canadian-controlled firms (French and English); 57% of total revenue
$539 million	Revenue from exports, of which $233 million are books and $306 million are rights
$220 million	Book export sales of Canadian-controlled firms; 94% of total export sales; 19% of total revenue of Canadian-controlled firms

Canadian domestic book sales

$1.5 billion	Revenue from Canadian domestic book sales; 74% of total revenue
$1.2 billion	Book sales by English-language publishers; 77% of domestic book sales
$803 million	Book sales by Canadian-controlled firms (French and English); 53% of domestic book sales (possibly approximately 8% less share than stated)
$761 million	Trade book sales (own and agency titles); 44% of domestic book sales
$578 million	English-language trade book sales (own and agency titles); 38% of domestic book sales
$381 million	Trade book sales by Canadian-controlled firms; 25% of domestic book sales; 50% of domestic trade book sales
$293 million	Estimated English-language trade book share if sectors match overall ratio ($381 million x 0.77)

"Own titles" in Canada (see definition above, includes buy-ins)

$940 million	Sales of own titles in Canada, all genres, all companies
$618 million	Sales of own titles in Canada, all genres, Canadian-controlled firms; 66% of all own-title domestic sales; (44% for foreign-controlled); 77% of sales in Canada by Canadian-controlled firms (23% for agency titles)
$684 million	Sales of own titles in Canada, all genres, English-language firms; 73% of own-title domestic sales
$417 million	Sales of own trade titles in Canada
$293 million	Sales of own trade titles in Canada by Canadian-controlled firms; 70% of the market share of domestic sales of own-title trade books
$304 million	Sales of own trade titles in Canada by English-language firms; 73% of the market share of domestic sales of own-title trade books

Source: Statistics Canada, Catalogue no. 87F0004X

A slightly different view of the industry comes from older Statistics Canada data that included both publishers and distributors. In 2004, the foreign-controlled sector was composed of nineteen firms (including some distributors) that published 710 new trade titles and reprinted 885, for a total of 1,595 trade titles. These new and reprinted title numbers can be compared to overall sales to estimate per-title average sales.[204] For the foreign-owned sector, per-title average sales were $58,047. The 311 Canadian-controlled firms surveyed in 2004 by Statistics Canada originated 10,351 new titles and reprinted 3,711 trade titles for a total of 14,062. Their average per-title earnings for new and reprinted titles was $14,176, while a calculated upper limit of average per-title earnings for trade titles is $19,258. These results are summarized in Table 5.3.

Table 5.3
Canadian- and foreign-controlled firms compared (French- and English-language), 2004

	Canadian-controlled firms	Foreign-controlled firms
Total revenues	$1,086,605,000	$949,906,000
Sales in Canada	$561,050,000	$808,418,000
Own-title sales in Canada	$429,653,000	$425,255,000
Own-title trade sales in Canada	$199,338,000	$92,591,000
Number of firms	311	19
Trade titles originated	10,351	710
Trade titles reprinted	3,711	885
Estimated average sales per trade title	**$14,176**	$58,047
Sales of trade imports	$37,953,000	$255,121,000

Source: Statistics Canada, "Book Publishing Industry," *The Daily*, July 10, 2008.

The differences between the data of Tables 5.1 and 5.3 are partly the result of data in Table 5.3 including non-publishing distributors. The particularly noteworthy figure, bolded for emphasis, is the estimated per-title trade sales of Canadian-owned firms: $14,176. Allowing for error and including wishful thinking, the average per-title trade sales of Canadian-owned firms might be as high as $19,258. Whatever the exact figure, Canadian-controlled publishers are crippling themselves with such per-title sales.[205]

Structural support

The main structural support for the industry is copyright law, to which, over the years, the federal government has introduced some special features that serve this country's particular needs, as well as measures that have precedent elsewhere. Other structural measures include the Public Lending Right, the Convention on Cultural Diversity, BookNet Canada (a quasi-structural agency), and support for a robust education system.

Copyright

The distinguishing feature of Canada's copyright law relevant to the issues being discussed is the extension of copyright to a distribution right. In general, copyright provides the foundation for a time-constrained, infinitely divisible monopoly that allows authors and publishers to benefit from their investments in the creation of a book. In keeping with the provisions of the Berne Convention, Canada extends national treatment to foreign authors — that is to say, it extends the treatment it provides to its own authors and publishers to those from other countries whose books are sold in Canada.

The distribution right is covered in Section 27 of the Copyright Act and is elaborated in a document appended to the act entitled Book Importation Regulations.[206] Section 27 grants sole supplier status to Canadian importers and distributors, which allows Canadian exclusive distributors to supply books to the trade (i.e., wholesalers and retailers) without competition. This status parallels the position of a publisher, but is conditional on the distributor meeting certain conditions, including timely delivery of the books for a surcharge of not more than 10 percent on U.S. books and 15 percent on books from other countries. Should the distributor exceed this surcharge, booksellers and other purchasers of multiple copies, such as libraries and schools, are free to import the titles directly from the originating foreign publisher. A similar sole supplier provision applies to the publication of Canadian editions of foreign titles. The publisher of a Canadian edition must notify industry members of the availability of that edition in specified ways. In return for doing so, the publisher establishes its right to be the sole provider of the title to the Canadian market.[207] Notwithstanding the distribution and Canadian edition rights, individuals can import books, and booksellers can import single titles on behalf of clients upon the client's request.

The distribution right addresses what is called parallel importation, which is the importation of a book that a Canadian publisher or distributor has the exclusive right to sell directly to Canadian bookstores and libraries. Parallel importation usually involves third parties, such as wholesalers and remainder houses, that are not party to the agreement between the original publisher and the exclusive agent. But parallel importation is not restricted to third parties. It also occurs when the original publisher dumps remainders (unsold, discounted copies of books) into the Canadian market by selling them, directly or indirectly, to a remainder retailer in Canada, when the Canadian distributor is still selling them at market price.

One issue the distribution right could address, but does not, is the ability of U.S. publishers to obtain North American rights inclusive of Canada. A stronger copyright act provision could require all books sold in Canada to be distributed by Canadian-owned firms, but the political will to require by law what is normal practice in the United States and the United Kingdom has been lacking. Speaking historically, the distribution provision improved what had become an untenable policy operating on the foundation of an ineffective customs tariff, referred to as Schedule C.[208] In 1987 the Canadian government incurred the wrath of the U.S. film industry by trying to require Canadian distribution of all films.[209] In the end, the government succeeded in enacting regulations roughly parallel to those of the book publishing industry; for example, preventing new entrants to the film distribution industry or takeovers of Canadian-owned distributors, and implementing subsidies to the Canadian-owned sector.[210]

The extension of the Copyright Act to control book importation shows both inventiveness and a recognition that Canadians must tailor their laws to the reality of the Canadian industry. The distribution provision is absolutely crucial for the domestic book industry. The elevation of importers and distributors to a legal status that approximates a rights holder's status helps sustain a cross-Canada distribution infrastructure that serves both imported titles and Canadian-originated titles, which are usually by Canadian authors and are often on Canadian topics. If that distribution infrastructure did not exist, Canadian publishers would likely have to sell and ship their books to the various regional warehouses of U.S. national wholesaler Ingram, which would in turn ship them back into Canada. As most Canadians know, the reason Sir John A. Macdonald caused a national railway to be built was to avoid the same kind of Canadian dependency on the U.S. for transcontinental shipping and travel that would result from a lack of recognition of Canadian book distributors. As well, given that agency titles account for approximately 50 percent of the market, the lack of a distribution right could easily cut the size of the Canadian book publishing industry by half. The fact that a distribution right is unique to Canadian copyright law underlines the role that the political environment plays in implementing laws that shape industry.

Canadian copyright law also allows for the formation of collective societies that ensure creator rights holders receive remuneration for use made of their intellectual property. To this end, the Canadian Copyright Licensing Agency was formed in 1988. Originally called Cancopy and now known as Access Copyright, until 2012 the agency issued two types of licences: the "comprehensive licence" covers occasional photocopying by members of an organization, while the "transactional licence" charges about ten cents per page for making identified photocopies, usually multiple copies of trade and scholarly books. In 2012, the agency stopped issuing transactional licences because, now that digital content can be placed on password-protected sites, there is no way for unauthorized copying to be monitored and held in check. Access Copyright collects fees from these licences and, after deducting administrative costs, distributes those fees to the creators (and publishers) whose work is copied. To provide a sense of the scale of its operations, in 2010 Access

Copyright collected just over $33.7 million and distributed $24.0 million to copyright holders. It used a percentage of the difference to cover expenses, seek compensation, expand its activities, and hold a surplus. The comprehensive licence is currently a source of contention, because educators argue that much of the photocopying they are paying for is allowed under the "copying for private study" element of the fair dealing clause and if it is not, it will be covered by a new "education" exception under fair dealing that was proposed in Bill C-11 in 2012.

A number of other elements of Canadian copyright law are notable. One is the existence of the Copyright Board of Canada, a statutory body that can set law — for example, tariffs for photocopying. Fair dealing is another element in the Copyright Act that is more restrictive for users than the U.S. equivalent, fair use. This restrictiveness results in a flow of funds to rights holders in Canada, especially from educational institutions, that exceeds the flow to rights holders in the United States. In 2010, proposed revisions to the act, contained in Bill C-32, and repeated in 2011 in Bill C-11, maintained this comparative restrictiveness, but included expansion of the exceptions within fair dealing to education, satire, and parody. The term of protection in the Canadian Copyright Act is also shorter than similar terms in the U.S. and E.U. acts. In those jurisdictions, copyright expires seventy years after the death of the author, whereas in Canada it ends fifty years after death. Moral rights in Canada are weaker than they are in, for example, France. In Canada, explicit moral rights are largely restricted to allowing an author to control the association of his or her work with a product, person, or institution.

These copyright laws and policies have been crucial to Canadian book publishing. Without them, the domestic book publishing industry would be a shadow of its present state and size.

The Public Lending Right

The Public Lending Right (PLR) program provides additional structural support that is directed at authors and indirectly benefits publishers. The PLR Commission operates a fund, created and maintained by the Canadian government, to provide payments to Canadian authors whose books are in the collections of public libraries. As of 2011, approximately 17,500 authors had registered over 81,600 titles, 70,865 of which received support. The PLR Commission surveys Canadian public libraries each year to determine how many times listed titles appear. Based on this survey, in the fiscal year 2010-11, the commission distributed $9.9 million, with the amount any one author could receive capped at approximately $3,400. In 2005–6, the maximum an author received for a single title was $340 and the average amount individual authors received was $566.[211] The fund obviously favours authors who write books that are likely to be purchased by public libraries, and it extends the time over which many authors receive remuneration for their work. It is not uncommon for authors to receive more in PLR payments than in royalties.[212]

The Convention on Cultural Diversity

Structural support extends into the agreements dealing with international trade. Canada is a signatory to the Convention on Cultural Diversity (CCD), and as such has committed to abiding by the articles of the convention. If its ninety-plus signatories embraced the CCD enthusiastically, incremental change could take place over the next two decades that would result in dramatic shifts in the structures of cultural markets, improving the balance between domestic productions and foreign imports. It remains to be seen whether that will happen and, indeed, whether Canada will continue to lead the effort to recognize the social and economic benefits of cultural diversity. The Conservative government of Stephen Harper has shown little interest in the CCD, and without the strong voice of a G20 nation bringing forward initiatives that maintain a focus on balancing domestic production and global products, the CCD is liable to founder. Canada has shown belated interest in joining the Trans-Pacific Partnership, which is promoting economic integration among Asia-Pacific nations. Involvement in this group would offer Canada a chance to make the influence of the CCD felt.[213]

BookNet Canada

According to its website, BookNet Canada "finds and creates solutions that make it easier for Canadians to buy books, sell books, reach new audiences and ride the ever-cresting wave of new technology."[214] The foundation of its services, and hence its powerful structural support, is the provision of machine-readable book metadata (product information such as title, author, price, genre, length, format) that can be used for a variety of purposes. The main function is to track titles anywhere in the supply chain, from publisher through bookseller to final sale. As well, data on the warehouse holdings of the publisher, the stock levels in any bookstore, and the weekly sales are available for every title. And all data are available to all participating members of industry.

These data facilitate inventory management, allowing publishers to see for the first time how many of the books they've shipped are in the hands of consumers, rather than sitting on bookstore shelves or in warehouses. Sales data also give publishers a sense of what is doing well in the market and where, either by region or by store. This allows them to manage reprinting with information, rather than educated guesswork. BookNet's coordination of this rich information base has the potential to diminish the number of returns, as both publishers and retailers are in a better position to track sales, understand demand patterns, and forecast future sales. Marketing campaigns and their impact on sales can also be tracked.

The same information base is used for streamlining sales between publishers and bookstores. The electronic data interchange (EDI) system put in place and overseen by BookNet is Internet-based and allows ordering by bookstores, and acknowledgement, invoicing, shipping documentation, and record keeping by publishers, all paper free. The benefits are in the immediacy of the interactions and the reduced transaction costs.

As well as increasing efficiencies in the day-to-day operations of publishers and booksellers, BookNet's services allow publishers to undertake extensive analyses of trends over time. Through this reorganization and effort, BookNet Canada is encouraging book publishers to make greater use of standards, databases, and machine-readable information. The combined services that BookNet makes possible have been valued at $80 to $100 million a year.[215]

Regional publishers also rely on data provided by software called BookManager, a stock-control and sales-tracking service that has been sold to four hundred independent booksellers, with high numbers using it in British Columbia and in the smaller population centres of Ontario.[216]

At its debut, a few small publishers saw BookNet as a Trojan horse. They reasoned that BookNet provides need-to-know information to large publishers about their own books and those of their competitors, but only nice-to-know information to smaller publishers, who already know their markets well. BookNet data encourages large operations, such as Chapters/Indigo, to treat all books as equal, rather than assuming that the sales profiles of books published by small publishers will be lower as a result of their limited marketing resources. It also allows large publishers to spot authors who are becoming successful, making it easy for them to woo such authors away from the smaller houses.

BookNet is a structural measure, but it can also be seen as industrial support. It requires government funding, and in 2008 the Harper government proceeded with a scheduled reduction of funding for the agency from $3.5 million to $2.5 million. The reduction was originally scheduled with the understanding that BookNet would establish itself, then move to a maintenance function. However, the project has changed the economies of publishing and bookselling and, hence, introduced structural change.

Summary

Federal, provincial, and municipal governments provide vast indirect support of the structural foundations of book publishing. At the forefront is the protection granted authors and publishers by copyright law and by Canada's extension of that law to protect distribution. Complementing copyright law is the Public Lending Right, the Convention on Cultural Diversity to help strengthen cultural markets, and BookNet Canada's provision of sales information on all titles to the industry as a whole. But the pinnacle of public support is actually the billions of dollars spent on education and literacy. These are obviously of benefit to book publishers, even if the direct beneficiaries are the individuals receiving education and, more generally, the society they live in. Funding for public libraries has also contributed immensely to ensuring that Canadians are literate and have a literary culture that is widely, if not universally, accessible. While publishers might argue that libraries cost them sales — and such arguments were put forward in the debates preceding establishment of the Public Lending Right — it is clear that without public access to written material, there would be many fewer readers in Canadian society. Public libraries can be seen as an extension of the education system, providing the backbone of literacy for the country.

Of course, Canada is not unique in investing in education and literacy, but it is easy to take such investment for granted. As suggested in Chapter 1, a country's organizational infrastructure has an enormous impact on the publishing and reading of books. That societal or organizational infrastructure begins with roads, transportation, and warehouses; it includes such basics as law and order, a postal system, and a tax system; and it extends to respect for learning and for the various professions that encourage knowledge creation and allow time for knowledge creators to engage in book writing. The value literate societies place on literacy and books, and hence authorship and publishing, reflects humanity's fundamental desire for knowledge and understanding, in both a personal sense and for their contribution to the development of a community of understanding; naturally, when these things are valued, book publishers benefit immensely. In reviewing the development of Canadian book publishing and governmental support, one can sense the determination of the Ontario Royal Commissioners and subsequent governments to build a nationally sensitive, but internationally referential, knowledge community in Canada, founded on the legacies of such notables as Egerton Ryerson, Lorne Pierce, and Vincent Massey. With their understanding of the nature of books, those involved in establishing a book industry in Canada would have been aware that they were also laying the foundation for the further authorial opportunities and, indeed, the national and international acclaim Canadian authors have received.

Industrial support

All levels of government have put in place direct financial support programs that contribute to populating the book marketplace in Canada with Canadian titles. At the federal level, the two main programs for book publishing, administered by the Department of Canadian Heritage (DCH) and the Canada Council for the Arts, provide, respectively, industrial and cultural support.

The DCH industrial support program, known for years as the Book Publishing Industry Development Program (BPIDP), was renamed the Canada Book Fund (CBF) in 2009 "to signal a shift in its focus from developing the industry to providing support to a mature industry that will help bring Canadian books to readers."[217] DCH also partially funds the Association for the Export of Canadian Books (AECB — now Livres Canada Books), which supports publishers' foreign-rights marketing. The Aid to Scholarly Publishing Program (ASPP), funded by an arm's-length agency, the Social Sciences and Humanities Research Council (SSHRC), complements these two main programs. In three provinces, British Columbia, Ontario, and Quebec, there is further industrial support, which in B.C. and Ontario is delivered in the form of non-refundable tax credits. Other provinces provide cultural support programs.

Book Publishing Industry Development Program/ Canada Book Fund

The following description of BPIDP programs is based on information available at the time of writing. It describes BPIDP, but according to the Department of Canadian Heritage website, the Canada Book Fund (CBF) will carry on in much the same manner, although the support is purposed somewhat differently: "The principal objective of the Canada Book Fund (formerly Book Publishing Industry Development Program) is to ensure access to a diverse range of Canadian-authored books in Canada and abroad."[218]

Until it was renamed, the Book Publishing Industry Development Program (BPIDP) provided direct assistance to over 220 publishers and also financed several projects designed to assist the entire industry.[219] The BPIDP application guide listed several criteria publishers had to meet before receiving direct aid, including requirements that they were primarily engaged in book publishing; were financially viable; had been in business for at least thirty-six months; were 75 percent Canadian owned and controlled, with a headquarters and 75 percent of employees in Canada; and had fulfilled their contractual obligations to authors (i.e., their royalty payments were up to date).[220] Eligible publishers could be private sector firms or university presses. In its "reference year," effectively a firm's previous year of business, a firm's annual sales could not exceed $20 million, and profits could not exceed 15 percent. The firm must also have published at least fifteen new, Canadian-authored trade books, or ten new educational or scholarly titles, during the previous three years, ending with the reference year. During the reference year, it had to have published at least one new Canadian-authored book. Normally a firm was required to have minimum annual sales of its originated titles of $200,000. There were limits on the percentage sales value of self-published titles and vanity titles. For scholarly publishers, support was provided for Canadian-authored titles or those translated or adapted by a Canadian.

Eligible publishers could receive funding for books that met the program's criteria. They had to have been written by a Canadian author or adapted or translated by a Canadian, clearly and publicly attributed to the author(s) or translator(s), at least forty-eight pages in length (except for children's books, which could be less than forty-eight pages), the publisher's own title, published under the publisher's imprint or under an imprint for which it had distribution rights, and printed in Canada (except for co-published books or books with an acceptable justification). There was also a list of ineligible title types, which included vanity titles, directories, calendars, games, catalogues, and the like; books underwritten by a political party; books that contained advertising other than the publishers' promotional material; periodicals; and books that contained undesirable content such as hate literature, pornography, and gratuitous violence.

The amount recipients received from BPIDP was determined using a set of coefficients based on eligible sales. (This method of allocating funds remains in use by the Canada Book Fund.) Sales are multiplied by the appropriate coefficient for each category of book. Table 5.4, taken from the CBF application guide, reports the 2010 coefficients. The table is best read by focusing initially on the third line, which shows that the first $400,000 of sales of Canadian-authored books (including learning kits) have a coefficient of three. This supports small firms. Sales greater than $400,000 have a coefficient of one — a substantial drop, based on the notion that above the $400,000 sales level, firms become much more viable. Note that there is no further drop, say to 0.5 or zero, for sales beyond a certain higher level. Line 5 also places emphasis on digital books and rights, with a coefficient of three for sales — up to 15 percent of all sales of Canadian-authored books, well above what would be normal for a publisher's program. Note also that translations prepared by Canadians from one official or Aboriginal language to another receive more financial support than original books. This extra funding directly reflects social policy and was put in place because publishers must pay both the author and the translator.

Table 5.4 also indicates that some support was available for foreign-authored books adapted or translated by a Canadian — although not a great deal, reflecting the assumption that many books falling into that category would be able to sustain themselves in the market. Non-print materials and digital files, distributed online or on physical media such as CD-ROMs, were also supported to some degree, as were book packs (books accompanied by complementary objects such as toys). These categories covered the practicalities of the book market for specific groups, such as computer users and children.

Support for organizations and associations

Following BPIDP, CBF continues to administer the program Support for Organizations and Associations.[221] Eligible activities include marketing, professional development, strategic planning, internships, and technology projects. In the category of marketing, CBF funds activities that advertise and promote "Canadian-authored books published by Canadian-owned and -controlled publishers ... [including] promoting/advertising to the public the work of nominees or award-winners."

Table 5.4
Aid to publishers sales coefficients

Canadian-authored books		
Eligible sales up to $400,000	Official or Aboriginal language translations	3.75
	All other Canadian-authored books, including learning kits	3
Eligible sales > $400,000		1
Foreign-authored books adapted or translated by a Canadian	Up to 25% of sales of Canadian-authored books	0.5
	Portion > 25% of sales of Canadian-authored books	0
Digital editions or digital rights	Up to 15% of sales of Canadian-authored books	3
	Portion > 15% of sales of Canadian-authored books	1
Physical non-print editions e.g., DVDs	Up to 50% of sales of Canadian-authored books	1
Portion > 50% of sales of Canadian-authored books	Portion > 50% of sales of Canadian-authored books	0
Book packs	Sales up to $400,000	1.5
	Portion > $400,000	0.5

Note: For official language minority publishers and Aboriginal publishers, the Canadian-authored books coefficient is 4.5 for eligible sales of up to $100,000.
Source: Department of Canadian Heritage, Canada Book Fund, *Support for Publishers Application Guide* (Ottawa: DCH, 2010.), http://www.pch.gc.ca/pgm/flc-cbf/sae-sfp/guide/102-eng.cfm.

Under professional development, it funds "projects that will help Canadian book industry professionals acquire skills and knowledge related to the book industry," including "formal training, workshops, mentoring and other innovative, information-sharing opportunities." Strategic planning funding is available for "projects for organizations and professional associations that assist recipients in adopting strategic approaches to key challenges." Through its support for internships,

CBF funds "internships that provide valuable on-the-job training, thereby contributing to the development of the next generation of Canadian book industry professionals," and through its technology projects funding it will support "collective technology projects that support the industry in becoming more efficient, viable and adaptable to constantly changing trends," including "development and implementation of a communication and distribution infrastructure; and, improvement to the collection, use and dissemination of information; or, training for and development of the respective human and technological resources necessary to adapt to industry changes."[222]

Association for the Export of Canadian Books/ Livres Canada Books

The Association for the Export of Canadian Books (AECB) has also been renamed. It now operates under the name of Livres Canada Books. It remains funded by the Department of Canadian Heritage, although there is nothing to confirm that fact on the website.[223] One of the more interesting facts that can be found on the website, which gives a sense of the program and its successes, is that whereas

> export sales of Canadian-controlled firms increased by 74% between 1992–93 and 1998–99 (not including rights sales) . . . between 1993 and 2000, total export sales for [the group of AECB permanent beneficiaries] increased by 231%, with distribution sales growing by 221% and rights sales by 381%! During the same period, this same group of Canadian publishers' rights sales increased by 767% in Germany, 545% in France, 368% in the U.S., and 26% in the U.K.! Also between 1993 and 2000, this group of Canadian publishers' distribution sales increased by 331% in the U.S., 214% in the U.K., 73% in Belgium and by 18% in France.

The two main AECB programs were the Export Marketing Assistance Program (EMAP) and the Foreign Rights Marketing Assistance Program (FRMAP). The objective of EMAP was to "assist Canadian publishers in their efforts to develop and strengthen their sales to foreign markets (rights and finished products) and to expand foreign markets for their Canadian titles"[224] by subsidizing export sales. FRMAP "provid[ed] funding for export sales trips and participation in international trade events." Under Livres Canada Books (LCB), EMAP has been transferred to the Canada Book Fund. LCB continues to produce catalogues, distributed globally, that list Canadian books available to foreign publishers for distribution or rights sales. It coordinates the Canadian presence at various international book fairs during the year, designing and staffing Canada's displays and providing subsidies to publishers wishing to attend. It also provides export market intelligence and funds professional development.[225] The market intelligence consists of "unique market insights to help Canadian publishers discover opportunities, monitor trends, develop strategies and hone their export marketing efforts." Included under this rubric are blogs by experts who address wide-ranging topics "from industry statistics and trends to buying patterns and book categories, key editorial sectors, supply chain issues (for example, those having to do with distributors, wholesalers and booksellers), market segments (such as book clubs, non-traditional accounts, libraries and more), mergers and acquisitions, book deals, events, awards, resources and more."[226]

Financial totals

The 2004–5 edition of *Printed Matters*, and its renamed sequel *The Book Report* of 2005–6, both of which are annual reports of the Book Policy and Programs Division of the Department of Canadian Heritage, reveal the level of support provided by all DCH industrial support programs in the context of the value of the Canadian book publishing industry.[227] For example, in 2004–5 and 2005–6, support provided through the Aid to Publishers program was $26.5 million and $26.7 million respectively, just under 10 percent of total sales of Canadian-authored titles. The Aid to Industry and Associations was $3.6 million and $4 million, while the Supply Chain Initiative amounted to $3.5 million and $2.6 million. International Marketing Assistance through the AECB was $4.8 million in both years. Total support provided was just over $38 million in both years and 14.1 percent of revenues for Canadian-authored books.

Spread over the total number of firms, this is approximately $175,000 per firm. In percentage terms, support provided can also be stated as 10 percent of all publishing revenue ($38.4/$393 million) and nearly 6 percent of total revenue ($38.4/$658 million). Given that the primary purpose of the support is to publish Canadian authors, 14 percent seems the most appropriate figure to use. Table 5.5 is a statistical overview of the DCH programs for the six years from 2000–1 to 2005–6. A much more limited set of figures can be found in the Annual Report of the Department of Canadian Heritage.[228]

Table 5.5
Aid to publishers: Statistical overview

	2000-1	2001-2	2002-3	2003-4	2004-5
Number of book publishers – total	213	217	217	221	218
French-language book publishers	105	105	104	102	100
English-language book publishers	108	112	113	119	118
Number of new titles published – total	5708	5874	6277	6270	6098
French-language book publishers	3368	3446	3856	3889	3376
English-language book publishers	2340	2428	2421	2381	2722
Sales of Canadian-authored titles – total (in millions of dollars)	260	281	288	281	272
French-language book publishers	130	137	144	146	143
English-language book publishers	131	144	144	135	130
Publishing revenue – total (in millions of dollars)	369	388	410	416	393
French-language book publishers	191	199	203	224	216
English-language book publishers	178	190	208	192	177
Total revenue (in millions of dollars)	627	615	635	671	658
French-language book publishers	258	267	280	304	301
English-language book publishers	369	348	355	367	357

Sales of Canadian-authored titles as a % of total revenue	41.6%	45.6%	45.4%	41.9%	41.3%
Profit margin (median)	4%	2.3%	2.9%	4.8%	3%
Number of publishers per total revenue category					
Less than $200,000	22	26	23	22	19
$200,000 to $999,999	101	104	98	102	96
$1 million to $4,999,999	60	56	65	62	70
$5 million and greater	30	31	31	35	33

Publishers and contributions per region in 2004-5	Publishers	% of total	$ millions	% of total
British Columbia	23	10.6%	2.4	9%
Prairies and North	26	11.9%	2.3	8.8%
Ontario	62	28.4%	7.3	27.6%
Quebec	97	44.5%	13.8	52%
Atlantic	10	4.6%	0.7	2.6%
Total	218	100%	26.5	100%

Notes: Total may not sum due to rounding.
Data are reviewed and revised on an on-going basis.
This can result in changes to data published in previous editions of the annual reports.
Source: Department of Canadian Heritage, Book Publishing Policy and Programs Printed Matters 2004-05 (p.50) and The Book Report, 2005-06 (p. 30)

Cultural support

As noted, the third type of government support for book publishing, cultural support, comes mainly from two Canada Council programs, Block Grants and Emerging Publisher Grants. The goal of the programs is "to offset the costs of publishing Canadian trade books that make a significant contribution to the development of Canadian literature."[229] Many of the same criteria used by BPIDP, and now by the CBF, are considered by these programs. To be eligible for the Block Grants program, firms must have published at least sixteen titles. Those with fewer titles in print, and all first-time applicants, must apply to the Emerging Publisher program. The guidelines note that "the Canada Council's mandate includes supporting production in the literary arts, and the study of literature and the arts," so it funds only titles in the following categories: fiction; poetry; drama; graphic novels (of more than forty-seven pages); eligible publications for children and young adults; and literary non-fiction, which is defined as

> narrative text about real events, people or ideas, where the writer's voice and opinion are evident and the narrative is set within a context and a critical framework. The work should be accessible to a general reading audience and cannot be intended for a specialized or academic readership. Eligible literary non-fiction titles make a significant contribution to literature, or to information about the arts or to the enjoyment of writing by Canadians. Titles within the following subjects are eligible, if they meet all other eligibility criteria: art, architecture, biography, history, literary criticism, nature, philosophy, politics, reference, social sciences, sports and travel.

There is also an extensive list of subjects that are not eligible for funding, including reference books, calendars, cookbooks, guidebooks, and publications destined primarily for an educational or scholarly market.[230] And there are minimum requirements for size of print runs, amount of Canadian-authored content, and number of pages.

Grant amounts are determined through an evaluation of the publisher's program in previous years, but they are awarded to offset future costs of production of eligible titles. Awards consist of two elements: a base component, which is calculated from the number of eligible titles the publishing company has produced over the previous three years (for multi-year grants) or two years (for annual grants), and a bonus portion determined by a committee of peers, who evaluate each applicant's publishing program in comparison with the programs of all other applicants. Table 5.6 shows the amount of support made available in 2005–6.

Table 5.6
Canada Council for the Arts book publishing support programs in 2005-6

Grant Program	Amount Awarded	No. of Recipients
Emerging Publisher Grants	$350,000	23
Block Grant Program	$7,885,000	156
Translation Grants	$666,400	99
Author Tour Grants	$316,900	126
Grants for Art Books	$109,800	16
Total	$9,328,100	

Source: DeGros Marsh Consulting, *Essential Support for Literary Publishing in Canada: A Review of the Canada Council for the Arts' Programs in Support of Book Publishing* (Ottawa: Canada Council for the Arts, 2006), 1, www.canadacouncil.ca/writing/.

Table 5.6 shows that a total of 179 publishers were eligible for grants from the Emerging Publisher and Block Grant programs. Normally only Canadian-owned firms may access Canada Council funding, but in the case of author tour grants, at least one foreign-owned company received support.

The Canada Council supports creative trade books and scholarly books. A sample of scholarly presses and titles indicates that in 2008, the average per-title grant received from the block grant program was approximately $3,300.[231] Scholarly titles, whether published by trade or university presses, are also eligible for assistance from the Aid to Scholarly Publishing Program, which is funded by the Social Sciences and Humanities Research Council (SSHRC) and administered by the Canadian Federation of the Humanities and Social Sciences. It provides $8,000 per title for costs of production, which include promotion as well as typesetting, printing and binding. In 2005–6, allocations to publishers under this program totalled $298,000.

International marketing programs

The Department of Foreign Affairs and International Trade (DFAIT) programs for culture, specifically those for literature and publishing, also benefitted publishers, though not by putting money in their pockets. Instead, these programs expanded marketing efforts, resulting in increased sales of books published in Canada and also of rights to translation and foreign publication. They can be viewed as additional industrial support to book publishers.

The objective of DFAIT's literature and publishing program, cancelled in 2008, was to increase international awareness of Canada and Canadian literature by supporting the promotion in foreign markets of recently published fiction and literary non-fiction. Financial assistance was provided for promotional projects by authors in direct support of a literary work, and for participation by writers or their agents in important international book fairs.[232]

DFAIT also complemented the work of the Association for the Export of Canadian Books through its Canadian Studies Program, and through the Cultural Personalities Exchange Program (also cancelled in 2008). The Canadian Studies Program awarded funding to academics and more than twenty-four Canadian Studies associations outside Canada for teaching, research, and publication about Canada in various disciplines, including English- and French-Canadian literature, culture and the arts, history, social and political sciences, geography, business studies, the Canadian political system, the economy, status of women, aboriginal issues, multiculturalism, social values, the environment, law, information media, and even regional development. The objective was to develop a greater knowledge and understanding of Canada, its values, and its culture among scholars and other influential groups abroad. Facilitating the sale of Canadian books to Canadian Studies courses in foreign countries was a high priority, as was funding travel for noted Canadian authors so they could mingle with the members of the Canadian Studies associations.[233]

Direct funding to these associations was supplemented by the Cultural Personalities Exchange Program, which provided exposure for outstanding Canadian achievements in the arts and scholarship. Applications from all cultural, social, or academic sectors were entertained, with support usually limited to airfare to priority countries involved in the promotion of Canadian Studies abroad.

Technology-inspired programs

At the federal level, the creation of Canadian Culture Online within the Department of Canadian Heritage, and its four programs — the Canada New Media Fund, the Virtual Museum of Canada, Culture.ca, and Culturescope.ca, the online service of the Canadian Cultural Observatory — were positive developments. The aim of the observatory was to inform and advance cultural development in Canada by giving cultural policy researchers, planners, managers, and decision makers specialized services that foster more responsive research, encourage informed decision making for policy and planning, and stimulate community debate and improved knowledge exchange. Culture.ca was a free, interactive online hub that connected interested Canadians to data about and discussions of government policy on culture. Most significantly, the Canadian Culture Online programs were completely in keeping with the view of cultural industries as contributors to economic growth. Unfortunately, they were all cancelled in 2008, with only the New Media Fund revived in 2009.

As the final draft of this book was being prepared, an initiative was being put in place at the federal level through the Supply Chain Initiative, the federal program under which BookNet was funded until early 2010. Basically, it amounted to training industry members for digital realities by funding technology internships. These internships are available both to publishers and publishers' organizations.[234]

Writing programs

Although aimed at writers, the Canada Council's writing programs are notable for their benefit to publishers, as well as for their extensiveness. They make it possible for authors to set aside time to write, enhance marketing by providing support for festivals, and underwrite some costs to foreign publishers for translations. They also provide support to literary magazines, which often serve as incubators for writers honing their craft before moving on to create books. The Canada Council also goes to considerable effort to include programs specifically addressed at Aboriginal forms of literary expression, such as story telling. And in keeping with the changing forms of creative development, a program called the Artists and Community Collaboration Fund has been created.

Finally, there is also a set of Canada Council–administered prizes for which writers and publishers are eligible, including the CBC Literary Awards, Canada-Japan Literary Awards, Governor General's Literary Awards, Japan-Canada Fund, Joseph S. Stauffer Prizes, and Molson Prizes.

Federal industrial grants summary

A conservative estimate of federal grants to book publishing can be made by totalling the values of the various programs. Unfortunately, the latest data available are not for the same year, thus the overall accuracy of the estimate is somewhat decreased. In 2005–6, BPIDP support was $38.1 million. According to 2005–6 data, the Canada Council added a further $9 million to bring the total to $47.1 million. If this is divided by the number of eligible firms, the result suggests that grants make up just over 18 percent of revenues. This calculation does not take into account DFAIT funding, nor ASPP funding. But neither does it take into account that only some firms are eligible for Canada Council support and, similarly, only some firms are eligible for BPIDP/CBF support.

Provincial support

Various provinces provide structural, industrial, and cultural assistance to publishers, both directly and indirectly (by supporting writers). This support assists publishers in maintaining viable businesses.

British Columbia

As mentioned, Ontario and British Columbia offer tax credits to publishers,[235] which are seen by many economists as grants by another name. But they can equally be seen as structural support — rewards, in the form of forgiving taxes, for taking on work unlikely to generate normal profits. British Columbia's system, instituted in May

2003, is simple. According to government documents, "The amount of the refundable tax credit for a taxation year is equal to 90% of the ... BPIDP contribution" received for the taxation year.[236] Since BPIDP represents 9.7 percent added revenue (as previously calculated), the B.C. tax credit adds a further 8.8 percent (0.9 × 9.7%) revenue for Canadian-authored titles. If a firm qualifies for the B.C. tax credit, BPIDP, and Canada Council funding, the percentage of added revenues could equal as much as 26.8 percent (18% + 8.8%).

British Columbia publishers also receive assistance from the B.C. Arts Council, which in 2006 provided $450,000 to nineteen publishers, an average of $23,600 to each. This adds another 1.9 percent of revenue for Canadian-published titles, bringing government support to a total of 28.7 percent of revenue (based on the national average firm size, which is undoubtedly greater than the size of the average B.C. firm). The intention of the support is to assist individual titles that contribute to provincial or national arts and culture. The amount awarded is calculated on the average deficit for the genre of the title, with reprints funded at 50 percent of the average deficit of an equivalent original title.[237]

Ontario

The Ontario Book Publishing Tax Credit (OBPTC), instituted in 1998, "is calculated as 30% of the eligible Ontario expenditures [pre-press, printing, and marketing costs] incurred by a qualifying corporation with respect to eligible book publishing activities, up to a maximum tax credit of $30,000 per book title."[238] Overall spending and assistance by title figures are unavailable, but it is reasonable to assume that they would be in the same ballpark as support in British Columbia.

The Ontario Arts Council also assists publishers with cultural support.[239] In 2006–7, it awarded $754,000 to thirty-eight publishers for 516 new eligible titles, which amounts to $19,800 per publisher and $1,460 per title. The council also provided some funds for reprints.

As well as providing direct support, the Ontario government is also continually developing policy analysis that leads to incremental changes. In particular, the programs of the Ontario Media Development

Corporation (OMDC) signal changes in keeping with the re-visioning of cultural industries that began in 1994. For example, in 2008 the OMDC facilitated a strategic study of the book industry that examined the opportunities and challenges presented by online and other digital technologies (including e-books).[240] Overall, the report argues that the industry requires the means to acquire new skills and knowledge to cope with coming technological change, as well as the resources "to explore and experiment with new business models."[241]

The OMDC report suggested the government establish a "Digital Transition Fund" and made fifteen recommendations, three of which, in whole or in part, embrace the digital world, with a fourth embracing cross-sector fertilization of initiatives. The remainder of the recommendations focus on the still-dominant print-on-paper medium, not surprising given that the steering committee was composed of existing book publishers.

Other provinces

Most other provinces provide some assistance to book publishers, though not quite at the same level as British Columbia and Ontario.

Quebec provides export assistance, business assistance, assistance to book fairs, loans, loan guarantees, and a refundable tax credit through the Société de développement des entreprises culturelles (SODEC). Other support comes to literature in general through the Conseil des Arts et Lettres du Québec and the Conseil des Arts de Montréal.

Through the Cultural Industries Guarantee Fund, the Alberta government provides loan guarantees that help publishers obtain bank loans. The government also sponsors a set of awards for publishers through the Alberta Foundation for the Arts. In 2004–5 the foundation awarded $255,000 to eleven publishers, $23,000 per firm. In 2006–7, $282,000 was awarded to ten publishers, $28,200 per firm, while in 2007–8, $360,000 was awarded.[242] In 2010, Alberta publishers were lobbying for a Book Publishers Operating Support Initiative (BPOSI), "a transparent and straightforward operating grant formula per eligible publisher of 30% of cost of sales (COS)."[243]

Saskatchewan has two sources of funding: the Cultural Industries

Development Fund, which provided approximately $85,000 to sixteen projects in 2005–6; and the Saskatchewan Arts Board, which supports Saskatchewan's three literary publishers with approximately $200,000 and provides funds to other literary projects, some out of province.

Through the Book Publishers Project, the Manitoba Arts Council provides grants to that province's publishers to encourage the publication of original editions of Manitoban and Canadian books of cultural significance. The maximum award per title is $3,000. The Manitoba Ministry of Culture, Heritage and Tourism administers the Manitoba Book Publishers Marketing Assistance Program, which helps Manitoba book publishers reach new markets and audiences and increase sales by providing grants for publishers' marketing activities. The Manitoba Book Publisher Project Support Program helps Manitoba book publishers afford new technologies of professional skills training, while the Industry-Wide Assistance Program helps the Association of Manitoba Book Publishers defray costs for projects that benefit member publishers, including *Prairie Books Now* (a newspaper that promotes books), Manitoba Book Week, and the annual writing and publishing awards, Brave New Words.[244]

There is no support for book publishing in Prince Edward Island, and support is limited in Nova Scotia, where eight publishers received $166,000 in 2006. In 2008, New Brunswick announced a program that would distribute $550,000 per year for three years. In Newfoundland and Labrador, the Publishers Assistance Program provided $200,000 to five publishers in 2009.[245] Overall, it appears that each firm in Nova Scotia, New Brunswick, and Newfoundland/Labrador receives approximately $40,000 annually in support of their activities.

Support-derived drawbacks and constraints

It would be a mistake to review current policy and support programs and treat them as simple ways of injecting needed funds into firms that are publishing important books that are insufficiently supported by the market. Support programs inevitably have drawbacks and constraints that distort the industry they assist. That said, virtually every industry is supported in one way or another by governments, be it through foregone tax revenues during an initial period of production, favourable tax regimes to encourage research and development, multibillion-dollar roads into the wilderness, the building and staffing of lighthouses, the provision of weather reports, or permission to exploit resources such as minerals and water flows. Such supports clearly advantage certain firms and certain industries and they are accepted as part and parcel of a free market.

The support Canada provides to its various cultural industries can be seen as having gradually evolved from after-the-fact subsidies to the structuring of a "free" market, in which Canadians have access to writing by Canadian authors on a variety of subjects for readers of various tastes and ages, attending to regional differences, in our various languages. The support governments provide can also be seen as directed towards the creation of a cultural milieu in which those with the potential to write are encouraged by society to do so, knowing that they must compete with others for significant audiences. At the same time, the support Canada provides to publishers can be seen as a contribution to an infrastructure that helps Canadians to know about their country and be motivated to contribute to it, rather than, for example, looking longingly south at greater apparent freedom and opportunity.

As mentioned, the notion that Canadian cultural industries are weak and thus require governmental crutches to survive, let alone thrive, is not a fact but an interpretation. This point of view fails to consider that all industries require structural intervention to make it possible for them to operate. For the cultural industries, that intervention consists largely of providing early and continuing opportunities for both creators and cultural industries to participate in production, to learn the nuances of cultural business operations, distribution, marketing, intellectual property, and so forth. The cultural industries can also be seen as making substantial contributions to social cohesion and social capital that pay off nationally and internationally. In the case of Canada, these industries require the fine tuning of the market infrastructure to counteract the size, shared language(s) with other countries, and cosmopolitan nature of our domestic market. As well, Canada's cultural industries can and should be seen as driving the country's participation in both domestic and international cultural marketplaces, encompassing generic entertainment products, culturally distinctive products, and, increasingly, online-mediated social networking.

It is this view of Canada's cultural industries as the foundation for participation in national and international markets that is endorsed and embraced by the Ontario government's *Business of Culture* document and the SAGIT-recommended international instrument for trade in cultural products that resulted in the Convention for the Protection and Promotion of Cultural Diversity, the CCD. Both emphasize the social importance of culture and of providing Canadians with opportunities to participate in the domestic and international cultural marketplace.

The Canadian challenge is to provide industrial and cultural support to maximize both creativity and market participation. This is difficult. For example, low profit levels mean that publishers have no retained earnings to invest in firm building and development. A history of low profit levels, combined with a plethora of small firms, means that there is no culture within book publishing for firm building.[246] The intent, heralded in *Vital Links*, of enriching financial support to book publishing and allowing large firms to diversify and, thereby, to be self financing, turned out to be empty rhetoric.[247] No such building took place.

Living within a break-even environment and being accustomed to receiving funds from government creates a culture of entitlement. This, coupled with a culture of poverty brought on by limited retained earnings, leads publishers to look first to government for funding when desirable initiatives such as adoption of new technology are being considered. Many small publishers believe that they have no funds to invest in training, technology, or firm reorganization. They feel unable to borrow money and to hire expertise.[248] As a result, they continue in operational modes to which they are accustomed, unable to take advantage of new developments. The culture of poverty can extend to a lack of responsibility in the spending of public funds. It is not unknown, but neither is it common, for firms to seek assistance for marketing initiatives with the full knowledge that the funds spent will never be earned back in revenues. The project may proceed because funds exist for that purpose. Were a greater percentage of funds coming from the applicant's retained earnings, greater rigour would be brought to the decision to seek assistance and the design of the project.

The attitude of Canadian-owned book publishers towards training is in contrast to those of other industry and professional associations, such as health, engineering, architecture, and teaching. While these professions involve a great deal of day-to-day common sense, all require considerable training in the disciplines upon which practice is built. Opportunities are created and rewards are given to those who pursue further training. In publishing, given that the industry employs highly skilled artists, editors, and marketing personnel, there is ample room for encouraging and taking advantage of advanced training. Editors seem to be the most keen in professional development, but there is little financial recognition of their efforts. On the other hand, the proliferation of continuing education and college level programs, rather than graduate programs that take more seriously the core concepts of publishing, tend to encourage continuity in the organization of publishing rather than the capacity for change — at a time when change is most possible and most advantageous.

Even though book publishing employees will accept relatively low wages, it is wasteful of human resources to start a university graduate with a Master in publishing and a second Masters' degree at a salary equivalent to an entry-level clerical position, let alone to assign that person clerical work. This is doubly the case when the same graduates can earn up to twice that salary doing publishing work outside the industry. Low wages are sometimes justified as part and parcel of a lifestyle choice that a person is prepared to make. Such thinking encourages mediocrity, feeds off a culture of poverty, and buys into a false notion of the nobility of poverty. It is a disservice both to those who are underpaid and society as a whole. Moreover, it is indefensible that government should be subsidizing an industry that does not compensate its employees with a living wage. Paying low wages not only drives people out of the industry but also encourages an inefficient organization of work.

Eligibility criteria for grants have also created barriers to entry in book publishing. Federal government policy now makes it such that a person or group of people who wish to begin publishing books cannot qualify for a full set of grants without having published for three to four years at a rate of four books a year. Consider these simplified calculations. If, as determined in the review of grants conducted earlier in this chapter, a new book publisher were to publish with average revenues and average costs, and have no access to grants, that publisher would have a percentage loss of 28.8 percent minus 3 percent equals 25.8 percent, or rounded down, 25 percent. If average per-title costs were $30,000 then the losses would amount to (.25 × $30,000 per title × 16 titles) $120,000. We also know that approximately one third of sales is required in working capital. This adds on another $40,000. Thus, without at least $160,000 in capital, it would be folly to start a new publishing firm. Even then, to earn back the $120,000 loss and earn out the $40,000 in working capital at 3 percent profit would require $5.33 million in sales. This is not taking into account compound interest.

Emerging Publishers grants at the Canada Council improve the situation somewhat, bearing in mind that Canada Council grants are for "creative" titles. Assuming they contributed the average of 4 percent, after four titles had been published, these grants would reduce the accumulated debt to $105,600,[249] requiring only $4.9 million in sales to earn back losses and working capital (again ignoring compound interest).

These figures mean that any grant-eligible company is worth at least $120,000, independent of its backlist. Most probably the least successful company, publishing only four books per year, would be worth a minimum of $150,000. The above reasoning and calculations mean that there is an effective barrier to entry of $150,000 plus working capital to any person thinking of founding a new book publishing firm. This certainly holds in B.C. and Ontario, and to a slightly lesser degree in some other provinces. An initial investment of anything less than $200,000 would be absolutely insufficient to start a book publishing company in Canada.

Cultural partners of the book industry

The supportive relationships that book publishers and authors share with various "cultural partners" are critical to the success of individual firms and of the industry as a whole. These are not the business connections that exist between author, publisher, and bookseller. Rather, they are the social and cultural interconnections with institutions that have the same general goals as the book publishing industry: to bring cultural content to Canadians. The conception of newspapers, television, radio, magazines, libraries, awards programs, and certain parts of the education system as cultural partners that share general goals with book publishing and each other facilitates an understanding of the nature of such partnerships and a recognition of how vulnerable they are in Canada's cosmopolitan environment.

A business partnership involves a close working relationship. When publishers supply books that sell well to retailers, both parties benefit. Retailers' sales efforts benefit themselves but also their suppliers. Reliability of supply is also mutually beneficial, as is proper display and promotion within a store. Cultural partnerships are different. Within a framework of complementary goals, each organization in a cultural partnership is free from obligation to any other partner and has the responsibility to choose whether or not to support any particular activity or product. Individual media outlets can decide which book, magazine, song or album, or movie is of material interest to its audience. A newspaper can decide that book reviews are unimportant and dispense with them altogether. Book review journals in Canada can turn the major part of their attention to writing outside Canada, something that would not happen in the United States, the United Kingdom, any European country, China, Japan, and so on — unless, of course, the publication touted itself as an international outlet. In contrast to business partnerships, where there is a direct sell-in, negotiation, and feedback on performance, interaction between cultural partners is more distant. Books are proffered. Reviews may or may not be commissioned. Those reviews that are commissioned may not come in, which means the book may be missed. Cultural partners have no moral obligation to devote time or space to individual titles, and they may decide to ignore Canadian productions altogether.

The history of cultural partnerships in Canada, and in many other colonial countries, is typified by an ingrained tendency of the partners to look to the outside world as much as the domestic world. It was really only in the late 1960s and early 1970s that Canadians challenged the notion that cultural products generated elsewhere were somehow of better quality and greater significance than those created in Canada. What was evolving in the years prior to 1995 and is still with us — and, again, Ontario's *Business of Culture* document signalled a turning point — is a greater cohesion among cultural partners that takes for granted the common cause of creating and disseminating Canadian cultural products to national and international markets. This is a major milestone in Canada's cultural history.

Newspapers

In 1993, a study assessing the cultural partnership between newspapers and book publishers showed that, in a sample of eight newspapers and one weekly newsmagazine over two four-month periods, 3,472 titles were reviewed and given total coverage of approximately 250,000 column lines.[250] A subsequent 1995 study, which made reference to the 1993 paper, reported that CBC Radio and commercial talk radio allocated substantial time to discussion of books and authors, and book talk had been extended over the years into public and commercial television.[251] This same study revealed that book coverage in big-city dailies had increased, that there was occasional coverage in community weeklies, and that the number of book reviews had increased in Canada's two national newspapers. For example, on Saturday, March 19, 1994, the *Globe and Mail*, following a typical pattern, included four full pages of books coverage, consisting of a quarter-page column focusing on publishing, nine full book reviews, a children's books column, two brief book reviews, a bestseller list, and book advertising.

Between 2000 and 2008, greater attention was being paid to books. On Saturday, March 17, 2007, the *Globe and Mail*'s weekly sixteen-tabloid-page *Books* section contained the following items:

- a one-tabloid-page picture-dominated cover with a Chapters/Indigo banner ad
- a contents page (tabloid size) with a 60 percent Chapters/Indigo ad

- ten full book reviews of various lengths on 6.6 tabloid pages
- five children's book reviews on 0.6 of a tabloid page
- seven paperbacks noted on 0.2 of a tabloid page
- 1.8 pages of other book-related material, including a best-seller list
- 5.1 tabloid pages of ads, including 1.7 pages of Chapters/Indigo ads, 0.9 of other bookstores, 1.4 of foreign-owned publishers, 1.0 of Canadian publishers, 0.1 of libraries, and 0.9 of awards organizations
- a one-page essay by a well-known author

In addition, in a separate section of the paper, there was one full broadsheet-page feature (with a picture taking up almost half of the page) on Jane Austen and her representation in books, television, and movies. Total coverage was nine full broadsheet pages, but with nothing on publishing. A review of subsequent issues of the *Globe* showed that this coverage was typical. Publishing news does appear on certain days — one reporter is generally assigned to the industry — when it shares space with further book coverage. Up to the end of 2008, books were being discussed by at least two other columnists, and a business reporter combined book coverage with business features from time to time. Such business coverage may be more productive for publishers, since books are their products, and industry dynamics are of interest to relatively few — perhaps, at most, half a million followers of the book industry (see Chapter 7).

The story changed in 2009. As newspapers lost readers to the Internet — to the point where the *New York Times* was in danger of closing its doors — they seemed to refocus on readers rather than the general population. Pitching the change in the most positive way possible, the *Globe* combined its books coverage with a commentary section called "Focus" to create a new two-part section called "Focus and Books." On January 24, 2009, the "Focus" section consisted of several pages of articles and columns, covering topics such as terror and death in Sri Lanka (2.5 pages), China's contribution to the ongoing financial crisis, and the inauguration and presidency of Barack Obama (two articles and a column). Another column offered bizarre quotes from around the world over the previous week; there was also a full page on poet Robbie Burns (with an ad for a book by former U.S. president Jimmy Carter), and a 1.5-page excerpt from a forthcoming book. The "Books" section consisted of seven pages — much less than the stand-alone *Books* section used to be — led off by a one-page review of two new biographies of Burns, accompanied by a poem and three 100-word related-reading recommendations. It also included a half-page review of two books about rivers; a one-third-page review of an anthropological book on linguistics and the nature of language; a one-sixth-page review of what John Polanyi, winner of the 1986 Nobel Prize in chemistry, happened to be reading; a column on a Canadian book club; one six-book review; brief reviews of nine books about dogs; a review of a potential Nobel Prize–winning author's latest book; two bestseller lists; a column on "the next Jane Austen"; a review of a fishing novel; short reviews of five children's books; six brief reviews of paperbacks; and almost one page of book or book-related ads.

Throughout 2009, the *Globe and Mail* did a certain amount of experimentation with its "Focus and Books" section, and it has been difficult to discern the newspaper's strategy for serving the readers of books and the book publishing industry. By the end of 2009, book coverage was to be found in the review section, in "Focus and Books," and online under the general category of Arts. If anything, book-related news and commentary seems to have expanded, both in its perspective and in the number of column inches it receives. Generally speaking, the book industry seems well treated and the editors of the *Globe* seem to believe their readership has a significant interest in books. Other major papers, such as the *National Post* and the *Toronto Star*, have positioned themselves in relation to the preferences of their readerships and in relation to the *Globe and Mail*. Most interesting is the resurgence of newspapers' cultural partnership with books and authors, in an environment of competition between media.

Libraries

Ever since Francess Halpenny's invention of the concept of Canadian collections, Canadian libraries have shown themselves to be fairly reliable cultural partners to book publishers and authors, insofar as they purchase Canadian titles and hold events focused on Canadian books and authors. Most libraries have built up extensive collections of Canadian authors; however, in 1998, in response to a significant downturn in library purchases, the Association of Canadian Publishers commissioned a study called *Collections: How and Why Public Libraries Select and Buy Their Canadian Books*.[252] While some of the findings showed a limited buy-in to Canadian books — for instance, only 53 percent of the 33 large urban public libraries surveyed held 21 or more of 30 adult fiction titles named in a sample of mid-list titles written and published by Canadians in the previous two years — on the positive side, the 33 libraries held 12,141 copies of the 100 mid-list titles surveyed. No quantitative statistics are available, but according to one board member of the Public Lending Right Commission, sampling appears to indicate that library purchases of newly published Canadian books are healthy.

Education

Post-secondary institutions have also partnered implicitly with the book community by developing courses on Canadian literature, geography, society, culture, ethnicity, politics, economics, business practice, art, communication, law and criminality, anthropology, and archeology, and the use of Canadian-authored and -published books has increased steadily. One company, Broadview Press, has specialized in creating Canadian-authored critical editions of Canadian and foreign classics that are traditionally studied in university-level courses. In 2008, Broadview split in half, with the humanities half remaining independent and the social science and history half merging with the University of Toronto Press (UTP), to become UTP Higher Education, specializing in post-secondary course books.

Schools are a different matter. While a number of studies were undertaken in the 1980s,[253] more recent data are difficult to obtain. In 2001, under the auspices of the Writers' Trust, Jean Baird found that Canadian writing is still not systematically introduced and explored with any commitment and vigour in schools across Canada. Of the teachers who responded to Baird's survey, 31 percent reported the existence of a Canadian literature course in their school. This does not mean that Canadian authors are absent in other literature courses, but Baird notes that "the majority of literature taught in Canadian schools is American authored."[254] In the conclusions to her report, she also notes the continuance of the attitude among some teachers that Canadian literature is substandard and lacking in moral fibre[255] and, thus, does not merit teaching in the schools. She observes that "community standards and fear of reprisal [have] a large impact on classroom teachers' selection of materials" and that it is difficult for publishers to market books to teachers because of "the lack of a network to share information about and promote Canadian literature to high schools."[256] The most effective device for such marketing is the writers-in-schools program administered by the Canada Council.

Baird continued to investigate the issue, and in July 2008 she announced that Saskatchewan was the only province to require graduating students to have read some Canadian literature.[257] Later the same month, however, she learned that the new English curriculum in British Columbia "requires B.C. high school English teachers to assign at least one Canadian [-authored] book per year."[258]

Magazines and scholarly journals

Other supportive complementary institutions include both literary magazines and scholarly journals. Today, as in 2000, approximately forty literary magazines operate in English and are funded by the Canada Council.[259] Literary magazines open the door of publishing to new authors and recognize and help publicize published authors by means of reviews. They give authors a chance to experience how their writing is received, first by reviewers and then by audiences. As authors add to their article output, they gain confidence and a sense of the rigours of authorial production. And either through publication itself or through the winning of prizes, these now-published authors come to the attention of book publishers. This may not take place until they submit a manuscript with an attached résumé, but then again it may happen when a publisher reads a story or article in a literary magazine. In either case, the knowledge that one is dealing with an author who has published gives a publisher a sense of confidence that otherwise would be missing.

Canada's more than 250 social science and humanities journals serve the same role, but in a different genre. A succession of articles may provide the foundation for a non-fiction book that is either a scholarly monograph or a more accessible trade title. Scholarly journals also review published books, bringing them to the attention of journal readers and now, by means of online publishing, Internet surfers.

Review publications also greatly assist in bringing attention to Canadian books. The leading publication, because it winds up in the hands of a great number of librarians, is *Canadian Materials*, while the *Canadian Literary Review* is also important and impressive. But more local publications also exist, such as *BC Bookworld*, *Prairie Books Now*, *Atlantic Books Today*, and the *Montreal Review of Books*. As well, professional publications such as *Quill & Quire* and the *Canadian Bookseller* help stabilize the industry and pass information to industry members.

Publishing and writing programs

Of further assistance to book publishing in Canada is the wealth of creative writing, journalism, and publishing programs that have emerged over the last three decades. The various writing programs, of course, train authors. Publishing programs directly train future employees. A variety of programs at universities and colleges across the country (including Simon Fraser University and Langara College in Vancouver; Mount Royal College in Calgary; and George Brown, Centennial, and Humber Colleges and Ryerson University in Toronto) prepare new employees and help existing staff members upgrade their skills for the publishing industry.

In 2006, the Cultural Human Resources Council (CHRC), an independent, not-for-profit organization that addresses workforce issues and workforce planning initiatives in the cultural sector, commissioned a study to examine the state of training in writing and publishing.[260] Researchers interviewed employees to tap their opinions about their skill competencies, gathered information on training program curricula, and assessed topic coverage in the curriculum. While this was not a strong methodology for assessing actual competence to undertake required tasks, the study identified two areas where employees felt their knowledge was not really adequate: information management and the management of rights and contracts. The study also reported two major findings. The first was that "book publishers are not using competencies to a large degree in their people management practices, such as screening, selection and training. They are much more interested in the perceived fit between the candidate and the 'culture of publishing.'"[261] And the second was that "there is a paradox in how training is regarded: on the one hand the book industry is very supportive of specific 'technical' training for employees, but on the other hand the passion for training (e.g., personal development) is less there. Importantly, marketing and technology are the two areas in general that they are interested in seeing staff learn more about."[262] This said, it is now common practice for book publishers to tell those searching for employment that they only hire employees who have completed a post-secondary program in publishing.

Awards: Cultural partners and government acting together — or at least in parallel

Perhaps the most direct and close interaction between the publishing industry and its cultural partners occurs in the celebration of books and authorship. It is not just cultural partners that fund book awards; governments are also active, and literary awards are many and varied, including the $50,000 Scotiabank Giller Prize for the best Canadian novel or short story collection published each year in English and the $100,000 Griffin Poetry Prize (divided between a Canadian winner and an international winner), both of which unabashedly celebrate literature. Indeed, a reasonable argument can be made that such generous prizes inspire other cultural partners to embrace more enthusiastically the literary output of the nation. Non-fiction awards are also significant. The $25,000 Charles Taylor Award for literary non-fiction and the $40,000 British Columbia National Award for Canadian non-fiction stand out. The awards themselves, and the gala events at which they are presented, bring Canadian books and authors to the "top of mind," as the market researchers say, of those who care about writing and publishing.

The Governor General's awards, sponsored by BMO Financial Group (otherwise known as the Bank of Montreal), are, as a group, the richest.[263] They are presented each year to the best English-language and the best French-language books in each of seven categories: fiction, non-fiction, poetry, drama, children's literature (text), children's literature (illustration), and translation (from French to English and from English to French). Winners receive $25,000 and a specially bound copy of their book. Non-winning finalists receive $1,000 in recognition of their selection as finalists. The total value of the awards, which are funded, administered, and promoted by the Canada Council, is about $450,000.

Just how many literary awards are there? There appear to be far more than a hundred in Canada alone. Besides the Governor General's awards, the lieutenant governors of at least three provinces now have their own literary awards. And beyond the Canadian awards are the international awards — the Nobel Prize for Literature, the Man Booker Prize, the Orange Prize, the Prix Goncourt, and the Pulitzer Prize for those born in the United States.

Conclusion

The combined structural, industrial, and cultural support programs of various federal and provincial government agencies have provided a somewhat stable environment of financial assistance for domestic book publishing. Canadian copyright legislation has addressed the needs of authors and publishers, particularly in terms of the importation and distribution of titles. Direct industrial support accounts for nearly 30 percent of publishers' incomes in Ontario and British Columbia (and in Quebec, although that province has not been analyzed here). And the Canada Council and provincial arts councils address the limitations of the literary marketplace by providing cultural support for literary titles.

Valuable support for book publishing has also been increasing from cultural partners in the private and not-for-profit sectors. Books have come to hold a place of pride for Canadians, who view their authors as talented and insightful creators who deserve support. The mass media, and other media, build on the groundwork laid by authors and publishers, getting the word out to all who might listen. The attention books and authors are receiving in contemporary Canadian society is warm, welcoming, and substantial.

Much of the support expresses, in a straightforward way, admiration for the contribution books and authors make to Canadian society and the world. But the reasons for the support run deeper. The colossal investment of resources that Canada, and many other countries, make in books and authors derives from the belief that language is not just a medium of representation, but also a means for knowing about the world. At one and the same time, language, and especially written text, is capable of explicit denotation and implicit evocation; of carrying a narrative and facilitating an explanation; of conveying feeling and facilitating and illustrating logic; of linking the senses to our cognitive abilities. Such power is far greater and more flexible than other media. This immense investment of resources assures an audience for books; encourages the emergence of authors, those desirous of turning their talents for working with words into a tool for making meaning; and underlies the activities of publishers, who desire to turn writing into public documents to contribute to the continuous evolution of the human condition.

CHAPTER 6

Operational and market realities in Canadian book publishing

Introduction   207

Cost structure   208

    Table 6.1 Cost structure of Canadian-owned, title-originating, book publishing companies   208

Market structure   208

    Book selection, imports, and run-ons   209
    Run-ons   210
    Remainders   211
    Title specificity   212
    Discourse communities   213
    Pricing   215

Demand structure   215

    Book purchasing   216
    Table 6.2 Percentage of mentions of reader-identified influential factors on book purchases   217
    The market for Canadian authors   218
    Table 6.3 Three measures of interest in Canadian authored books   219
    The role of bookselling and booksellers   220

Book retailing: Chapters/Indigo and the selling of books to Canadians   224

    An alternative Chapters   226
    Kobo: A hopeful, then dashed, initiative   229
    Restructuring Chapters' retail environment: Clouds on the horizon   230

Market functioning   231

    The dynamics of foreign competition   232

Net benefits and acquiring and retaining authors by
   Canadian-owned publishers   233

The role of authors' agents   236

Authors and publishers' nationalities   237

Sales, marketing, and distribution   237

Returns   241

Copyright reform: Market opportunity or threat?   242

A concluding reflection   244

Introduction

Operating a book publishing firm is no easy task, despite government support, cultural partnerships, and the push for new rules governing international trade in cultural products. It is certainly more difficult in Canada than it is in the larger markets of the United States and the United Kingdom, or in various European countries such as Germany and France, which have their own languages. The overall challenge derives from the normal dynamics of international competition as they play out in Canada, and the curious, as well as the usual, economics of cultural products and markets outlined in Chapter 4. Books originating in Canada, usually written by Canadian authors, must compete with run-on imports exported from more populous countries where economies of scale prior to export dwarf those in Canada. Canadian books must also compete in an environment replete with celebrity marketing machines that diminish greatly any significant cultural discount (see Chapter 4) for U.S. products flowing into the Canadian market. And support from the publishers' mass media cultural partners for Canadian books and Canadian writing is continuously compromised by their need to maintain their mass audiences. These dynamics become particularly important in the absence of international trading agreements regarding cultural products that are designed to be beneficial to all parties.

Imports dominate the domestic market for Canadian books, and, in general, foreign production structures the market by strongly affecting sales patterns and the prices and character of titles. The nature and operations of the retail sector, particularly the dominance of the national bookstore chain — variously called Chapters, Indigo, or Chapters/Indigo — with its emphasis on mainstream titles, and the more recent rise of discount merchandisers such as Costco and Walmart, also influence demand patterns, as do publishing practices that favour large-scale enterprises. As we shall see, one bright spot has been the development of BookNet Canada, which has brought increased efficiencies in the supply chain and, hence, better firm survival. As we review these marketplace and operational realities in this chapter, perhaps the most surprising revelation is that while, structurally,

Canadian-owned publishers are at a considerable competitive disadvantage, and are barely profitable after grants, they manage to survive frequent dramatic market disruptions.

Cost structure

The major categories of costs to book publishers provide an introductory sketch of the challenges of meeting those costs. The figures reported here are rough averages and include data from Canadian-owned companies as small as those generating $100,000 in revenue per year, and as large as those generating $5 million per year.[264] As Table 6.1 indicates, what is normally termed "cost of sales" — all the costs associated with the creation of the book, including editing, layout, cover design, and manufacture — are in the range of 40 to 50 percent of all costs. Promotion and marketing absorb about 15 percent. Distribution, which includes the cost of fulfillment and returns processing as well as damaged books, is often underestimated, and at 14 percent accounts for almost as much as promotion and marketing. All other operating expenses, such as general acquisitions, paying the publisher (who often is not involved in individual titles), running the office, and so on, consume just under 30 percent.

Table 6.1
Cost structure of Canadian-owned, title-originating, book publishing companies

Cost categories	Low range	High range	Approx. avg.
Development and production	40%	50%	44%
Promotion and marketing	10%	18%	15%
Distribution	10%	16%	14%
Administration	28%	30%	29%

Market structure

The origin of titles (i.e., Canadian or imported), the ownership of participating publishers (i.e., foreign or Canadian), and the price range offered to the consumer are also key variables in the structure of the marketplace.

Book selection, imports, and run-ons

Imported titles have always come as blessings and burdens, and these remain. On the one hand, imports provide distributors and some publishers with a relatively low-risk method of "publishing." In acting as sole distributors of foreign works, they build a business and market presence. Distributing imported titles within Canada also creates well-travelled distribution pathways, along which Canadian titles may also flow. Without such pathways, Canadian publishers might have found it necessary to use U.S. distributors and wholesalers to distribute books across Canada. On the other hand, imports increase the competition for display space on bookstore shelves and, even with distributor markups, set price points that are difficult to match with titles that are originated in and for the domestic market. Considering all factors, barring the politically unpalatable and practically impossible alternative of forbidding them, and hence impeding the free flow of ideas, imports will remain a fact of life in the Canadian book market, and their market share is growing.

Looking at the positive side of imports, in the print era, the book community liked to claim that Canadians had access to a greater selection of books than book purchasers in any other country.[265] The basis for this small conceit, most often put forward by small Canadian-owned book publishers trying to find space on bookstore shelves, is that many independently owned bookstores, before the turn of the last century (when their numbers were decimated by Indigo and Chapters), were accustomed to carrying *Canadian Books in Print*, *British Books in Print*, and *American Books in Print*. These three catalogues listed over 90 percent of all books published in English, sometimes including books from India, South Africa, Australia, New Zealand, and other countries of the Commonwealth, as a result of rights acquisitions by publishers of the three titular nationalities. "Available" was, of course, a euphemism, for though a title might be listed, certainly the vast majority were not on a shelf or in the warehouse of any bookstore or publisher in Canada. "Available" meant that many such stores were prepared to place an order for the book and knew where to place the order. But the likelihood of timely delivery was small. Backlisted titles could take anywhere from six weeks to twelve months to arrive. Even delivery of current titles was tardy — three months was not an uncommon waiting time for an order to be fulfilled.[266]

Despite these problems with timely delivery, the potential availability of a wide variety of titles, as well as the actual availability of a large subset, was undeniable. Beyond the fact that many bookstores were willing to order any title from any of the three editions of *Books in Print*, there were four main reasons for this wide selection: some titles were run-ons, some were remainders, bookstores were independent, and used bookstores were common. Today, Internet-based selling has at least maintained, if not enhanced, title availability. First to the run-ons and their market impact.

Run-ons

When a British or U.S. publisher with either a branch plant or an exclusive agent resident in Canada is determining the print run of a title, it has the opportunity to add an extra few hundred copies to cover off the Canadian market, and thus reduce the overall per-unit cost slightly. Given that the cost of printing and shipping is in the order of 20 percent of the retail price, and the distributor will pay at least 30 percent of retail for copies, producing the extra copies would seem to make sense — especially to British publishers, in that the British have the fifty-three nations of the Commonwealth for their export market.[267] If Canada is good for five hundred run-on copies, then so, probably, is Australia, the European Union, and maybe India; add one hundred or so to New Zealand and South Africa, five hundred more for the rest of Africa, five hundred for various Asian markets, a certain number for other countries where English is strong as a second language, and the latest publication on the life of Darwin is a going proposition. (Of course, this all presumes that the originating publisher has not sold off rights, rather than exporting books.) The export market of the U.K. publishing industry is regularly 30 to 40 percent of a total print run, compared with 10 percent in the U.S. British and U.S. run-ons, made available at a price reflecting the lower per-unit cost of the total print run, fill the shelves of virtually every Canadian bookstore not specializing in discount books.

Over the years, exports of titles by U.S. publishers have been evolving towards the same sort of business structure as Britain has, but in countries within the international sphere of influence of the United States. Larger firms have also been establishing wholly owned subsidiaries around the world, while smaller U.S. publishers, alongside authors' agents, seek out foreign publishers and distributors to represent their titles and authors. While book export statistics show lower percentages for the United States compared to those of Canada and Britain, it may be that rights sales, combined with simultaneous publishing by a parent company and various of its branch plants, camouflage U.S. "book exports." Simultaneous publishing, like agency distribution, allows the lowering of per-unit costs by amortizing set-up costs over the whole of the print run.

Remainders

Extended print runs, mainly from the U.S. but also from the U.K., supply Canada with only some of the low-priced books that compete with Canadian-originated titles. Another source is remainders, which are basically unsold leftover print copies. It is a constant challenge for publishers to supply just enough books to enough bookstores for there to be a reasonable chance of a sale to a person who wishes to buy a particular title. There are hundreds of bookstores, millions of people, no way (until recently) of tracking copies once they are gone from the warehouse, and thousands of new titles each season. (In 2004–5, according to Statistics Canada, the industry published 16,776 new titles; Library and Archives Canada processed 22,972 new titles that same year ending March 31, 2005, including self-published and non-commercial titles).[268] Obviously, some are going to be left unsold after the initial demand dies down. Many such books become remainders. They are shipped back to a warehouse and sold by volume or weight or shipping container to the highest bidder, with the buyer knowing little of the actual contents. Prices regularly drop below one dollar per book, and such books, in perfectly good condition, with perfectly good production values, written by perfectly good authors, on perfectly good subjects, can be found on the discount tables of nearly every Canadian (and foreign) bookstore. They take up the vast majority of space in all discount bookstores. One practice that seems particularly unsound is the remaindering of hardback copies in anticipation of the publication of a trade paperback edition. The avid consumer can often find remaindered hardback copies at a lower price than the trade paperback copies that appear in the following months. Remainders also represent a rather haphazard extension of the number of titles available to Canadian readers. But, like run-ons, remainders represent competition with Canadian-originated titles for the dollars of consumers.

These two book sources can be viewed in a favourable light, especially from the point of view of the reader. They mean that a wider range of titles becomes available at low prices. Ironically, because it is important to meet a passing but high demand for bestsellers, as much as any title, bestsellers can be found as remainders a year or less after their release. In part this is because the market must be 20 percent oversupplied to maximize sales; hence, even in a perfect world, publishers get 20 percent returns. In the more usual imperfect world, returns rise to 35 and 40 percent. As well, it often seems that publishers are overly optimistic in their printing of important titles. The serious reader benefits, because such titles can be found in the discounted sections of bookstores, or on the Internet, through AbeBooks and other online bookstores. From the point of view of publishers, however, run-ons and remainders are simply discounted competitors for book purchasers' money. As well, there are three other less-than-favourable effects of the wide range of title choice that run-ons and remainders provide to Canadians: First, such wide-ranging choice leads to expectations of title specificity that are difficult for Canadian publishers to manage (an issue discussed in the next section). Second, they set price expectations that Canadian publishers ignore at their peril. And third, run-ons and remainders place a premium on bookstore shelf space that makes it difficult for Canadian books to compete.

Title specificity

The issue of specificity and its impact on sales can be illustrated with a success story. Generally speaking, the greater the specificity of a title, the smaller the market. *Birds of the Stikine River Valley* is likely to be less successful than *Birds of Northwest British Columbia*, which is likely to be less successful than *Birds of the Raincoast* (including Washington and Oregon). A certain level of generality is required to attract a large enough number of consumers to make the book viable. Canadian readers who are not conscious of this sometimes expect an unrealistic degree of specificity in domestic titles. A political figure, a newsworthy event, or an educational issue that is worthy of a book in the United States or the United Kingdom is, in principle, as worthy of a book in Canada. If U.S. presidential and U.K. prime ministerial memoirs are *de rigeur*, Canadian prime ministerial memoirs should be too. The trouble is that while such a formula may work at the level of national leader, at some point, not very far down the ladder of noteworthiness, the market in Canada is too small to support such an equivalence; the subject is too specific. Publishers know this. But Canadian readers, generally speaking, do not. When Canadians see, for example, multiple biographies of important figures in other countries, they complain of the lack of a similar selection of biographies of equivalent figures in Canada. When they consider likely reasons for this deficit, many Canadians are too quick to engage in deprecatory commentary about their fellow citizens as unimaginative, lacking entrepreneurship, and afraid of risk. "Realistic evaluators of the Canadian market" is a more appropriate epithet.

Lone Pine Publishing has turned the dynamics of generality versus specificity on their head. Based in Vancouver and Edmonton, and recognizing that title specificity was attractive to readers, the firm produced a range of highly specific titles — *Birds of Vancouver, Victoria, Seattle, San Francisco, Los Angeles, British Columbia, Washington State, Alberta, Calgary, Edmonton, the Pacific Northwest Coast* — by using slightly adjusted content for each. Lone Pine has even been able to extend its formula to ghost stories, an indication that this strategy has more potential for publishers than one might think.

Discourse communities

Consumer expectations with regard to title specificity are a challenge. There is no law regarding the appeal of specificity that applies to all titles; most often, this appeal arises from the individual character of the geographic, social, and political community about which the author writes and with which readers identify. It applies more weakly to creative writing — fiction, poetry, short stories, and some drama — in which the specific is often meant to represent the general, rather than itself. The value of such titles comes from their creativity, which is beyond their specificity. Yet even in creative writing, readers look for concrete elements to which they can relate. To some extent, the notion of title specificity operates within the context of what might be called discourse communities — communities made up of the like-minded or the like-concerned. Consider that books of poetry by unknown poets tend to sell about as many copies in Canada as they do in the United States: in both countries, sales of 250 copies are considered respectable. In the 1970s, sales were usually somewhat larger in both countries. In the late nineteenth century, even with its smaller population, they were far higher. In the case of new poetry or even poetry in general, the notion that such books should sell ten times more in the United States than they do in Canada, because the U.S. population is ten times that of Canada, does not hold. One might also surmise that per-title sales levels for poetry are equal in Canada and the United States because more books of poetry are published in the United States (as many as ten times more), and they must all share the market. It also seems likely that individual titles tap into a discourse community, composed of those with sympathy for the author, the subject, and the genre, and that for many subjects and authors, the size of such a community is unrelated to the national population size. On the other end of the scale, Barack Obama's autobiographical works easily sold ten times the number of copies in the U.S. than would ever be expected of a Canadian prime ministerial autobiography, in spite of the lower reading levels of U.S. citizens. Obama's book sales reached the level they did not only because the community of people interested in Obama is directly related to the size of the population of the United States, but also because that country's position of power in the world propelled the book into the hands of an international discourse community interested in powerful men and the politics of the era, and including many readers with an interest in trying to predict the actions of the United States. In short, in this case, world power status combined with population size to determine the size of the market.

Pricing

The challenge of topic specificity is exacerbated for publishers of Canadian-originated titles by the price brackets established by the print runs of imported books. As noted, imports dominate both selection on shelves and customers' choices. On average, the number of Canadian-authored books purchased on any day in any bookstore in Canada varies between 10 percent and, at most, 25 percent.[269] While book buyers often claim that price does not make a great deal of difference to their purchases, and they also report that cover design is immaterial, the experience of publishers differs. Indeed, in the latest Canadian national reading study, respondents were queried about price, discount price, and book cover as factors influencing the purchase of a book. In respective order, 26 percent, 32 percent, and 10 percent cited those three factors as influences on their book purchase.[270]

Although this may be changing, the mindset (with regard to price) of the Canadian book buyer appears to be this: Why should a book by a Canadian author cost more than any other book, especially a book by a more famous foreign author? Only in very recent years, with increased emphasis on organic food products as well as economies of scale, have the realities of production costs — which, in the case of books, means the size of print runs — become persuasive. Prior to those developments, try as they might, Canadian publishers had not been able to persuade the book-buying public that Canadian-authored and Canadian-published books are the gold-riveted designer jeans of the market. In short, in spite of a substantially higher per-unit cost, the price differential for books created solely for the Canadian market remains a disincentive to purchase, as does a poorly designed cover.

Demand structure

Demand structure addresses the demographic characteristics of book purchasers (age, gender, education level) and their purchase preferences, such as format and genre, frequency, price, for whom they purchase books, and at what time of year, as well as the retailers they tend to use. For instance, women buy more books, at a lower average price per book, than do men[271], while teenagers ignore any book-buying advice from teachers, much preferring to select books under ten dollars recommended by friends.[272]

Book purchasing

Reading and Buying Books for Pleasure, a 2006 study commissioned by the Department of Canadian Heritage, reveals a host of factors that influence book purchasers.[273] In 2007, Lorimer and Lynch undertook a second analysis of the data, with publishers in mind, entitled *The Latest Canadian National Reading Study: Publishers' Analysis*. The data clearly show the factors identified by responding readers as influencing their purchases.[274] Respondents were asked to identify all influential factors, and percentages were calculated based on the number of respondents identifying a factor. Table 6.2 shows that when readers whose primary language is English are divided into light, moderate, and heavy buyers, the rank order of influential factors remains consistent across all categories, with a single insignificant exception (bolded in the table). In rank order, the top five factors are topic/type of book, author, discount price, price, and title. The next six factors, in rank order, are Canadian author, size of print (for children and older people), length of book, illustrations and graphics, book cover, and, finally, publisher.

To some degree, the obviousness of the leading factor, topic/type (i.e., what a non-fiction book is about and whether a fiction title is a novel or genre fiction), make it easy to dismiss. But it shows that, for book buyers, content in non-fiction, and whether a book is fiction or non-fiction, share the pinnacle of factors that influence purchases (at 70 percent, topic/type leads by almost a factor of two over any other variable).[275] Two other factors in the top five can also be seen as oriented to content and form (i.e., topic and type). They are author (mentioned by 41 percent) and title (23 percent) — author, in the sense that the purchaser has some sense of how the author writes, and title, as a short expression of the content of the book. Book purchasers

Table 6.2
Percentage of mentions of reader-identified influential factors on book purchases

Factors	All Buyers %	Light Buyers %	Moderate Buyers %	Heavy Buyers %	English Buyers %
Topic / type of book	70	65	71	70	69
Author	41	33	36	50	41
Discount price	32	29	32	34	32
Price	26	28	24	25	26
Title	23	24	26	20	21
Author is Canadian	15	19	15	12	13
Size of print	14	17	14	12	13
Length of book	12	14	11	11	11
Illustrations and graphics	11	13	11	9	10
Book cover	10	12	9	9	10
Publisher	7	8	3	6	6
Mode of purchase					
Impulse	42	42	36	46	46
Planned	56	56	63	52	52

Source: Lorimer and Lynch, *Latest Canadian National Reading Study, 2005: Publishers' Analysis.*

also tend to be open minded, as the lower part of the table indicates. Nearly half their purchases are on impulse (42 percent), while slightly over half are planned (56 percent).

As Table 6.2 indicates, three other influential factors identifiable in the data refer to readability but also relate to marketing: size of print (14 percent), length of book (12 percent), and illustrations and graphics (11 percent) affect readability in a broad sense, and marketability to a particular purchaser. A book's cover is another significant marketing factor (10 percent). The final two influential factors identified in the table are the author's being Canadian (15 percent), and the publisher. The fact that 15 percent of readers attend to whether the authors of the books they purchase are Canadian has some significance. This share is in the same general vicinity as the percentage sales of Canadian-authored books in the market, as reported in a variety of studies. The low ranking of publisher, last on the list at 7 percent, can also be read as positive for most publishers. While it is difficult to build loyalty with so few book purchasers paying attention, it also means that the dominant publishers have not so captured the market that small publishers must break through brand loyalty.

The market for Canadian authors

The 2006 data suggest a somewhat increased interest of readers in Canadian-authored books, compared to the levels identified in two previous reading studies carried out in 1981 and 1995.[276] As Table 6.3 indicates, while only 4 to 5 percent of respondents described themselves as "very familiar" with Canadian authors, 31 percent reported that they were somewhat familiar with Canadian authors. Also, 77 percent of the total sample reported reading a Canadian author at some point in their lives, while 83 percent, 86 percent, and 80 percent of regular readers, heavy readers, and mainly literary readers, respectively, reported having read a Canadian author at some point in their lives.[277]

Table 6.3 also reports the average number of Canadian-authored books read over the previous year. The overall average for all readers was 2.4 books, with 1.4 for light buyers and 5.5 for heavy readers. Notable is the low average number of books for moderate buyers (1.6) in comparison with regular readers (3.1).

How do these data translate into book-oriented behaviour by Canadians? Even such low percentages as 4 and 5 percent suggest that there are enough Canadians who care about Canadian titles to make publishing on Canadian topics a viable proposition. If 5 percent of readers follow Canadian authors, and if we assume that 50 percent of Canadians aged fifteen to sixty-four constitutes the whole population of those who read books, this means that Canadian authors and publishers have a following of just over half a million readers (566,615 to be exact, or $22,664,600 / 2 \times 0.05$). Using the same 50 percent assumption and multiplying that number by 2.4 (the average number

Table 6.3

Three measures of interest in Canadian authored books

Familiarity with Canadian authors	Total readers %	Regular readers %	Heavy readers %	Mainly literary readers %	Total buyers %	Light buyers %	Moderate buyers %	Heavy buyers %	English buyers %
Very familiar	4	4	5	4	4	4	2	4	4
Somewhat familiar	31	37	41	33	33	28	33	37	30
Has ever read a book by a Canadian author	77	82	86	80	79	74	79	85	72
Number of books by Canadian authors read in the last year									
Percentage of readers that have read 1 to 5 books by a Cdn. Author in last 12 mos.	40	43	38	41	41	33	46	44	34
Percentage of readers that have read zero books by a Cdn. Author in last 12 mos.	39	31	24	37	39	53	39	27	48
Average number of Canadian-authored books read in last 12 months	2.4	3.1	5.5	2.6	2.4	1.4	1.6	3.8	1.9
Interest level in reading books by Canadian authors									
Very interested	16	18	23	16	16	16	14	17	13
Somewhat interested	55	55	56	57	52	47	55	55	50

Source: Lorimer and Lynch, *The Latest Canadian National Reading Study, 2005: Publishers' Analysis.*

of Canadian-authored books read each year) gives us an approximation of the number of books written by Canadian authors and read by Canadians each year: about 27 million — somewhat more than one book for every man, woman, and child between fifteen and sixty-four. Such data are rough estimates, but they strongly suggest that there is a reasonable-sized community of somewhat interested to avid readers of Canadian authors and Canadian topics. Reaching that community with appealing titles is the challenge, which leads to two final factors that deserve mention in this description of the general dynamics of the book market, bookselling and the media.

The role of bookselling and booksellers

Bookselling, both past and present, provides a certain insight into the nature of the book market and the dynamics of the marketing of books. According to one well-known bookseller, the overlap in inventory from one independent bookstore to the next used to be extremely small — only about 10 to 20 percent.[278] This low overlap reflected the eclectic tastes of Canadian book buyers, revealed in two marketing studies conducted in 1996 and 1998.[279] In both studies, the percentage of titles purchased only once during the study period, in the sample of book purchases from across the country in a wide variety of bookstores, was very high. In the 1996 study, of 304 Canadian-published books purchased, 250 titles were purchased only by one person, only fourteen appeared in at least one published top-ten list of the time, and just two titles were purchased more than four times. In the 1998 study, of 1,003 books purchased, there were 900 different titles; fifty-eight were purchased by more than one person, eight were purchased more than four times, and the title most often purchased was bought by six different people. The laudable element in bookseller stocking patterns, which was reflected in book buying patterns, was that they provided genuine choice in the market, which allowed readers a choice of favourite bookstore. However, from the point of view of publishers selling their wares, such a cornucopia of choice often interfered with a reader's ability to purchase a just-published book that he or she had heard reviewed, or found out about otherwise.

In short, before Chapters/Indigo attained its dominance, the likelihood a customer would be able to complete an intended purchase of a particular title beyond a very narrow range of bestsellers was not high. Given limited shelf space, booksellers' independence of mind, and the media's frequent failure to give publishers or booksellers prior notice of a coming review of a title, the chances of a timely match of consumer desire and in-store availability were slim. The industry was rarely able to ensure timely delivery of titles before consumers' desires waned, except around Christmas time. Six weeks was the normal waiting period for delivery of a title request. The creation of Chapters and Indigo, with their larger store formats, increased the chances of a title being in stock, if only because there was a greater selection of titles available in the larger stores. Both firms had employees whose primary job was to coordinate availability with demand and, generally, anticipate the wishes of consumers. While Chapters/Indigo's performance was very far from being perfect, the chance of finding a recently released book garnering good reviews increased, especially when the publisher paid for premium space.

The price Canadians have paid for the emerged dominance of Chapters/Indigo is the loss of booksellers whose lives were made meaningful through the intrinsic value of buying, reading, and recommending books and authors and intuiting the enjoyment of individual book readers. Chain bookstores focus on counting, not beans, but the number of books that go out the door, some with the apparent personal endorsement of the owner and her staff. Specifically, chains are most interested in generating the highest possible level of sales per square foot of store space. To do this, they present a wide variety of products in a spacious store to potential purchasers. To increase the probability that customers will make repeat visits, the chains add to the illusion of an all-but-infinite selection with a constant churn of product, forever offering new choices. It is industry wisdom that if a book does not begin selling well within six weeks in national bookstore chains, it is removed from the shelves and returned. For certain categories of mainstream titles this is true, but for certain limited numbers of titles, even in Chapters, it is definitely not the case.

Where does this approach to bookselling leave titles from new authors, or books from small Canadian-owned publishers? Mainly, their books are relegated to the back area of stores. Amidst the hundreds of new books being churned through on a six-week trial, and the slew of imports that inevitably will sell less than five hundred copies in Canada, new titles from new Canadian authors can be found. In most subject areas, Canadian titles are intermingled and all but lost amidst these also-rans. And although the calculations have not been done, there is every possibility that Canadian books are purchased at no better than a chance selection.[280] Remarks made by Larry Stevenson, the founding owner of Chapters, suggest that in his opinion the whole supply of books from Canadian publishers, with the exception of books destined for the high-sales area of his stores, could have dried up at any time and cost him not a penny in profit.[281]

While Chapters/Indigo has increased the number of titles available and the likelihood that readers will find a publicized book they are looking for when they visit a megastore, the loss of the knowledgeable bookseller has affected Canadian book publishers and authors rather dramatically, especially in the case of children's books. Throughout the 1970s, '80s, and '90s, the publishing of high-quality children's books was an ever-building Canadian success story. The leading publisher, Kids Can Press, was even sold for the unheard-of sum of $7 million in the heady days of the first dot-com boom. But the good news turned bad with the rout of the independent bookstores. Women (who purchase over 80 percent of children's books[282]) shopping at easily accessible chain stores were left on their own, with no trusted advisors to turn to, and often needed to make purchasing decisions in a rush — say, on a Saturday morning en route to a birthday party. The result: a growing triumph of brand, in spite of readers' generally low ranking of the importance of the publisher. When retail service quality fails, a recognizable brand such as Disney or Golden Books will do.

Bookselling exists in a social nexus surrounding books, an established finding that the purveyors of e-readers are attempting to reproduce in the digital world. (See Chapter 7.) "The Latest Canadian National Reading Study" found that "friends' recommendations" is the leading source of awareness of titles in general, while the choice of individual titles is determined primarily by "interest in topic." In short, as the study says, "while we listen to friends, our interest in a topic or the type of book determines whether we actually read or purchase a book, having heard a friend's recommendation."[283] Bookstores are powerful contributors to that social nexus, in that impulse buying accounts for over 40 percent of purchases and 60 percent of book buyers purchase a title with no prior intent to do so before they entered the store (according to other reading and purchasing studies).[284] In this context, the selection of stock by bookstores becomes important — to readers, for the breadth of choice offered; to publishers of books, especially those that are not potential bestsellers; and to Canadian society, for how the stores shape the reading public's knowledge and awareness.

In the context of this social nexus surrounding books, publishers try to stimulate word-of-mouth publicity, some through guerrilla marketing,[285] to feed friends' recommendations. And then there are what could be called "special friends," whose word becomes gospel in the choosing of book titles. The titles chosen by Oprah Winfrey for her television-based book club catapulted to celebrity status, their sales increasing, in some cases, more than twenty times. In his time, in Canada, and without the trappings of a book club, Peter Gzowski had a like impact on book purchasing behaviour. Booksellers can be seen both as extensions of friends' recommendations and as neighbourhood Oprahs: accessible, sincere, and trusted figures who help to engage their customers with books by providing them with a reason to read.

Book buyers have traditionally relied on recommendations from well-read booksellers. Repeat customers were able to take advantage of their independent booksellers getting to know their individual tastes through their purchases and their reactions to those purchases. In the chain bookstore environment, where young people work part time for low wages — not that booksellers were ever able to pay themselves handsomely — the advisory role of the bookseller is lost.

The emerging replacement for knowledgeable booksellers is the computer that tracks both search behaviour and actual book purchases from online and brick-and-mortar bookstores. The online environment also offers book reviews by an assortment of people, links to books that others who bought your selected title have purchased, and so on. There may be a significant opportunity for a service sector to be carved out here, a personal reading recommender helping fully employed, double-income families make book choices. Such a service would save parents, mainly mothers, from having to do the "shadow work" (in the terms of modern family dynamics) required to maintain an awareness of appropriate books to meet the ever-changing needs and interests of their children.

Book retailing: Chapters/Indigo and the selling of books to Canadians

With publishers such as Kim McArthur of McArthur & Company claiming that Chapters/Indigo accounts for 70 percent of her company's sales,[286] and Douglas Gibson agreeing to the same figure for McClelland & Stewart when it was offered to him in conversation,[287] it would be remiss not to devote space to the retail environment that Canada's dominant bookstore chain creates. Chapters/Indigo is a publicly owned company. The controlling shares are held by Trilogy Retail Enterprises L.P., the named principals of which are Gerry Schwartz, chairman of Onex, a Toronto-based investment firm that does $16 billion in business annually (making it nearly eight times the size of the Canadian book market), and his wife, Heather Reisman, who is chair and CEO of Chapters/Indigo.

When Reisman and Schwartz purchased Chapters in 2001, book publishers hoped that their relations with what the industry now calls "the chain" would improve. Under the leadership of founder Larry Stevenson, Chapters' dealings with publishers were something of a disaster, and the bookstore's actions arguably put a key distributor of Canadian titles, General Distribution, and associated companies controlled by Jack Stoddart, out of business.[288] Given Reisman's avowed interest in making the chain into a cultural department store[289] (something that may have changed), Schwartz's controlling ownership of the Cineplex and Famous Players theatre chains (through Onex), and various culture-friendly public statements the couple made, some imagined Chapters/Indigo would evolve to become a firm serving the educational, cultural, and entertainment interests of Canadians, elevating Reisman and Schwartz to queen and king of Canadian culture.

This has not happened. Schwartz gives the impression of total non-involvement. Reisman has maintained many of the business practices that Stevenson developed, and extended some, such as returning unsold copies outside publishers' terms of sale, delaying bill payment well beyond the terms of sale, and refusing to pay interest on late payments.

In approving the merger of Indigo and Chapters, which gave the firm a monopoly position as the single national bookstore chain, Canada's Competition Bureau imposed, with Chapters/Indigo's consent, an agreement that the firm would improve its performance with respect to timely payment of bills and would abide by other performance measures. Those measures required Chapters/Indigo to comply with agreed-upon percentages of returns; return all books in resalable condition; work within the terms of sale set by publishers; ship all returns to the correct source; and, in general, cease its unilateral setting of terms of sale.[290] Subsequent interviews with publishers indicate a continuing level of dissatisfaction over the years with Chapters/Indigo's adherence to those conditions. Complaints or attempts to assert terms of sale result in orders diminishing or disappearing for a period of time, after which they slowly begin to flow again. Many of these actions increase publishers' costs. Given that publishers are in receipt of government subsidies to counterbalance their chronically low profitability, increased costs to publishers translate into Chapters/Indigo finessing benefits from government grants.

The company's size makes possible other initiatives that indirectly place public funds in its hands. In January 2007, the *Globe and Mail* reported that six medium-sized, Canadian-owned publishers were asking the government for about $100,000 to undertake a book publishing "pilot project" to market books in Chapters/Indigo stores during February. A later article said that the publishers had reached a deal with the Canadian Booksellers Association to include them in the project.[291] While the project was initiated by publisher Kim McArthur, the participation of Chapters/Indigo in government-supported projects that benefit its operation amounts to an indirect subsidy of already wealthy owners. February is the slowest month of the year for booksellers, so it is difficult to make the case that the proposed project was an efficient use of public money.

Size also allows Chapters/Indigo to benefit in ways that are potentially harmful to publishers. Publishers normally sell to booksellers at a discount from the suggested retail price. Each publisher advertises its terms, which generally start at 40 percent. Increased numbers in a single order lead to free shipping or higher discounts, which can climb up to 50 percent. Chapters/Indigo has insisted on higher discounts based on its market share, without regard to publishers' costs. No other bookseller except for Amazon, which demands 55 percent discount, has that ability. A negotiation based on the cost savings publishers could realize by dealing with a larger entity would have created a much more positive business environment, but Chapters/Indigo has been more inclined to demand its terms than negotiate them.

An alternative Chapters

The performance and the possibilities of Chapters/Indigo are a study in contrasts. With its size and the original upscale orientation of Indigo, the chain could be both highly successful and greatly admired by both publishers and consumers. But in its purchasing methods, merchandising, and absence of connection to what should be its natural communities (e.g., publishers, libraries, authors), Chapters/Indigo seems to spurn cultural partnership in favour of the dogged pursuit of self-interest.

For example, one Chapters/Indigo buyer indicated, in a presentation to publishers-in-training, that a sales rep has no more than thirty seconds to make a pitch for an individual title.[292] This is absurd. In such a situation there can be no pretense that the quality of an individual book is assessed, or even matters. Consistent with the general wisdom on branding, there is little doubt that Chapters/Indigo relies on the publishers that supply it to maintain quality, since the publishers have a vested interest in maintaining their good reputations by means of high sell-through ratios. However, this method of buying stock does little to ensure good books achieve a market presence and reach appropriate buyers, nor does it favour talented emerging authors or discerning buyers. The thirty-second buy-in might better be replaced with publishers becoming jobbers — that is, supplying what they consider to be the appropriate mix of titles for Chapters/Indigo stores. Alternatively, Chapters/Indigo could rent space to publishers outright.

The thirty-second buy-in may be another symptom of the chain's drift towards the mainstream and its migration away from the upscale orientation that differentiated Indigo from Chapters. This migration is apparent in the similarity from store to store and has also been confirmed by a statement attributed to Reisman, in which she said she wanted Chapters/Indigo to be as effective in its operations as Walmart.[293] While such a goal is based on merchandising theory, it is mismatched with the curiosity book purchasers demonstrate in their selection of books and in the pattern of Amazon's online sales, a pattern described in Chris Anderson's book and essay, *The Long Tail*.[294] Alternative retail strategies would engender reader interest and spur purchasing. Such strategies might include setting up sections in selected stores for books from Quebec, the Maritimes, or the west; designing a section for French books, in French or in translation, with the trappings of a French bookstall; or maybe stocking a section with titles from India, the third-largest publisher of titles in English in the world. Chapters/Indigo could also set aside managers' sections for books with special appeal to the clientele of each location, and it could incentivize the management of the section by handing out bonuses for sales levels that exceeded sales elsewhere in the store.

Chapters/Indigo has been publishing ads promoting "Heather's Picks," which portray Reisman as a trusted source for book recommendations. However well this marketing idea is working, the difficulty is that Reisman is not in the same position as Oprah, or even Peter Gzowski. Reisman is simply the owner of the single national bookstore chain. Chapters/Indigo would do books and authors an enormous service were it to create a public platform and build up trust for several persons whom Canadians might accept as reading advisors. An insightful and erudite Canadian literature professor; an informed, book-reading techie; someone in the thick of politics; an environmentalist — all would feed a liberal pluralism that would be welcome in Canadian society and would generate a tremendous amount of good will for Chapters/Indigo among book purchasers, authors, and publishers. In June 2010, Chapters/Indigo headlined a one-page ad in the *Globe and Mail* with "INDIGO RECOMMENDS." This may be brand building. It certainly does not approach the effectiveness of friends' recommendations.[295]

It would be fair to say that there is an unfortunate absence of a warm cooperative relationship between Chapters/Indigo and the book community — authors, book publishers, libraries, and book reading communities. The chain's dominance also creates another issue, of a much different type. There is a tendency in Canada to allow concentration of domestic ownership, which is protected, formally or informally, by means of policy or business practice. This leads to a small group of companies — or one, as in the case of Chapters/Indigo — commanding a fairly unassailable position of dominance. The general result is that Canadians pay more for products and services than they would under more open competitive conditions. Eventually Canadians object, or, in these days of free trade, foreign companies mount a challenge to the protective barriers. Because the domestic companies are powerful, their foreign competitors must also be powerful. Enter U.S. companies of the same size or larger, and the eventual buyout of the Canadian giant by a foreign company with the strength that derives from a market ten times the size of Canada's. With such a buyout comes roughshod treatment of the distinctive demands and patterns of consumption of Canadians. In the end, priceless cultural distinctiveness is sacrificed at the altar of minimal consumer savings.

While this book was being written, the federal government approved the establishment of a distribution centre in Canada for Amazon.ca. Reisman responded by noting that she wanted to be sure that she could bring U.S. investors into her company to level the competition.[296] Then in the fall of 2010, the federal government began laying the foundations for a loosening of the restrictions of national ownership in the industry by calling for a consultation on the matter.[297]

Clearly, the Harper government's general welcoming attitude to foreign investment in the culture and communications sector, and the exact wording of the policies governing foreign investment in cultural properties, affected the federal government's approval of Amazon.ca's plans. But there is also no doubt the federal decision on Amazon.ca has enhanced competition in the online selling of books and, hence, has weakened Chapters/Indigo's effective monopoly position as a national chain. It has done little for Canadian authors, book publishers, and readers. In this developing context, and the more recent interest of Walmart and other general chains, if Chapters/Indigo created two divisions — say, Indigo and Chapters — organized on quite different principles, in order to provide consumers with a choice and to create internal competition, it would change the brick-and-mortar bookselling environment very much for the better. Internal competition would not prevent integrating certain non-competitive elements, such as warehousing and shipping, to take advantage of economies of scale. With both divisions achieving profitability, one could even be sold to a Canadian buyer to abide by foreign investment regulations, just as Thomson Reuters sold Nelson Education, even though it was perfectly profitable. Alternatively, such a differentiation could enhance profits by better serving the market.

Kobo: A hopeful, then dashed, initiative

In the summer of 2009, Chapters/Indigo launched an e-book sales service called Shortcovers. Originally intended to sell content rather than devices, the service offered consumers a wide variety of titles in formats compatible with such reading devices as computers, Apple iPhones and iPods, Sony eBook readers, BlackBerrys, Google Android mobile devices, and Palm Pre. By May 2010, Shortcovers had been renamed Kobo (an anagram of "book") and the product line was expanded to include an e-book reader at $149, also called Kobo. A pre-Kobo presentation on Shortcovers by Michael Tamblyn, vice president of content, sales, and merchandising (and former CEO of BookNet), indicated that the system was set up to log critical data to create a history of purchases by all readers; it can also track how quickly a person reads a book, when that person reads, if he or she downloads only items that are free, and so forth. This may eventually raise concerns about privacy, but at the time, the sharing of these data with industry members seemed to be the harbinger of more cooperative relations between Chapters/Indigo and the full range of Canadian book publishers and distributors. For a time, it even looked possible that Chapters/Indigo could have learned from the apparently more respectful partnership Kobo seemed to be pursuing with publishers. Alas, it was not to be. In November 2011, Chapters/Indigo sold 100 percent of Kobo to the Japanese online international retailing giant Rakuten,[298] part of the process of increasing non-book items from 25 percent to 50 percent of Chapters/Indigo's sales,[299] granting the chain the right to return unsold copies of books a mere forty-five days after receiving them — without necessarily even putting them on shelves.

Restructuring Chapters' retail environment: Clouds on the horizon

On April 9, 2011, the *Globe and Mail* ran a cover story entitled "Less Tolstoy, more toys."[300] The story spun the drop-off in traffic at Chapters stores in a positive manner, attributing it to an increasing digital market, including both online purchasing and the purchasing of digital copies from online sources. The article talked of an expanded non-book product mix that was increasing traffic into the "bookstore," a spin that had been in existence for at least a year.[301] Following Reisman's lead, it likened the anticipated turn-around to that achieved by Starbucks when owner/operator Howard Schultz "closed 900 underperforming stores, fired thousands of employees, added new products and slowed dramatically the hunt for new locations . . . and achieved a sales boost of 10 percent."[302] In a rather curious gesture, one week later, in a full page ad in the *Globe and Mail*, Reisman encouraged readers to purchase Schultz's autobiographical account of Starbucks[303] by offering them 1,000 extra points on their Plum rewards card.[304]

A more realistic assessment of the role of Chapters/Indigo is this: The large-format bookstore has not proven itself as a viable model, particularly in Canada. It displaced independent bookstores not by performance, but by its access to capital. True, it attracted some investors in its build-out. More stores led to increased revenues, and the slow collapse of the independents led to more increases. The centralization of decision making allowed Chapters to coordinate title availability with the marketing campaigns of bestselling titles, which increased the sales levels of those titles and perhaps the revenues of both the bestseller publishers and Chapters/Indigo. The uncertainty comes from not knowing the overall impact of discounting bestsellers. But faced with the migration to digital, having failed to build the level of loyalty that made customers cry at the closing ceremonies of their local independent bookseller and caused publishers to forgo legal action at the closure of at least one city chain (Duthies in Vancouver), and being beholden to generating decent returns on investment, storm clouds are clearly visible on Chapters/Indigo's horizon. Those clouds are certainly not being dissipated by talk of opening bookselling to foreign competition. And it is also important to recall that Chapters/Indigo's sales patterns don't appear to match the reading patterns that existed prior to their entry into the market — when, for example, bestselling titles accounted for less than 10 percent of book purchases.[305] In addition, Chapters/Indigo gives no hint to readers that it is aware that different titles have different sales patterns, and that it is the place where reader demand will be met in whatever form that demand arises. To be concrete, while digital sales of bestsellers, genre fiction, and reference titles seem to be healthy, digital sales of regional and other niche market titles can be puny: a recently released (2011) regional title targeted at readers in their twenties and thirties sold 8,000 print copies and only eight digital copies.[306] Interviews with publishers in the course of writing this book have revealed that some markets have simply vanished, not necessarily because readers have vanished, but because Chapters/Indigo is an inadequate delivery vehicle to those readers and the independents that could serve that purpose have been decimated. The accumulating losses of Chapters/Indigo in the summer of 2011,[307] the sale of Kobo, and the announced reduction of time of display to forty-five days speaks of a firm in market, if not financial, trouble.

Market functioning

While concerns about business-to-business relationships with Chapters/Indigo loom large for many publishers, there are other systemic operational realities that have a substantial impact on book publishing firms, including the competitive position of foreign-owned firms and titles, some finer points of publishing policy, the role of literary agents, and the differential impact of returns.

The dynamics of foreign competition

In many, but not all, industries, importers and distributors evolve into domestic manufacturers by learning about the nature of the products they handle and the markets they serve and substituting locally manufactured goods. However, when the product is books, this import substitution is impeded, because titles are not interchangeable in the minds of book buyers, and the dramatic economies of scale in book production place title origination in small markets such as Canada's at a permanent competitive disadvantage to imports. Thus, both imports and original Canadian titles remain in the market and compete for book-purchasing dollars.

Currently, the importation and distribution of general trade books in Canada is carried out mainly by four foreign firms — Random House, Penguin (a division of Pearson), HarperCollins, and Simon & Schuster. This is not to say that no Canadian firms are involved. The very substantial H.B. Fenn and Thomas Allen, along with the smaller Firefly and Raincoast, are Canadian-owned firms that are primarily importers and distributors, although it should be noted that Fenn filed for bankruptcy protection in 2011.[308] However, a quirk in Canadian policy has turned three of the four foreign-owned firms into the main publishers of leading Canadian authors. The general role these foreign-owned firms play in the market can be appreciated, to some degree, by noting the number of "agencies" they represent in Canada. Many such agencies are imprints within larger publishers, but a good number are also separate publishing firms. In terms of size of operation, 90 percent of these agencies would probably qualify as BPIDP/CBF recipients were they owned and operated in Canada and publishing qualified titles. As of 2010, Random House Canada represented 169 agencies, Pearson represented 30 and Penguin represented 135, HarperCollins represented 161, and Simon & Schuster represented 52.

In total, the foreign-owned firms represented 547 agencies. On the other hand, the Canadian-owned firms represented 223, with Fenn representing 138, Firefly 14, Raincoast 55, and Thomas Allen 16.[309] There are other agency operations, but in 2010 the above-named firms were bringing the publication programs of some 770 entities into the Canadian market, entities comparable to the 282 Canadian-owned publishers identified by Statistics Canada in 2006. Actually, expressing the comparative performance of fewer than 300 Canadian firms (at 53% domestic market share) in comparison to the 770 foreign imprints (at 47%) suggests that Canadians publishers are at least 250 percent more successful in serving the Canadian market than their foreign competitors.

Net benefits and acquiring and retaining authors by Canadian-owned publishers

The Baie-Comeau Agreement (introduced in Chapter 3) was set up in 1985 as an attempt to patriate Canadian book publishing. It failed as a policy because it required foreign firms to shut down their Canadian operations and form partnerships with Canadian owners. The only firm that had any interest in trying to conform to this policy was Bertelsmann, which found ways to work within at least the letter of the law, just as it did for a period of time in its operations in China, where foreign firms are prevented from owning book publishing companies.[310] As it evolved into its current manifestation as part of the Investment Canada Act (and ceased to be called Baie-Comeau), the policy required the federal government to conduct a review when any foreign-owned Canadian subsidiary was purchased by another foreign firm, usually as part of a sale of its parent company to a new owner. The government would examine such transactions to determine if they were of net benefit to Canadians. In the takeovers that did ensue, one of the net benefits the acquiring companies proposed, which the government encouraged, was that they would increase their publishing of Canadian authors. Not only does this seem logically to be a net benefit, but also, given that the foundation principle of policy in this area is cultural, and that is what exempts it from the full application of free trade principles, its consistency with policy principles is relatively unchallengeable.

However, the result has been negative for the Canadian-owned sector generally, and especially for the so-called medium-sized Canadian firms that attempt to compete in the general market for lead authors — firms such as D&M or McClelland & Stewart.[311] The early 2012 sale of McClelland & Stewart to Random House Canada highlights the inherent instability of Canadian ownership of book publishing firms and is discussed in the concluding chapter. The essential difficulty is that as well as having substantial economies of scale in terms of production, the foreign-owned triumvirate — HarperCollins, Penguin, and Random House[312] — have the scale of operations necessary to bid on those titles most likely to become bestsellers and blockbusters. Indeed, if needed, they can operate their Canadian origination programs at a loss.[313] This freedom comes about because they can cross-subsidize their Canadian publishing, not only with the distribution of run-on imports (just as Canadian firms did traditionally), but also with simultaneous publication (in effect, co-publication) of titles originated in the United States. Legally, co-publishing amounts to title origination — it is a splitting of rights by territory; practically, it is importation and distribution. MacSkimming describes a version of the comparative competitive advantage between Canadian and foreign-owned firms in his account of an anonymous executive from the foreign-owned sector who admitted that only one-third of his unnamed company's Canadian-authored and -originated titles were profitable,[314] a statement that suggests the foreign-owned sector has been winning over authors by offering advances against royalties that the authors need not earn out.[315] Given that there is a certain marketing advantage inherent in the splash of a big advance, this may have been true at the time the comment was made. Unearnable advances have diminished, because all publishers can now pay to access weekly sales figures for all titles from BookNet Canada. Any publisher can now calculate which titles earned out the advance given to the author.

The cushion provided by simultaneous publication and importation/distribution is unavailable to those publishers who are unwilling or unable to engage in such activities. The only competitive counteradvantage of Canadian-owned companies is access to grants, for which foreign-owned publishers are ineligible. In a way, grants attempt to allow Canadian firms to compete. In the end, net-benefit-induced Canadian publishing programs by the foreign-owned sector create a paradox: while the expansion of publishing programs by foreign-owned publishers benefits Canadian authors, it exacerbates the competitive disadvantage of Canadian-owned publishers.

Apart from their influence on the relative competitive advantage of domestic and foreign firms, arguments can be made in favour of net benefits. At the time of the Ontario Royal Commission and through the 1980s, when branch plants showed a profound lack of interest in title origination, Canadian book publishers were intent on wresting sufficient market share away from the branch plants to allow the Canadian-owned sector to predominate in both title origination and agency publishing (i.e. importing and distribution). With the foreign-owned sector now engaged in title origination and working with authors to publish new titles, a good many talented editors, marketers, designers, publicists, and production and sales personnel have risen to the top positions in these Canadian subsidiaries. Their commitment to their companies' Canadian publishing programs is understandably far different from the interests of their predecessors in the 1960s through the 1980s, which were purely in importing and distributing their parents' titles. Having whetted their appetites on title origination to meet their obligations to Investment Canada, the Canadian employees of foreign-owned book publishing companies have become almost as engaged by title origination as the Canadian-owned sector.

Still, a basic problem exists that was brought forward in May 2010. The Canadian Booksellers Association lobbied Heritage Minister James Moore to abolish the distribution right (in the Copyright Act) on the grounds that consumers were complaining about the 10 percent surcharge Canadian importers and distributors are allowed to put on books.[316] The removal of the distribution right would undermine the Canadian importation and distribution system and, with it, the rationale for the existence of the foreign-owned sector and several Canadian-owned distributors. Without that system in place, it is likely the Canadian subsidiaries, with their Canadian publishing programs, would close up shop. While such an eventuality would open up the market for Canadian-owned firms, it would also decrease publishing opportunities and the chance for international rights sales for Canadian authors. It would also place a slight downward pressure on the price of Canadian books.

The role of authors' agents

Agents are another matter to consider. Taking their lead from London and New York, the large foreign-owned firms, with their deep pockets, have outsourced a good part of manuscript acquisition to agents. They rely on agents to filter manuscripts, find good ones, and bring appropriate titles and authors to their attention, while understanding that they will have to bid for the rights to publish manuscripts that other publishers may want.

There is no simple answer to questions about the source of the money that pays the agents. On the surface, the agent's fee is a percentage of the author's royalties. But the agent is in a better position than the author to bargain and, hence, drive up the price (in advances and royalty rates) of a manuscript. This view of an agent's position implies that the money comes from the publisher. There is no definitive answer, and there is another aspect of the involvement of agents that is of even greater importance. The use of agents has meant that, increasingly, both Canadian-owned and Canadian-based publishers are offered Canadian rights — not world rights, not English-language rights, but Canadian rights only. This restrictive sale means that the originating publisher, who usually pays the bulk of manuscript development costs, gains no return for that work outside the Canadian market. In an effort to share origination costs, Canada's foreign-owned sector appears to be increasing its determination to make joint offers with its U.S. parents. Such deals are often brokered by New York agents, even for unproven authors. This was the case for a three-title contract signed by B.C. author Susan Juby,[317] and also for Andrew Davidson's $2.5 million contract.[318] It is more difficult for Canadian-owned firms to develop the close relationships with parallel publishers in other countries that would allow them to pursue this alternative, partly because of the dynamic discussed earlier — Canadian-owned publishers rarely have the chance to represent foreign authors in Canada.

There is little doubt that agents have done a better job for authors in international sales than publishers did prior to the increased use of agents.[319] Agents have also made the acquisition of lead authors more difficult for Canadian-owned firms and have made it more difficult for small firms to compete with larger firms. This is because, typically, successful small firms do not outsource acquisition, but rather seek out and invest in new authors. As a result, they overlap the agent's function and are less able to pay for such services than are the larger firms. The September 2009 announcement by Random House Canada that it was turning over its international sales to the Cooke Agency, a Toronto-based literary agency, confirms the value of literary agents to the large companies.

Authors and publishers' nationalities

The role of authors (and by extension their agents) in choosing publishers is a curious one. Few — with some significant exceptions, such as Margaret Atwood, Alice Munro, Rohinton Mistry, Jane Urquhart, and Mavis Gallant — have a sense that there is any reason for them to choose a Canadian-owned publisher over a foreign-owned one. Their reluctance to choose a publisher on the basis of nationality of ownership is understandable when it is clear that foreign-owned publishers are in a better position to invest in a title and to market it. From the publishers' perspective this is problematic, as their ability to invest in and develop an author in anticipation of title profitability in the future is restrained by the freedom of the author to choose a different publisher with any new book. Loyalty may make authors less willing to move from house to house, but, of course, money talks, especially when an author feels underappreciated.

Sales, marketing, and distribution

Also crucial to market functioning are sales, marketing, and distribution and the handling of returns. MacSkimming quotes a Canadian publisher as quasi-biblically intoning, "I saw revealed before me one of the truths of Canadian publishing: we can never compete against the Random-Penguin-HarperCollins model."[320] This was in reaction to the revelation by an employee of a foreign-owned publisher that the firm's typical overall costs for sales and distribution were 16 percent of expenses (6 and 10 percent respectively). For a Canadian-owned firm, the "costs for those services typically hit 25 percent or higher."[321] Such a difference is astronomical in terms of making a title work in the marketplace.

The notion that foreign-owned firms are not only able to carry out sales, marketing, and distribution more cheaply, but are simply better at them, bubbles up in the same chapter of MacSkimming's book. But the inherent competitive advantage provided by economies of scale and the associated market power hides, rather than reveals, reality. With respect to distribution, foreign firms do not in fact deliver any faster than some Canadian firms, such as Raincoast. They do not have more effective sales reps than the Canadian-owned companies. But because of their willingness to purchase prime retail space in Chapters/Indigo, because they are a continuous source of bestselling titles, and because they represent so many different imprints, Random House, Penguin, and HarperCollins have formed a greater, more equal, mutual dependence with Chapters/Indigo than exists between Chapters/Indigo and Canadian-owned firms. It would not be a surprise to learn that Chapters/Indigo buyers are more open to negotiation on entire lists with these three (and with Simon & Schuster) than with smaller firms. Such negotiations would make that process less vulnerable to the imperfections of the thirty-second sell-in and the passing idiosyncrasies of a national buyer. Because it is in neither the publishers' nor the retailer's interest to ship titles only to have them returned, and because the large foreign-owned publishers and Chapters/Indigo are much closer to one another in their thinking, it is reasonable to conclude that this compatibility results in "greater effectiveness" in sales, marketing, and distribution.

Erin Williams has documented the vulnerability of the Canadian-owned sector with respect to distribution and returns, using as her case study the collapse of General Distribution Services in the context of Chapters' massive returns in early 2001.[322] When Jack E. Stoddart took over his father Jack's company, he inherited an agency business that imported and distributed well-established, mainly British, publishers. Like Jack McClelland, Jack Stoddart Jr. (henceforward referred to here as Stoddart) had different ideas of publishing than his father. While continuing the agency business (some of which went to his sister Susan and her company, Distican, when the two parted company as business partners), Stoddart was ambitious to become a real publisher. He did so partly by building up his own titles and partly by acquiring other companies. He also took minority ownership in several companies that wished him to undertake distribution on their behalf, and he contracted to provide distribution for other smaller companies. Stoddart was seen as a white knight acting on behalf of many small, often literary publishers, and charging very fair prices for the services he provided.

By the late 1990s, all was not well with General Publishing and General Distribution Services (GDS). The firm ended up with financing from an organization called Finova, which itself went into bankruptcy protection, causing GDS to lose its access to credit. In 1998, in what was a bold and, in the end, ill-advised move, Stoddart moved from an 85,000-square-foot warehouse to one that was 300,000 square feet. He appeared to believe that he could become the main supplier, perhaps even a wholesaler, serving Chapters. Perhaps he thought he could position his firm for sale to Chapters, which was expanding its operations vertically. Whatever his intentions, his efforts did not bear fruit. Publishers were not being paid and not receiving records of transactions. By the time the company went into receivership, many in the industry quite reasonably assumed the business was insolvent. According to Williams, *Quill & Quire* editor Scott Anderson reported, "In a letter to publishers in February 2000, Jack Stoddart explained that in January [of that year],[323] Chapters debited back returns of $4.3-million rather than paying its bills. At the same time, returns from independents exceeded 'any year in memory.' In total, returns 'have been nothing short of astonishing.'"[324] Williams notes that Stoddart filed an affidavit which showed that even as late as 2001, the firm was profitable.[325] However, there was insufficient cash flow to run the company; many publishers felt the firm's staff was too small to process returns efficiently, by getting them onto warehouse shelves so they could be shipped back out to Chapters or other booksellers. The bizarre element of this tragedy was that, given its marketplace power, it was possible for Chapters to return unsold books, debit the amount it owed, and reorder the same books, thereby granting itself, at the cost of shipping the books back, a further 90 to 120 days to pay for the one that ended up sold.

In the end, the books Chapters had "borrowed from publishers through GDS" to fill its shelves and present itself as a new kind of book-buying experience were no more than extremely low-cost product displays. Chapters solved its cash-flow problems in part by returning stock for credit, but this meant Stoddart's business, in the midst of transition and built on the assumption of normal sales volumes and a normal level of returns, was suddenly flooded by returns at a time of vulnerability, throwing the rest of the system into chaos.

With the collapse of GDS, thousands of books were caught in bankruptcy proceedings. The courts deemed the books were owned by GDS rather than the publishers, even though the publishers had not been paid. Chapters' presumed cash-flow problems became the cash-flow problems of GDS and the publishers. So extreme were the problems stemming from the downfall of GDS that the British Columbia government commissioned a report to determine how badly B.C. firms were suffering.[326] The government concluded that it needed to address the problem, and did so by means of a special grant. It is difficult not to conclude that this was an indirect cost to the taxpayer caused by the manner in which Chapters conducted its business.

In the late summer of 2011, New Star Books publisher Rolf Maurer provided an analysis of the difficulties for small publishers inherent in the current distribution and retailing system.[327] The arrival of computer-based inventory systems and consolidation of distribution in the mid-1980s promised three improvements. They were: something closer to timely delivery to the right reader; sales feedback that would allow market growth; and, hence, greater profitability. What has happened? According to Maurer, industry-wide returns are about three times what there were in the mid-eighties, resulting in publishers printing, shipping, warehousing, ordering, picking, packing, receiving, and shelving three books for every two sold. Average discounts to retailers have increased steadily, giving retailers half of the book buyer's dollar. Initial orders from bookstores are, on the whole, smaller, and returns are higher and faster, with few re-orders. Market demand, rather than distinctive publishing programs and significant literature, is regarded by many as the primary interest that should drive what gets published. And the fit between many small publishers' programs and the development of a literature on the one side, and distributors' and booksellers' sales emphases on the other, is increasingly difficult. All in all, an unfulfilled promise that may be workable for large publishers, but certainly has not resulted in either efficiencies or profit for small publishers.

Returns

In 2001, Tom Woll of Cross River Publishing Consultants undertook a study of returns for the U.S.-based Publishers Marketing Association.[328] At the time, returns cost the U.S. publishing industry (both trade and college markets) $7.1 billion, an amount that was made up of approximately 32 percent of hardcover trade books, 23 percent trade paperbacks, 18 percent juvenile hardcovers, 19 percent juvenile paperbacks, 17 percent scholarly hardcovers, 21 percent scholarly paperbacks, and 23 percent higher education textbooks. The study was commissioned in response to an increase in returns over the years 1995 to 1997, when large-format stores were being rolled out and attempts were made to introduce just-in-time inventory systems. Returns dropped somewhat after these three years, but they remained, overall, about 3 percent higher than their 1994 level.

In his report, Woll made nine recommendations, beginning with the advice that publishers should sell to retailers for a set amount and free the retailer to price as it pleased. He argued that new short-run-capable digital printing forms allowed for sales on a non-returnable basis. Perhaps his most effective recommendation was that booksellers make greater use of markdowns in place, just as cheese is marked down as it gets closer to its "best before" date. Woll also advised publishers to make greater efforts to use the Internet to sell directly to end users and establish percentage returns limits. (This change would likely have zero impact on Chapters/Indigo practices, as that company would insist on its own terms.) He also advocated that publishers become more knowledgeable about the nature of markets for books and distribution methods and channels, and that publishers (and booksellers) analyze sales and returns and take advantage of point-of-sale data. (Such analyses are now generated in Canada by BookNet.) Woll suggested that small publishers seek out non-traditional sales venues and sell into those venues on a non-returnable basis — common practice in Canada — and that large publishers should plan for returns (most do).

All these recommendations have value, and publishers have explored many of them, at least tentatively. The three key ideas that would dramatically change the U.S. (and Canadian) market are moving to net pricing — which means that publishers would sell books to bookstores for a set amount, rather than selling at a discount on a set retail price; converting to a system where price markdowns of already delivered books are *de rigueur*, leading to eventual full non-returnability; and increasing non-returnable sales using other venues and technology. Unfortunately, dramatic change has not occurred in the U.S. industry in the aftermath of the study. Nor has there been appreciable change in Canada. The sole, and dramatic, change has been a move to the agency model, which has the publisher commissioning the bookseller to sell for a defined percentage of the revenue. This does nothing for returns.

Copyright reform: Market opportunity or threat?

As was discussed briefly in Chapter 5, copyright forms the main legal foundation of book publishing. Copyright law grants to authors a time-constrained exclusive right to make their intellectual work available to the market. Authors may assign or license their property rights, but not moral rights, to other legal entities, usually publishers, in return for editing, layout, marketing, distribution, and royalties, usually a percentage share of earnings. Four distinguishing features of copyright attempt to balance the exclusive (or monopoly) right granted to authors to control the dissemination of their work with the social interest derived from the circulation of ideas. First, copyright grants protection only of the expression of an idea, not of the idea itself. Secondly, author rights are time constrained; in Canada they extend fifty years after the death of the author. Third, in part because written material is not destroyed by the act of reading, both people and institutions may lend or give their copy of a work to others. However, the Public Lending Right fund compensates authors (but not publishers) for lost sales resulting from this social service. Fourth, photocopying certain amounts of copyrighted material was legalized by placing an obligation on photocopy service providers to remit a fee to Access Copyright, to be passed on to authors and publishers. As mentioned earlier, in addition to these four features, under provisions for fair dealing, Canada's Copyright Act (as of March 1, 2012) included a clause that allowed the copying of parts of a work for private study, review, and criticism.

The digital world has changed the dynamics of both copying and fair dealing. Digital copies are far easier to make than photocopies, and digital copies are easy to circulate out of public view. Digital copies can be posted to password-protected sites, and users, such as students, can access the files and make either digital or print copies themselves. To add to the changes wrought by digital technology, the Canadian government, through Bill C-32 and again through Bill C-11 (Bill C-32 died on the order paper), suggested that fair dealing should be extended to include education, parody, and satire. Bill C-11 also suggested that if a work was brought to market in a digital form that included a digital lock, it would be illegal for anyone, under virtually any circumstance, to break the lock — in pursuit of making a copy of the work, or for any other purpose. The inclusion of education as a "fair dealing" certainly seems to contradict the very spirit of fair dealing. Allowing thousands of students, or institutions acting on their behalf, to make copies of works without compensating the authors or other rights holders deprives creators, especially of educational materials, of fair compensation. It is akin to asking teachers to teach for no salary. However, it could be argued that offering publishers iron-clad digital lock protection gives users and creators equal weaponry to work out a mutually acceptable *modus operandi*. Perhaps the greatest surprise in the debate about digital copying is the vehemence with which both educators and librarians support the right of users to access the works of creators. A prime example of this vehemence occurred in the presentation of the Council of Ministers of Education of Canada to the hearings of Bill C-32, in which the group made clear that they saw copying for educational use under fair dealing as giving them the right to make classroom sets free of any compensation to authors and publishers.[329]

While Bill C-32 died with the call of the election of 2011, its successor Bill C-11 maintained education as a permitted exception; all parties were evidently wooed by the political attractiveness of supporting education by making it a fair dealing.

A concluding reflection

The foregoing examination of book publishing over a span of thirty-plus years in this and preceding chapters reveals what appears to be an anomaly that speaks to market and operational realities. From time to time, there are major disruptions in the book market. When General Distribution declared bankruptcy, many of the firms distributed by GDS lost all of their stock that was in General's warehouses — based on a legal interpretation that it was owned by GDS. When Duthie Books, the foremost independent bookstore chain in Vancouver, closed, publishers lost millions of dollars. When Edwards Books and Art, a bookstore in Toronto, folded after the owner called a loan — quite an unusual maneuver — the publishers lost all their stock, which was a major setback. When Mike Harris's Conservative provincial government instituted its "Common Sense Revolution" in Ontario between 1995 and 2002 and removed government guarantees on publishers' loans, a number of firms had to scramble to cover their debt. At least one bought the debt from the government for twenty-five cents on the dollar, imposing on taxpayers a hefty expense that was completely unnecessary and utter nonsense. From the late 1970s and through virtually all of the 1980s, including a couple of years of significant recession, the Association of Canadian Publishers seemed continuously to describe its members as being in a state of crisis. Yet with rare exceptions, book publishers carried on. General/Stoddart and Coach House Press closed their doors, but Coach House Books was soon up and running again, due to the efforts of its original founders and owners.

There are times when one wonders how, exactly, publishers manage to persevere in face of these tsunamis of losses, when they show few profits year after year after year. The basic answer to the riddle is twofold. First, profits come after wages, and publishers can pay themselves whatever wage they deem appropriate after profits diminish to zero. Second, publishers are smart: they have assets in the warehouse in the form of a backlist; they can borrow from printers by paying their bills slowly; they can let many of their staff go, reduce personal expenditures, and pull through a winter of disruption; and they are small enough not to have continuing, committed, large expenditures they cannot back out of. Third, few are beholden to the banks. In other words, many firms have considerable short-term resilience. Canada, authors, readers, and the publishers themselves are lucky for that.

This resilience represents a real strength in the face of the considerable challenges of the marketplace for Canadian book publishers, and it is founded on their cultural commitments and their determination. As this chapter documents, those challenges exist in the nature and structure of the market, in cost structures, in the range of competitive products, in demand dynamics of book purchasers, in retail's favouring of mass market products, in the competitive disadvantage of Canadian-owned firms compared to their foreign-owned counterparts, and in the custom that allows the return of unsold copies — and they are not disappearing. The nitty gritty of the survival of Canadian book publishing is recognition that these realities are systemic and unlikely to be addressed by structural intervention. Hence, the need for financial support to address competitive disadvantage is continual. But rather than questioning the evolved organization of the Canadian-owned sector and hence, in a backhand fashion, the value of the titles that a heterogeneity of firms brings to market, governments and the book community can follow the lead of the 1989 Ontario paper, which said, "Because there is no question as to the quality of Canadian books, the problem now is how to continue financing production." The evidence on quality is clear: international recognition of Canadian fiction authors, and the interest of cultural partners in Canadian non-fiction, confirm the cultural value and success of book publishing as a cultural enterprise. The sustainable rationale for continuing financial support is to be found in the concepts of creative economy and social capital. Continuing financial support is investment in infrastructure for the generation of social and economic wealth, rather than a necessary palliative.

As we will see in the next chapter, the view of continuing financial support as infrastructure investment is complemented by a whole new set of technology-based changes that are imposing themselves on book publishing. The challenge for the industry is to make such technologies work in favour of the cultural goals that book publishers espouse, and to maintain the heterogeneity, in terms of size, location, and genre orientation, that is the foundation of the cultural nature of the industry's production.

CHAPTER 7

Trajectories of technology-incubated massive change

Introduction 249

Technology trajectory 1: The virtual reorganization of publishing production 253

    Pre-press production and achievements and limits of WYSIWYG 254
    Database-driven publishing 257

Technology trajectory 2: New media: The changing nature of books, representation, knowledge, and understanding 263

    The book as a medium of representation 263
    Ultra libris: Beyond the book 266

Technology trajectory 3: Interactivity: social networking and creation 271

    The social dynamics of writing and reading 271
    Experiments in interactivity 273

Business applications 1: O'Reilly Media as a manifestation of the three technology trajectories 276

Business applications 2: Maximizing sales from market-based print-technology choices 278

Business applications 3: Service publishing, a.k.a. self-publishing 281

    Table 7.1 The top ten print-on-demand publishing companies operating in the United States and the number of titles psublished in 2009 283

Business applications 4: AbeBooks and digitally mediated bookselling 283

Business applications 5: E-books 285

Business applications 6: Apps 286

Overview and conclusion 289

Introduction

Book publishers are producers and purveyors of mind-transformative goods: cultural artifacts that, by carrying ideas, are exemplars of human creativity and enlightenment and may remain in our minds and inform our spirits for the span of our lives on earth. Books are also a medium of record, and, as Alberto Manguel so aptly points out, personal libraries can express the nature of our being.[330] The contribution of books to civilization is non-trivial, and book publishers are often drawn to their profession by the desire to add to the contribution that information and ideas make to society.

In the West, beginning after 1450 and carrying through to 2000, printed books reigned supreme. After 2000, trade publishers, in Canada and elsewhere, were dubious that a different technological form could ever replace the simple and wondrous materiality of the printed book. In spite of new developments in electronic reference materials, it seemed impossible that some kind of electronic device would replace beach-, bath-, and bedroom-friendly, variously bound, inexpensive print on paper.

Also in the early years of the present century, book publishers became increasingly aware of the emerging competition for consumers' leisure time and expenditures from the television killer, the Internet. While they were concerned, Canadian trade book publishers carried on more or less as they had done over recent years, publishing printed books and shipping them off to bookstores. Then, courtesy of a Canada Foundation for Innovation grant to Canadian libraries, industry members were offered the chance to participate in a bulk sale of digital copies of their backlists to more than seventy Canadian libraries through the Canadian Research Knowledge Network (CRKN).[331] As they deliberated over whether to participate, the offer seemed to convince doubters that the digital market was a real market and, more importantly, that it was not dependent on a perfect e-book reader for its existence.[332] For most, the sale represented a stimulus to create and organize electronic files of their titles so that they could respond to emerging online sales and service opportunities requiring digital content (for example, providing files to Google for its Partner Program, or to Amazon to allow electronic browsing).

Following this stimulus came an emerging awareness of the Google digitization project. After signing an agreement with five major libraries — the university libraries of Harvard, Stanford, Michigan, and Oxford, and the New York Public Library — but without obtaining permission from rights holders, Google set out to create a twenty-first-century universal library modelled after that of Alexandria in ancient Egypt. By default, authors and publishers were included and offered approximately sixty dollars per title to allow users to browse Google's electronic version; by taking action, they could opt out. Publishers were also offered the opportunity to participate in three other revenue models if they allowed Google to make their titles available in full. By mid-2009, the Authors Guild and the Association of American Publishers had signed an agreement on behalf of writers and publishers, and negotiations continued that would legitimize Google's annexation of the record of civilization in book form written in English[333] — all for the princely sum of $125 million,[334] or 13.9 percent of what Google bid for scooping up the intellectual property and patents of Nortel dealing with 4G wireless technologies and semiconductors.[335] Thankfully, on March 22, 2011, Circuit Judge Chin, of the United States District Court of New York, denied the motion for final approval of the Google Books Amended Settlement Agreement ("ASA"), concluding that the ASA was not "fair, adequate and reasonable." The court noted:

> While the digitization of books and the creation of a universal digital library would benefit many, the ASA would simply go too far.... [It] would grant Google significant rights to exploit entire books, without permission of the copyright owners. Indeed, the ASA would give Google a significant advantage over competitors, rewarding it for engaging in wholesale copying of copyrighted works without permission, while releasing claims well beyond those presented in the case.
>
> Accordingly, and for the reasons more fully discussed below, the motion for final approval of the ASA is denied.[336]

Other signs of a lively electronic book market were appearing by the end of 2009. For example, Harlequin was reporting increasing acceptance of e-book formats, and sales levels in the order of 5 percent.[337] The October 2009 release of *The Lost Symbol*, Dan Brown's follow-up to the mega-bestselling *The Da Vinci Code*, sold more copies in e-book format than in print during the first few days of its release. Notably, however, e-book sales fell off precipitously in the following days, as readers ambled by their favourite brick-and-mortar book outlets to pick up a copy of the printed volume.[338] By the end of 2009, various announcements revealed the imminent release of a host of smartphones that would be able to access books using Google's Android operating system.[339] As well, a slew of e-book readers began to appear on the market in early 2010. On July 28, 2010, Amazon claimed that Stieg Larsson had become the first author to sell more than one million e-book downloads.[340] And in China, it has been reported that 600,000 titles now exist in digital form and that 30,000 e-book readers were being sold each week in early 2011.[341]

These developments drew publishers' attention away from the material qualities of the book, which had caused them to emphasize the relative crudeness of e-book readers and to dismiss electronic books as serious rivals to print on paper. Now, they refocused on the book as a cultural object. As Dutch technology-and-society theorist Wiebe Bijker might explain it, the book publishers' concern with the inadequacy of e-book readers reflected their vision of the printed book as an immutable socio-technical ensemble.[342] Of course, publishers' focus on e-book inadequacy is far from surprising, given that when bookstores order books, they expect to receive bound volumes of printed paper, not any random socio-technical ensemble a publisher feels like transmitting. That said, it appears currently that a high and growing percentage of consumers are aware that the cultural object they seek is potentially available in alternative technological forms. Moreover, well-capitalized players, led by Amazon and Sony, followed by Barnes & Noble, and Kobo, are in on the ground floor with the devices. No doubt Amazon and Sony had early dreams of a possible market monopoly — or at least a dominant market share. As events unfolded, hot on their heels was a proliferation of devices — including the most popular in China, the HanWang reader — which suggests that e-book readers are well on their way to a pattern of development paralleling other electronic gadgetry: a market filled with a whole variety of brands, all accessing standardized, non-proprietary e-book formats, in which e-books may vie with or surpass printed book sales.

A few words regarding this milieu of innovation are in order. To some degree, the devices that have been brought to market define the nature of the reading experience. Whereas, for example, the Kindle and the Sony e-book reader are the student's companions in that they offer users the ability to mark up a text, Kobo defines reading more along the lines of how heavy readers and readers of genre fiction approach books. The text is available, pages may be turned, one can mark one's place. And, for some time, in 2011, that was about it. In dramatic contrast, Apple's iPad defined book reading in a more modern context, in colour, with connectivity and, hence, live links — even via phone, if you like — on a larger screen, and with the capacity to handle sound and moving images.

But by the end of 2011, both Kobo and Kindle had released colour versions of their readers with connectivity — the Vox and the Fire, respectively — pricing them in the $200 range, rather than the iPad's $500 range. Readers' tablets, one might say, rather than full-function tablets like the iPad. As well, drawing on their roots in the book market, Kobo and Kindle have both been trying to drive sales by tying into the sociality of the reading experience, specifically the reading communities associated with their devices. Kobo boasts two distinctive value-adds, in addition to its established practice of allowing users to access any Kobo purchase on any device from anywhere in the world. The first is the integration of Pulse, a free news feed that is also featured on the Kindle, with Kobo's Reading Life and Facebook. That integration allows Kobo-using readers to share book comments to Facebook friends as well as Kobo users, rather than, by default, being restricted to a separate Kobo-based community. This, of course, would help promote reading among friends and thus stimulate book and Kobo sales. The second value-add is the opportunity to share perceptions with a five-million-member international community of readers, rather than a predominantly U.S.-based community of readers.[343]

Budziak summarized the issue as follows: "Whereas Amazon's success is based heavily on their sophisticated recommendation system, Kobo's approach is to continually embrace and encourage their customer base to generate feedback, embrace word of mouth, develop a community structure, and of course, join forces with [Facebook,] the most prevalent social networking platform in the world."[344] Where the market will go, or, more accurately, what markets of what size will emerge, is not yet apparent. But clearly both hardware and functionality are key.

Taking an analytical approach to predicting the future, this chapter introduces and explores three technological trajectories that are changing the nature and dynamics of book publishing. It then explores associated business applications. The first technology trajectory is the reorganization of production and operations through centralization with constant, location-independent access. This trajectory is manifest in content management systems (CMS) and utility-based computing, a phenomenon some call cloud computing. It is founded on electronic databases. The second technology trajectory is an expanded use of media. The third technology trajectory is interactivity, the increased opportunity for readers to become associated with or involved in the creative process of authors and the dissemination process of publishers. The simultaneous influences of these three change trajectories, in combination with hardware advances, evolving copyright law, and concomitant changes in patterns of social interaction, are bringing forward changes in business organization and practice that can be seen in the new activities of traditional industry participants, the appearance of new participants with new methods and manners of book publishing, and the appearance of new product forms. In combination, all these factors foretell some elements of the future of books and book publishing.

Technology trajectory 1:
The virtual reorganization of publishing production

The industrial revolution saw the harnessing of technological power and the mechanization of practices that utterly reorganized the manner in which goods and services were created and, hence, the architecture of human societies. One of the main offshoots of the Industrial Revolution was the standardization of parts and, eventually, the sequencing of mechanical tasks, now performed by information-technology-based robots, into an assembly line. The advent of information technology (IT), which many seem reluctant to call the information revolution, is having the same impact. To pick one example, the creation of databases as central repositories of content, and of metadata (otherwise known as tags) indicating the status of the content, constitutes a fundamental restructuring that is so intrinsic to IT-based organization that the word "database" has all but disappeared from general use. As much as, if not more than, any other type of operation, because they deal almost purely in information, book publishers are major beneficiaries of a transition to IT. To appreciate the nature of this benefit and the nature of the change IT has brought to publishing requires some background on the development, and current modes, of printing.

Pre-press production and achievements and limits of WYSIWYG

Printing was an early form of manufacturing. Its precursor was the alphabet, which transformed sound into visual symbols by dividing the continuum of vowel and consonant sounds found in words into atoms — that is, discontinuous basic units that it depicted visually and standardized as the component parts, or building blocks, of language. The result was that any new word or set of sounds could be depicted via a combination of letters. And as Lewis Carroll so amply demonstrated in "Jabberwocky," the alphabet could also be used to create evocative non-words.

Building on the foundation laid by the alphabet, printing with movable type represented a second application of atomization and rebuilding language. It mechanized the scribal craft inherent in the writing of individual letters and the creation of a page. [345] First, in the West as distinct from Asia, standardized metal letters were created. Then they were assembled to form a flat surface, with reverse letters as a pattern. Next they were inked and re-inked, to easily create a page of words many times over. And lastly, the page was re-atomized by breaking the letters apart, so that they could be recombined to create yet another page. The wisdom was in seeing past the first copy. Obviously, printing one page by assembling movable type was far more work than a scribe would take to make the same page. But printing with movable type rendered the making of more than ten copies relatively trivial. By transforming labour-intensive scribal craft into a manufacturing process, it became financially feasible for someone other than an institution with access to a vast labour pool (e.g., the church) to create multiple copies of such books as the Bible. As well, with enough type to set several pages, one person could be printing while another broke up type from previous jobs and reassembled it for pages to follow, a process that could continue until the book was complete.[346] By adapting a wine press, founding metal type in precise sizes, creating a form to hold the type in place on a level plane, inking the type, and bringing the evenly inked face of the type against a piece of paper, high-quality printing could be achieved by means of a one-time-only casting of letters.

While the printing press, combined with movable type, was a major and significant invention — some say the most significant invention of the past 600 years — it had four limitations. The first was the cost of setting up, correcting wrong letters, inking, and otherwise finetuning a block of type to allow a high-quality transference of ink onto paper. These set-up costs made short runs expensive.[347] The second limitation came with the need for reprinting. Going back for a reprint meant beginning anew, typesetting the pages again. Printing new titles was no more expensive than reprinting titles already in existence, with the unanticipated result that the manufacturing process encouraged originality. In contrast, with block printing, no such recomposition of letters was needed. All one had to do was access the storehouse of old blocks, ensure that all pages were accounted for, fill the cracks and worm holes and otherwise refurbish the blocks, ink, and a new printing was underway. With block printing, a new title represented a major new investment; thus, originality was discouraged by the manufacturing process. The third limitation was that printing was based on standardized parts. Anything that required originality beyond new combinations of letters and words — for instance, distinctive decoration — meant the printer had to resort to a hand-carveable surface for the creation of the distinctive image. The fourth limitation, which emerged as time passed, was that the more sophisticated the printing system, the greater the expense in correcting errors, if only because the power of the manufacturing process was put on hold while the system was being prepared for a new job.

By the late nineteenth century (1884), printing technology had evolved from movable type to linotype machines that could form letters more or less instantly by melting and casting soft lead (with the unfortunate side-effect of emitting poisonous lead fumes).[348] By the 1970s, printing had moved to film and phototypesetting, and by the 1990s to computerized plates. Printing formats became more flexible, but set-up costs were reduced only minimally. Hence, scale economies remained in place, and small print runs remained more expensive per copy than large ones.

The possibilities of print production, and the efficiency of the set-up processes, changed dramatically with the development of computer desktop publishing in the 1980s. The creation of the personal computer, particularly the Apple Macintosh with its graphical user interface (developed at Xerox PARC [Palo Alto Research Center]), and page layout programs such as the early Aldus Pagemaker (1985), were significant because the new hardware and software allowed the operator to create and see a reasonably close approximation, both in look and colour, of what was destined to be printed on the page. This development, captured in the term WYSIWYG (what you see is what you get), was effectively the birth of desktop publishing. While desktop-based WYSIWYG technology eliminated the requirement of transferring copy to a very expensive-to-change format prior to a final edit — i.e., typeset bluelines — it did not eliminate all costs of last-minute revisions. Once copy is flowed into a page-layout program, it is often not trivial to make changes, as they may require reflowing the text and fine-tuning it yet again.

Desktop publishing has served both publishers and the general public well, and both were quick to embrace it. Publishers pushed for ever-greater functionality. Non-professionals and professionals alike were able to create and adjust a page on a screen and then, after final edits, print it on laser printers at ever-increasing dots per square inch (dpi). In terms of design flexibility, the whole world changed. Images could be superimposed on each other or on the text, and the text could be formed around objects, all without the clumsiness of scissors and paste.

However, WYSIWYG-based printing has a major limitation. To some degree, it is a reversion to craft. Like the pasted-up page, desktop publishing is designed mostly for one-version use. If a smaller paperback version of a hardcover is to be issued, the work must be laid out anew: text, titles, headings, paragraphs, page breaks, and cover material must all be reformatted. As well, a WYSIWYG document is unstructured, insofar as it is composed of text and images rather than atomized elements with distinct identities, such as titles, authors, keywords, summaries, pull-out quotes, captions, footnotes, citations, and headings. Human beings recognize the structure — we know that the first non-sentence in large type is not a mistake but a title. But such recognition is difficult and sometimes impossible for machines. In short, WYSIWYG renders the meaningful components of a page — titles, quotes, captions, headings, footnotes — machine-opaque, in that it is a complex matter for a computer to discern the identity of the elements.

Database-driven publishing

Database-driven publishing eliminates machine-based opacity because all content is tagged as what it is — title, author, heading, quote, author bio, author photo, chapter headings, subheadings, text, and so forth. Tags and templates, which define what content goes where in what form for, say, a journal article or, for that matter, an entire book, define how the elements for one particular publication of the content will be displayed. The layout professional then calls up the content, fine-tunes its appearance and, *voilà*, the file for displaying the finished product in print or online is ready to go.

Database-driven publishing involves much more than defining the structure of a document and retrieving tagged elements from a database. It also can operate at the enterprise level to facilitate the organization of the publishing of a book. When a publisher receives a book proposal and decides to offer the author a contract, there is crucial information for the publisher to capture in an accessible form, even at this early stage: the proposal itself; the correctly spelled name of the author; the working title of the book; the projected number of words, pages, illustrations; the projected date of manuscript delivery; the basic terms of the contract; and the contract itself. As the project moves forward, the publisher will generate other information, again worthy of capture. A projected profit-and-loss statement for the title will have determined not only the royalty rate that can be offered, but also the size of the advance against royalties. This, in turn, will be based on a projected print run, including costs of development, editing, design and layout, printing, sales, marketing, publicity, returns, as well as potential revenue from rights sales, from softcover editions following a hardcover, and so on.

Once the book is in development, publisher and author can agree on a schedule, which will be entered into the database and will trigger an alert when portions of the advance should be paid. An integrated email system can provide the editor with a pre-written but customizable email, complete with the author's name, address, book title, and a phrase describing what is expected one month or so prior to the scheduled date of delivery, with a positive tone of anticipation. Any changes in the timeline that the editor might enter into the system can generate an internal email or other form of notification to colleagues, if it appears the change might affect the projected publication date or the activities of others in the publishing company.

The advantage of such a database, which in this case is the foundation of content management system (CMS) software, is that it creates a living record or digital trail of all relevant transactions as an adjunct to work being done. Unlike a paper trail, which is what would have been created in the past, authorized users can readily access a digital trail for adaptation and reuse — pulling out addresses for reminder letters, the title and author name for announcements of forthcoming publications, a record of work for review of workloads, information about the title category for list balancing, and so on.

The records described in the previous two paragraphs are the metadata, metatags, or just plain "tags": information that describes the work, rather than the work itself. Other functions can be hooked on to such a content management system. Version-handling software can help ensure that editors work on the latest version of the manuscript while earlier versions are preserved. An alert system becomes particularly useful once the manuscript has been edited. At that point, or even before editing is completed, other employees (such as designers and publicists) who are about to become involved with the manuscript can be given advance notification using a calendar and/or email system. At a designated time, other employees can be given access to this new title as part of the firm's database of titles, with auto-notification and information access presented on their desktops. If needed, those involved can be presented with only those fields of data that they will need for their work. Once they have material to add — for example, cover art — it, too, can be loaded into the title file. With a database serving as a single authoritative source, it is not unusual to find various departments accessing the server continually throughout the day.[349]

A single database also improves file organization for both primary and secondary uses. For example, in preparation for print on demand, a ready-to-print PDF (portable document format) file can be stored and made readily accessible either to users, or to printers anywhere in the world for sale to a single user or to a group. Potential users can be given partial previews of the file, either by creating separate excerpt files or by allowing access only to certain pages of the main file. Access to key sections of the work can be sold at a premium as breaking news items. Individual chapters or sections of the book can be made available for marketing purposes or for direct sales. Other forms of information — for example, available rights — can be made accessible to potential business partners.

Another major spin-off use of a database is populating a website. The enormous advantage of databases in this regard is that as soon as a decision has been made — for example, the page count or price of a title — the publishing house can input that information once, and it will appear throughout the website in that form.

Databases automate information handling, particularly in situations where the amount of information is beyond the carrying capacity of the human mind. This is an asset when it comes to succession. Effectively, the operation of a company can be at least partially coded into the database, rather than existing only in the memories of current staff. For example, all appropriately backed-up files relevant to a particular position can be lodged on the central server and organized with a common naming system, freeing file retrieval from the idiosyncratic file organization on personal hard drives.

Building on the foundation of a database is increasingly important in a marketplace in which flexibility of product format — a whole book or parts, printed or online — is a consideration. Databases are the only way to serve the long tail of demand, where occasional sales of backlist titles can account for an appreciable business, and they also facilitate the sale of rights to part of a manuscript for minimal cost. The ability to respond quickly, efficiently, and economically to what traditionally have been peripheral opportunities is particularly important in a market where multiple use of content prevails and where what seemed a secondary market — permitting uses in addition to making the printed book available in bookstores — can generate more income than the traditional primary market. This is especially the case when the expense of participating in the primary market is high, as it is for book publishing. Not only do book retailers swallow up nearly 50 percent of book revenue, but they also, with returns, constitute a very expensive business model. Scholarly monographs and articles are two types of content whose subsequent use in textbook and courseware compilations can earn more than original sales. The involvement of Pearson, Canada's second-largest educational publishing company, in custom publishing confirms that this is a profitable venture. An accepted set of tariffs for copyrighted content, which encourage a lively trade in second publication rights of whatever portion of a text is of interest to a market, is clearly of universal benefit, as is provision for combined print and electronic editions to allow for searching and referencing.

In short, databases are to modern publishing what the printing press was to scribal culture. They transform acquisitions, development, production, sales, marketing, distribution, publicity, and re-use through rights sales. What's more, looking outwards from publishing, all its business partners are being likewise transformed by databases. Writers can now access large information storehouses with a few clicks from their workstation. This eases research for writing, and writing itself. They are equally able to have their own corpus of work at their fingertips for modification and repurposing. And with largely the same hardware, but different software, a growing number are combining images and sound with writing. Printers are also transforming their business model from a simple service to strategic management over the (extended) life of each title. (This transformation is outlined in the sections on digital and service printing later in this chapter.) Retailers the size of Chapters/Indigo, or even mini-chains like McNally Robinson, depend on databases of product information to conduct their business. BookNet's weekly sales reports provide a foundation for stocking decisions for all bookstores, as well as for understanding patterns of consumption. Amazon's online bookselling operation is also based on the foundation of a complex database of ever-increasing information about books (and other materials).

Consumers, too, are in a position to benefit from databases. In the pre-database era, the only sure way of purchasing a desired title was to buy it in hardcover format amidst the first rush of publicity. Combined with changes in printing technology, the databases and operations of AbeBooks and Amazon have substantially expanded the window of buying opportunity. Virtually all titles can be accessed at any time, new or used. While Amazon presents almost a full selection of titles, AbeBooks, as we will see later in this chapter, taps into the inventories of used and new-title booksellers and small publishers throughout the developed world. As database-facilitated content browsing becomes increasingly established, the bookstore experience of sampling and reading cover blurbs and endorsements is being more faithfully replicated; word-of-mouth publicity, created by reviews that other readers post, helps consumers make choices.

But databases are merely one manifestation of the IT-based reorganization of publishing. Related developments, which may at first seem clumsy, atomize and reassemble publishing elements and thereby introduce substantial economies. Word processing itself separates authorship from the creation of a printed (or typed) document by inserting the following between the striking of a key on a keyboard and the end product: a computer, enabled by software, connected by software to a computerized printer, that in turn, via software, causes printing to appear on a page. Other specialized programs abound. For example, among other functions, the blogging software WordPress atomizes elements of a written document by transforming common writing commands, such as a hard return, into html tags denoting, in this instance, a paragraph break. Such a document can be immediately imported into a database and retrieved either for access or re-purposing. On the basis of such functionality, companies such as Leanpub.com offer to turn a compiled blog into a book in five minutes.[350] Even more significantly, they offer authors an unheard-of royalty of 75 percent, rather than the usual 10 percent offered by publishers.

Perhaps the most significant development is a phenomenon most often called cloud computing. Nicolas Carr makes the case for cloud computing accessible by a general reader in his book *The Big Switch*.[351] Carr's version of cloud computing re-atomizes the creation, processing, and dissemination of information, as follows: While every personal computer (PC) currently has sufficient computer processing power, by virtue of its hardware and software, to be a stand-alone device, most users vastly underuse the software programs and the hardware available to them. Hours go by without the computer being active, and even when it is being used, hackers can commandeer a person's computer as a proxy server without the knowledge of the user. The alternative is to have both hardware and software at a distance, equipping users with a terminal to access both on a service basis, just as we access electricity from a central utility when we need it. Similarly, there is no need for gigantic personal hard drives to store our information, complete with back-up hard drives. This too can be done in a cloud — i.e., in some other location that makes sense, perhaps where electricity is cheap. Carr's argument is that the organization of demand would save vastly, not only on the cost of PCs, but also on the maintenance and deployment of processing power, software and their upgrades, and storage. In the Google universe, the company's provision of such services furnishes Google with an incredibly vast database to facilitate its ultimate goal, which is nothing less than world-encompassing artificial intelligence.

Interestingly, cloud computing derived from a previous atomization in which mainframe computing was atomized by PCs, which took some of the capacities of the mainframe and, by means of software, created the specialized devices that we all now use. In fairly short order, people were calling for networked PCs, which emerged in a number of different forms — local area networks, wide area networks, and finally the Internet, the last leading to the greatest change in social architecture — all of which paved both the technological and organizational way for cloud computing.

Book publishers do not disappear in this revolutionary scenario. As always, their role in selecting, filtering, refining, marketing, and disseminating will remain. Indeed, as making content public becomes easier, publishers may find that their services are increasingly needed in a world bombarded continuously by "information" from all sources. Indeed, it may be that publishers' brands increase in importance as users begin to need a way to differentiate the credible from the non-credible, a speculation that brings us back to socio-technical ensembles.

Technology trajectory 2: New media: The changing nature of books, representation, knowledge, and understanding

Clumsy though the words may be, the concept behind Bijker's "socio-technical ensemble" is an aid to gaining a full appreciation of the social construct we tend to call the book. Books can be distinguished by their intrinsic character, the use of language, their extended nature, and the capacity to evoke ideas beyond what is written. They can also be distinguished by a comparative analysis.

The book as a medium of representation

Like other pieces of technology, printed books arose out of the social intentions of their developers, which in the case of printing in the western world means innovators such as Johannes Gutenberg (~1400–68) and Aldus Manutius (1450–1515). Mainly, their intentions were similar to those of today's e-book hardware developers — to increase the limited circulation of written text. Printing turned out to be an enormous success. In its ability to spread knowledge and ideas to large numbers of people, the printed book reconfigured social relations and made the pursuit of knowledge — not to mention questioning accepted conceptions of the human spirit — common activities for far more people than the privileged few members of the religious and academic institutions of the day. Together with libraries, the printed book encouraged widespread literacy and, in doing so, contributed to the Renaissance and thereafter to the Reformation, the Enlightenment, Protestantism, empiricism and science, and the Industrial Revolution, in which the libraries of Mechanics' Institutes figured with some prominence.

The widened circulation of books, and the widening opportunities for authorship, had further consequences and even changed the very notion of "the author." Prior to the eighteenth century, authorial creativity was seen as a God-given gift to reveal divine truth; in other words, God spoke through authors. Scholars of copyright note that during the eighteenth century, authors came to be perceived as the rightful owners of their own expression of ideas, with the ideas themselves in the public domain, belonging to society as a whole.[352] The secularization of knowledge made possible by the economies of printing (as opposed to scribal copying), which in turn encouraged wider literacy, opened classical Greek texts to view by freeing them from the fetters of the church-controlled knowledge system. With printed books circulating outside state and religious institutions, the body of canonical (and also more ephemeral) texts that form the basis of modern civilization — or, as McLuhan and Innis would have it, literate society[353] — was gradually transformed. The socio-technical ensemble that is the printed book reigned supreme in Western society for 550 years (1450 to 2000).

Such social dynamics were ultimately dependent on book content. Books, then and now, contribute to the affairs of humankind by drawing on language, humankind's most powerful and flexible form of symbolic representation. They increase the power of spoken language by fixing expression in a permanent form. By replacing an ear with an eye (as McLuhan describes it), the fleeting moment of oral speech is replaced by fixed written text in material form, rendering the expression of ideas available to intense, thoughtful consideration and reconsideration. Such consideration knows no bounds; it can begin with the logical structure and consistency of the expression, or its imaginative dynamics, or its comprehensiveness, or its narrative form, or its evocative power, or its range of applicability, and so forth. Once fixed, text exists independent of the author and, hence, must stand scrutiny on its own, and not merely within a passing rhetorical moment. More than any other medium, books provide an opportunity for the creation of insight and the generation of meaning through representation. In this context, "generation" contains both a transformative element (that is, description makes us see things differently) and an extension function (we see things beyond the text in a different manner).

In contrast to language and books are other media. The power of music, for example, is not its ability to reveal a deeper appreciation of, or further insight into, an object, event, or philosophy, or even a scene. Music is valued for its inherent sound invention, architecture, and life, for its re-expression of form, feeling, or movement. Figurative painting, sculpture, and other visual arts are tied closely to what they represent, rather than to their stimulation of reflection. Relatively few members of society are able, with any confidence, to appreciate a rich signification in visual art — that is, the meaning beyond the visual depiction of the subject matter. Similarly in contrast to writing, the historical tardiness in granting copyright to photographers illustrates a reluctance to accept this particular class of visual artists as creators. (The proposed revisions to Canada's copyright act proposed in Bill C-32 in June 2010 finally encompassed setting photographers on a par with other creators.) That lack of recognition of the creator-photographer can be read into Roland Barthes' *Camera Lucida*.[354] He speaks of the photograph as an image captured by the camera rather than as a composition of a creator.

It is not that it is impossible for other media to generate understanding and an appreciation of creativity that audience members can carry away from a work. A re-viewing of Federico Fellini's *La Dolce*

Vita affords an immersion into the imaginings of this genius auteur, who captured the existential angst of certain types of people in a particular time and place. *La Dolce Vita* is an intimate journey every bit as evocative as a book. At the same time, Fellini's work represents a phase in cinematic history, a largely unpreserved opportunity for individual creativity. While the much-reduced production costs of video, in contrast to film, are encouraging a return to the possibilities of auteur-driven productions, film is currently a slave to mass audiences. Movies are created by teams, and for each team, power relationships are worked out explicitly among producers, directors, editors, and lead actors to determine who has control over what and to what extent. In the end, the movie is most often a compromise of forces, with the primary object being to make back, through audience maximization, the vast investment that various institutions and people have put at risk. Secondarily, its purpose is to serve as a vehicle for those involved — the actors, directors, producers, editors, scriptwriters, and technicians — to accrue sufficient prestige to allow them to move on to even greater projects.

In contrast, books are usually written not by a team but by a single author, and each book stands, to some extent, on its own. Books are written for individual readers, each of whom encounters the book alone and in silence, usually with some implied claim to solitude. Unlike a movie, with its aural and visual bombardment, the book is a medium for achieving a peaceful interlude within the control of the reader.

In summary, a book is far more than descriptive words on pages. In the compositional elements — the language itself, but also the story, characters, the flow of ideas and concepts, and structure — is found an encouragement of the reader to generate meaning beyond the realities of the book's content.

Ultra libris: Beyond the book

Current information technology offers significant numbers of the general public the opportunity to engage in the kind of meaning generation that has previously been mainly the domain of written language, most successfully exploited in books. Various combinations of media, readily accessible to both creators and audiences, are becoming true competitive alternatives to the printed book, allowing users to generate meaning beyond the denotative, just as books do with language. These new media are capable of transmitting information and, by deft handling, nuancing and transforming the presentation of information while, simultaneously, communicating connotative elements that are unambiguous. The result is that they encourage in the audience/users an understanding beyond what is seen and heard, which they can extend beyond the "text." As much as, or more than, digital text files and the e-book, these media combinations are the digital book.

What are these media? They are the presentation technologies available to most computer users. They include the digital sound, images, and moving images now at the fingertips of the emerging generation of initiating authors. Creative work in these new media is not unlike what is found in a video documentary. The form demands skill and sensitivity in the selection of images and sound, determination of exposure times, foregrounding and backgrounding, and sequencing: a range of media competencies that is truly impressive, but nothing that is beyond the skill level of creative young people, or outside their realm of experience as consumers of movies and documentaries, recorded music, event theatre, slide shows, and other media presentations. Most importantly, this combination of media affords the opportunity to display authorial creativity in such a way that audience members or users can extend the portrayals beyond the context in which they were presented.

From a sociological perspective, the relatively low cost of the required technology decentralizes or, as U.S. commentators would say, "democratizes" production, dramatically expanding opportunities for people to participate as both consumers and producers. The technology and content is not without cost, but since about 2000 it has become commonly available and used. In this way, too, the new media are similar in accessibility to book authorship and printing.

The use of new media is not confined to young people. In specific areas of scholarly discourse, as the media used for representation change, knowledge fundamentals are also changing. Geographers have always used maps, but their ability to present them has been constrained by the costs of both creation and reproduction. In online presentations, it is now possible to include computer-generated maps that are not restricted by the expense of print reproduction. Moreover, the ability to depict phenomena in layered presentations — now there, now removed — is vastly increased.

In health science, moving images are being used both in the presentation of research and in clinical practice. For example, sound imaging allows medical professionals to view a beating heart inside a live human being in real time, so that they can examine the flow of blood and the operations of the valves for malfunctions. Because the video stream is a quick succession of digital stills, the moving image can be stopped and any single frame viewed with some clarity. The impact of a leaky valve can be seen and a prognosis developed. Because sound can be digitized and transformed, on the fly, to images, as with radioactive x-rays, we can see what otherwise is invisible. Many other applications exist, not the least of which is sending visual robots into human-life-threatening environments, not only to see what is happening but to locate, using GPS, exactly where something is happening.

At the level of corporate production, opportunities have already been anticipated. In 1991, Marsha Kinders undertook an analysis of *Teenage Mutant Ninja Turtles*.[355] A report on Kinders' work describes how the combined television show, action figures, comic books, feature films, and video games functioned as a "transmedia supersystem" that, as the report says, "saturates the market with each piece of the overall . . . story that spans different media forms [and] initiates a cascade where the reader is attracted to more than one form of the content as a result of their initial exposure."[356]

Further examples abound, but their significance — their contribution to a fundamental change in meaning making and, particularly, in accessibility — generally goes unrecognized. Museums and archives are creating three-dimensional representations of artifacts and displays and animating them so that distant users can explore them from all angles. Universities in Alberta have developed the Campus Alberta Repository of Educational Objects (CAREO) project, which "aims to create both an online repository of educational objects for post-secondary educators and a community that both creates and supports those objects."[357] The "objects" were assembled from a variety of online sites and at one point included a night map of the world showing light intensities, now available from NASA.[358] Also referenced was an extensive information resource containing visuals and text about Nova Scotia, to be found on the provincial website.[359] Likewise, one can find a three-dimensional reconstruction of a block in Victoria, B.C., at a particular point in history, which uses archived building plans, censuses, directories, and business records (including the names of employees) to provide insight into neighbourhoods and work places.[360] Digital humanities scholars are finding ways of presenting particular editions of primary texts, along with the complete body of secondary analytical literature derived from them — and making it all available at a scholar's work station. (See, for example, the work of Ray Siemens.[361])

Modes of production lead to product forms. McLuhan, Eisenstein,[362] and many others have documented how movable type fundamentally changed the publication process and, thus, the accessibility of knowledge. In a parallel manner, the book itself created a materiality that encouraged permanence, the organization of collections, and the presentation of visible metadata on the outside of the book. It also encouraged variations, such as Aldus Manutius's octavo, which extended reading venues from reading rooms in private homes to gardens and terraces, and, by doing so, expanded the market.

Similar dynamics will accompany the e-book. A 150-plus-page book printed on long-lasting paper, priced at $20 and up, is a marketable commodity. Its sales to a sufficient number of readers and reader representatives (such as libraries) support an infrastructure in which books are written, acquired, edited, designed, printed, packaged with designed covers, marketed, publicized, warehoused, wholesaled, retailed and displayed, and sold. Books work financially in the marketplace at the macro-economic or societal level, even if some are returned. The display technologies of brick-and-mortar bookstores also work, although not so well in the children's section, where the variously shaped and sized items fit poorly onto regular shelves. The economies of lower-priced mass paperbacks demand different forms of printing, distribution, retailing, and, indeed, of acquisition, editing, design, and marketing.

When the materiality of a book disappears, or becomes optional, as it does in the digital world, the whole supply chain is open for reconfiguration. For e-books, production stops at the point of going to press, at the creation of copy one. There is no manufacture. Production allows infinite product forms — browsable excerpts, works in progress, meaningful chunks, teasers, executive summaries — all repurposing the original content. Distribution can be instant, payment does not depend on location, transaction costs plummet. Only the social habits of consumers demand the existence of a retailer. As both Trajectory 1 and Trajectory 3 indicate, the author/publisher interaction can change from its very beginnings. As well, availability can be perpetual. And the range of media that can be included expands vastly.

In short, the book as a medium is in flux. As the electronic book takes shape, so will development, production, marketing, and distribution processes that may lead to surprising results. Already, we are seeing prices on bestselling trade e-books that are far below expectations. The prices of early 2010 — as low as $9.99 in the U.S. and $11.99 in Canada, and then $1.99 — reflect the presence of early entrepreneurs attempting to build the market and command market share. By April 2010, publishers had grown fearful that the low prices were going to become a long-term consumer expectation, and that once retailers tired of swallowing losses on e-books, they would pressure the publishers to lower their net price. Because of the bargaining power of outlets such as Amazon and Chapters, publishers would have been forced to do so, and without an immense increase in the volume of sales per title, a great decline in title choice would have resulted. To prevent this, many publishers redefined their relationship with booksellers, turning them into sales agents offering books at prices and margins set by the publisher.

What the dominant identity of the e-book will be is difficult to predict. It may become the mass-market paperback of the twenty-first century, with other enriched forms augmenting print as tied purchases, or perhaps replacing print in certain instances.

Technology trajectory 3:
Interactivity: social networking and creation

A third major change afoot is affecting the manner in which books are created by authors, brought to market by publishers, and read by readers. This change is being brought about by the ubiquity of technology-facilitated interactivity, mainly on the Internet, that has caused television viewers to rise from their couches and migrate to their work stations, an odd designation for a home computer. It has spawned an appetite for continuous contact and interaction in a host of ways, including conversation via smart cellphones, real-time broadcast commentary of ongoing events (Twitter), person-to-person business networking (LinkedIn), as well as for maintaining access to databases (the BlackBerry and other data-service-capable phones) and opening one's life to the whole of the online world (Facebook).

The social dynamics of writing and reading

The interactive dynamics of book reading begin when, fresh into the world as toddlers, we engage with printed books in a social setting by listening as someone reads aloud. We mimic that social form when we learn to read aloud ourselves. The books we are introduced to contain text accompanied by still images; images that some children, even as they listen, try to crawl into. Gradually, maturing readers re-enter the silence of listeners. Gradually, readers achieve solitude to commune with the text. And again gradually, images disappear from books as readers come to embrace pure text, or at least that is how it has worked in the past.

Just as the reader is alone, so, at least until recently, was the writer. Perhaps in keeping with the times, and perhaps reflective of the purpose for which the statements to follow were compiled — to establish creators' property rights — the solitary nature of book writing is emphasized in the literature on the history and philosophical underpinnings of copyright.[363] Two examples will suffice. One has the philosopher Denis Diderot, following in the tradition of Locke, Edward Young, Goethe, Kant, Schiller, and Fichte, saying: "What form of wealth *could* belong to a man, if not the work of the mind ... if not his own thoughts ... the most precious part of himself, that will never perish, that will immortalize him?"[364] The other has Keats reflecting upon Shelley and Wordsworth by saying: "I never wrote one single Line of Poetry with the least Shadow of Public thought."[365]

The author creates a composition, a work of art similar to a symphony or a sculpture. The task is one of invention. Until recent years, interactivity in literary creation came in the form of engagement in the world, inclusive of the perceptions and ideas of others. For the most part, this engagement preceded and continued into the literary creative work — the act of putting pen to paper, as they used to say. Following the author's turning that engagement into a literary form came the publisher's supplying editorial, design, and marketing. On the reading side of the equation, interactivity in reception of literature derived from shared meaning frameworks that, at the level of the solitary reader, encouraged a more subtle version of the very young read-to listener crawling into the story. Whether those meaning frameworks derived from something as foundational as narrative or character, or whether they derived from common social concerns or interests, the reader set aside the time and mental space to follow the creation of the author.

Stepping back from the author and the solitary reader, we can see that other forms of interactivity focused on the book abound. At the level of small groups — reading clubs, for example, or followers of literary mentors such as Oprah Winfrey — interactivity is in the exchange of perceptions drawn from a common text, a bonding exchange not unlike the parent-and-child bonding that derives from one reading aloud to the other. Such an exchange provides insight not only into the text, but also into the reader. At the level of society, as illustrated by the example of Hogarth Press and the Bloomsbury Group, interactivity came in the form of a sea change of opinion and behaviour that was nothing less than a re-articulation of being in the world. Recall that Bloomsbury itself was a group of creators, and it was, to some extent, defined by social interactivity, an idea we shall return to shortly.

In overview, until nearly 2010, the following appeared to be a common model in book writing: Out of a social nexus came engagement and reflection, followed by solitary creation. Proceeding through selection, fine-tuning, and release to the public, books were usually consumed in solitude, from which there was a return to social interaction in the writing and reading of reviews, in book clubs, and in reading mentorships that led to interpretive communities of many different kinds, from reading groups through to religions. But how is interactivity a major, non-trivial change trajectory in book publishing? The answer comes through examples.

Experiments in interactivity

We are near the beginning, and there have been failures. On September 8, 2009, which, in the time warp introduced by technologically induced change, seems like an eon ago, Anthony Zuiker (creator of the television program *CSI*) and Duane Swierczynski, in collaboration with the publisher, Dutton, released *Level 26: Dark Origins*, the result of a million-dollar investment in the development of a multimedia interactive book, dubbed a "digi-novel." When readers reached certain points in the thriller (occurring about every twenty pages), they were encouraged to go online and view a three-minute film clip that connected the textual content already read to the textual content to come. The producers' goal was to combine these written and visual elements with an interactive community website that would allow readers to talk about the digi-novel and contribute to the story. Fifty days after the book's release, a Google search for "Level 26" produced, first, twenty-one promotional items. The twenty-second item to appear was a review of the digi-novel. The reviewer followed the creators' instructions, reading, then watching the film clip, reading the next section, then watching, and so on.[366] He had two main reactions: going back and forth between media was very disruptive, and the video portrayals did not mesh with how he imagined the characters and events. Two of the next ten items in the Google search were reviews; one was a somewhat vacuous positive review from a person who described herself as a crime writer,[367] while the other was neutral. *Level 26* was an experiment. It is certainly not the future of interactivity. It disrupted the reading experience and missed a workable formula by a substantial margin. Eighteen months after the release of *Level 26*, Anthony Zuiker's video promoting it had received only 2,241 views.

An obvious contrasting example is Wikipedia, an undeniably successful, crowd-sourced resource that, on the whole, is the most broadly useful and used site on the Internet. The site is so powerful that it decimated a whole genre of publishing: commercial encyclopedias. Are the dynamics of Wikipedia repeatable? Yes, and much smaller but powerful examples exist, particularly in the creation and maintenance of open source software. Open source software is usually developed by an individual or group and, when substantial enough, as in the case of Drupal and WordPress, is adopted by a community of programmers. However, in open source programming, crowd sourcing does double duty. As well as developing the software cooperatively, the programmers develop techniques to facilitate cooperation. Nothing but their discomfort with technology stands in the way of a group of non-programmers copying this model for the creation of a new kind of book — a much more dynamic and integrative form than its historical predecessor, the anthology. The cooperative creative process can also be seen as an extension of contemporary author interactions with their fan clubs,[368] but this process is far more intense and involving. Some might call it web-mediated conversation. It can equally be called interactive creation; creation in interaction with a referential community, involving varying levels of co-creation. Academics are already experimenting with serious interactive writing by posting drafts for comment and incorporating the useful comments of their peers. The writing and sharing of fan fiction is yet another example. Acknowledgements do not diminish the credibility of the author; rather, they encourage partial intellectual ownership of the resulting book as testament to one's contribution.

Tweeting, of course, represents interactivity at its apogee. Tweeting and other forms of social media hastened the fall of Hosni Mubarek of Egypt in early 2011. In China, social media forced the government to back down from trying to cover up a high-speed train wreck. More generally, in later 2011, that government was constrained in its actions by the explosion of information by means of social media.

One result of the development of books along a trajectory of interactivity appears to be the emergence of authors, not as voices in the tradition of the Romantic poets, who saw themselves as just one step removed from a God who spoke through them, but as voices of communities of interest, concern, aesthetics, location, or knowledge. Such interactivity will have consequences that are far reaching. It may lead to authors as community champions as well as creative heroes, or an amalgam of the two roles. It may also create further challenges to copyright, especially in visual media that require reproduction of images to properly reference the past.[369] A persuasive argument in favour of this form of interactivity comes from interviews in which members of the management teams of Alberta magazines were asked to describe the role of their publications. Without consulting supporting documents, the senior staff of virtually all publications were able to describe both the dynamics and the nature of their target audience, and they stated that their magazines provided interactive leadership by articulating the issues and interests of that audience. [370]

Interactivity of a far different sort is emerging in educational publishing markets. Motivated in part by an attempt to maintain their marketplace position, and in part by their vulnerability to unauthorized copying in a digital world, educational publishers are looking to interactivity as a means of enhancing learning and thus preserving their markets. The first steps can be seen in right-answer subjects. Rather than merely making electronic versions of their textbooks available, textbook publishers are creating interactive engagement to help students master the subject matter. For instance, students are asked to resolve a chemical equation. The customized feedback that is drawn from a database is contingent on the answer the student provides. Random answers result in the student being advised to re-study the unit. As well, since tests are self-administered, students can proceed at their own pace. Understandably, developments in the humanities and social sciences lag behind the right-answer science subjects.

Combined with the expanded opportunities for generative multimedia expression of Trajectory 2, the dynamics of interactivity have the potential to reconfigure authorship and strengthen the connectivity between creative genius and the social nexus within which creation occurs. Perhaps it is too bold to claim that the potential exists for an emergent re-visioning of the nature of humanity that may nurture new forms of bonding, creation, and expression, as humanity begins to gain insight into the neuroscience of co-creation.[371]

Business applications 1: O'Reilly Media as a manifestation of the three technology trajectories

A company that demonstrates the power and possibilities of the three trajectories is O'Reilly Media, which began operations in 1978 and has evolved into a highly respected chronicler and catalyst of technological innovation. Described as a vertically integrated niche publisher,[372] O'Reilly's success in the marketplace combines transmedia dynamics, which some call vertical niche publishing, with the use of databases, cloud computing, new media, and interactivity. The following description is taken from the O'Reilly website.

> O'Reilly Media spreads the knowledge of innovators through its books, online services, magazines, research, and conferences. Since 1978, O'Reilly has been a chronicler and catalyst of leading-edge development, homing in on the technology trends that really matter and galvanizing their adoption by amplifying "faint signals" from the alpha geeks who are creating the future. An active participant in the technology community, the company has a long history of advocacy, meme-making, and evangelism.
>
> Publisher of the iconic "animal books" for software developers, creator of the first commercial website (GNN), organizer of the summit meeting that gave the open source software movement its name, and prime instigator of the DIY revolution through its *Make* magazine, O'Reilly continues to concoct new ways to connect people with the information they need. O'Reilly conferences and summits bring alpha geeks and forward-thinking business leaders together to shape the revolutionary ideas that spark new industries. Long the information source of choice for technologists, the company now also delivers the knowledge of expert early adopters to everyday computer users. Whether it's delivered in

print, online, or in person, everything O'Reilly produces reflects the company's unshakeable belief in the power of information to spur innovation.[373]

Some of O'Reilly's more notable innovations include its digital electronic reference library for programmers and IT pros, which is part of its online bookstore, Safari. As well as processing simple purchases, Safari allows a fee-paying user to access more than 4,000 often-referenced books for a defined time period and to access relevant materials online immediately. Either chapters or entire books can be downloaded, and in addition to O'Reilly publications, Safari carries the publications of about twenty other presses. It also offers Rough Cuts, a service that provides early access to cutting-edge books as they're being written. Short Cuts are short papers provided in PDF form dealing with subjects that require limited treatment, such as *How to Build an RSS 2.0 Feed and Automating InDesign with Regular Expressions*. Among other categories of content are pocket guides, hacks (for programmers), and theory-in-practice books on such topics as databases and project management. O'Reilly has invested in software development and a number of other technology start-ups. It also has a number of affiliated bookstores that work with the publisher to stock items in anticipation of demand. In Canada, they include the University of British Columbia, University of Waterloo, University of Toronto, and University of Calgary bookstores.

With these elements in place, O'Reilly has begun to emphasize large-scale events. Like rock concerts, these events bring users and fans of the company together to celebrate the worldview, actions, and products of O'Reilly Media and the insights of the community the company represents. Unlike rock concerts, where attendees walk away in a half-deaf buzz, attendees at Tools of Change or other O'Reilly conferences leave with a buzz of new possibilities for innovation and serving emerging markets. O'Reilly also runs a set of online courses, mostly focused on programming, for which participants can earn credit from the University of Illinois.

A number of elements contribute to O'Reilly's success. Its innovation and energy have certainly given it a certain status in the IT community. Its involvement as a purveyor of information in a variety of forms confers power and allows it to exploit content in multiple ways. Its myriad product forms complement the traditional book format. Its varied terms of sale — permanent/temporary, wide access/single-item purchase, free download or paid-for print version — convey implicitly its interest in its customers. In short, the very existence of its constellation of services and products to advance knowledge, skills, and understanding communicates relevance, demand, and value. Moreover, they exemplify developments embracing new media, interactivity, and databases put to the task of serving a targeted market.

Business applications 2: Maximizing sales from market-based print-technology choices

From a book publisher's perspective, printing used to be quite simple. A publisher would estimate the print run needed for a title, obtain quotes, select a printer, place an order, review proofs, and give approval for the run. Printing was carried out with film-based offset presses that were either web-fed (from a paper roll) or sheet-fed. Small runs were expensive, and publishers often produced more copies than they needed because the per-unit costs for the last 10 percent of a run were so marginal that it made little sense not to print them.

Then came print-on-demand (POD) technology. POD made it possible for a book to be written, edited, and laid out on a desktop computer, thereby eliminating the cost of tying up expensive equipment. A computer file of the laid-out page could then be sent to a combined laser printer-photocopier that ran off from one to 500 or more copies. Various companies, including Xerox, IBM, HP, Kodak, Ricoh, and, more recently, Perfect Systems, saw a market for such machines in academic institutions as well as for book publishers, and they developed integrated high-speed printer-photocopiers capable of printing a book in minutes. Combined with colour printers, trimmers, and binders, these machines were able to produce instant or "print-on-demand" books. At their debut, the speed of the machines was somewhat limited, but their main stumbling block was lack of quality, most obviously on the cover and binding, and also in the fineness of the print, which was initially 300 dpi. Since then, laser printers have achieved a resolution level (up to 1800 dpi) that is ample for most book printing and is inadequate only when there is a need to reproduce images with extreme precision.

The success of POD, which is now in use by a wide variety of publishers, is based on its minimization of up-front set-up time and costs, accomplished to some significant degree by off-loading them from the printer to the publisher. Digital printing's major limitation, besides the quality of printing, is that it yields negligible economies of scale: the per-copy printing and binding costs of a book using digital printing technology, excluding set-up costs, remain more expensive than per-copy offset costs. This means that digital printing makes sense for short print runs — less than 500 — but offset begins to make sense for runs of 500 to 1,000, at which point press set-up costs, amortized over the full print run, are reduced to less than the difference between digital printing costs minus offset manufacturing costs. (There are additional factors that this does not take into consideration, such as the fact that the ink used in POD is less environmentally friendly.)

Available computer memory has increased to the stage where digital image resolution has surpassed the crispness of analog film, although digital printing processes have not yet caught up. Moreover, page-layout programs have become adept at handling and integrating massive image and text files. Printing from computer files reduces costs for materials and for work required by various printing professionals. Even with the reduced time for set-up that has been achieved over the past decade, hourly costs of machine access remain high.

Where offset machines shine is in speed, combined with quality. Some can produce as many as 50,000 impressions per hour, with the potential for more than one page in each impression. This process exemplifies modernist industrial processes, whereby major investments are made in the development and perfection of the manufacture of prototypes, and then thousands of copies are made based on the customized output of these machines.[374]

In summary, digital printing speeds up and simplifies the set-up process, so even though the output speed is reduced to about 100 to 280 pages per minute (6,000 to 17,000 per hour),[375] it makes short runs far more feasible for publishers. Digital printing also makes it possible to print and bind small quantities to order, thereby reducing warehousing and handling costs, and it allows for personalization or other content adjustments as the print run progresses. Each copy can be slightly different.

Digital printing has opened printing to technical and organizational innovation, encouraging offset press manufacturers to eliminate film by developing techniques to allow direct printing from computer to printing plate, stimulating the development of new lines of digital printing machines that will meet the needs of various clients,[376] and driving a reorganization of print management that makes maximum use of database and digital printing technologies.

At a first level, a print file can be kept in a database on a server and readily accessed from a distance for printing at any time. But with a richer database that contains not only the print files but also the sales pattern of the title, much more is possible. Evidence of the strategic value of print management is provided by the Life of Title service, a trademarked system provided by the U.S. printing company Edwards Brothers. Life of Title allows publishers to analyze demand for titles tracked and stored in databases, and decide which technology to use to meet that demand with the least expense. For example, a publisher might initiate a 5,000-copy offset print run. This might be followed by one or two top-quality digital short runs, spread out over a few years, of 300 copies each. In turn, these printings might be followed by smaller orders for individual customers, fulfilled by printing for immediate delivery — true print-on-demand. Edwards Brothers emphasizes how it can bring a publisher revenue from OSI (Out of Stock Indefinitely) titles. For the University of Chicago Press, the company was able to generate $1 million in additional revenue in 2004 from the management of 8,800 titles with average print runs of twenty-five copies; it generated $500,000 in a year for Rowman and Littlefield from 4,500 titles with average print runs of eight.[377]

In Canada, Webcom offers a similar service, called Title Value Management, and Friesens and many other major book printers have followed suit. As Webcom's website makes clear, the idea of using multiple technologies to manage titles over their lifespan is establishing itself in the publishing marketplace.[378] The challenge for printers is not mastering the technology, but organizing production to take full advantage of that technology. Thus, a self-serve model, where the publisher loads the print files and requests a proof from the printer, makes far greater economic sense than having the printer handle the file directly.[379]

Many publishers are using digital printing technology; however, publishers, like printers, have yet to exploit the full flexibility of the technology, combining print-on-demand, short-run digital printing, and longer-run offset printing. It is certainly true that copy-by-copy POD rarely makes sense for an initial print run, because publishers need author copies, review copies, prize submission copies, sample copies for reps, advance reading copies, and so on, as well as copies for sale. Such needs may eat up as many as 200 copies. But conservative initial print runs followed by short-run printing and then print-on-demand could make sense for many of Canada's small trade publishers.

Business applications 3:
Service publishing, a.k.a. self-publishing

As the previous section indicates, digital printing has led to an expansion of services by established printers whose primary clients are publishers. It has also led to the emergence of new printer/publishers whose primary clients are individuals and self-publishers. Trafford Publishing of Victoria, B.C., was an early entrant and achieved market leadership before it gradually lost ground to offshoot companies such as Amazon's BookSurge and CreateSpace, Ingram's Lightning Source, and Lulu, at which point it was sold to a U.S. company based in Indiana.

Trafford did not risk its own investment in title development; rather, it sold publishing services to authors, including educators, public speakers, older people who had written memoirs, the parents of children who had created wonderfully charming stories and illustrations, previously published mid-list authors, authors with out-of-print titles, and the list goes on. As well, Trafford's business model was truly print on demand. When an order came in for one copy or ten copies, the books were printed and shipped out, usually on the same day.

Trafford clearly laid out its publishing services and the fees charged for each. It offered various levels of service, from the entry level Legacy, which published the book and gave the author a proof and ten copies, to Best-seller, which added on bookmarks, postcards, and other marketing tools; established relations with Amazon; offered discounts to libraries (15 percent), bookstores (40 percent), and wholesalers (48 percent); and sold books on a non-returnable basis. Over its first twelve years of existence, Trafford published 750 titles per year — just over three new titles per working day. Its website at the time noted that between 1995 and 2006, it published 9,000 titles for 7,000 authors.

Trafford was a bellwether company that identified a market for its competitors, highlighted the unserved market of authors (many with credible manuscripts), and tore down the wall that had preserved traditional book publishers as the sole legitimate gatekeepers of market entry. Self-publishing is now a growth activity of enormous proportions. Hugh McGuire has reported that in 2002, 215,000 new titles were published in the United States, along with 32,693 self-published and short-run books. In 2008, U.S. book publishers published 275,000 titles, while non-traditional publishers produced 285,000 titles. But the really dramatic change was yet to come. In 2009, book publishers published 288,000 titles, while non-traditional publishers accounted for 764,000 titles — a 182 percent increase over the previous year, which in turn was a 200 percent increase over 2007. The breakout year for non-traditional publishing was 2007, when the number of titles rose from 21,936 in 2006 to 123,236. In the same two years, the number of traditionally published book titles rose from 274,416 to 283,370.[380] Equally interesting is the number of titles published by the top ten U.S. companies involved in self-publishing, who hold 74 percent of the market share.[381] The data are shown in Table 7.1.

While book publishers do add value in a variety of areas — including editing, layout, market planning, marketing, and distribution — they may not be as valuable to an author who is also a successful travelling lecturer and whose sales are mainly to people who attend those lectures. With still other companies, such as Leanpub.com, entering the marketplace, the low cost and flexibility of self-publishing is breaking down publishing barriers even further. As well, many organizations have their own successful publishing programs that bring in revenue, where a traditional book publisher would merely be a revenue drain.[382]

Table 7.1
The top ten print-on-demand publishing companies operating in the United States and the number of titles published in 2009

Name of Company	Number of titles published
BiblioBazaar	272,930
Books LLC	224,460
Kessinger Publishing, LLC	190,175
CreateSpace	21,819
General Books LLC	11,887
Lulu.com	10,386
Xlibris Corporation	10,161
AuthorHouse	9,445
International Business Publications, USA	8,271
PublishAmerica, Incorporated	5,698

Source: Bowker, "Bowker Reports Traditional U.S. Book Production Flat in 2009," April 14, 2010, http://www.bowker.com/index.php/press-releases/616-bowker-reports-traditional-us-book-production-flat-in-2009.

Business applications 4: AbeBooks and digitally mediated bookselling

Digital, database, and associated technologies are also transforming the bookselling marketplace. Change has been dramatic in some areas. Online bookstores are certainly challenging brick-and-mortar stores for sales. When respondents were asked to name all sources of acquired books in a Canadian study undertaken in 2005, brick-and-mortar bookstores were by far the dominant source for the acquisition of books, at 81 percent for purchases (including books for others) and 62 percent for books the purchasers had read themselves. Public libraries were at 32 percent for reading. Big-box stores such as Costco were far behind, at 12 percent for purchases and for reading, while the Internet was at 8 percent for book purchases and 4 percent for reading.[383] By 2008, online booksellers were claiming a trade book market share of between 21 and 30 percent,[384] and in 2009 Amazon had worldwide sales of $24.5 billion, an ever-decreasing percentage of which were books.[385] The success of Amazon indicates that it has achieved a substantial market position, and it confirms this by continuously adding new features to its websites. But there are inherent inefficiencies to Amazon's operations. A substantial portion of its business — Chris Anderson claims it was 25 percent just prior to 2006 — is in the long tail of demand,[386] and the process of shipping single copies of rarely ordered books to Amazon to be redirected to customers is costly and inefficient. Those costs are borne in part by the publisher (shipping to Amazon) and in part by the consumer or Amazon (handling plus shipping costs to the consumer). The thorniness of the issue is exemplified by the continuous incentives Amazon offers consumers to buy in sufficient volume that Amazon will make enough on the sale to forgive the shipping costs.

Like eBay, AbeBooks developed a smarter, low-touch model, setting itself up as an order processor. The company attracted used-book sellers to its Internet presence and, hence, to its worldwide audience by offering an inventory management tool that let buyers know where a title was available. It then linked buyer to seller and processed the order. After AbeBooks expanded its supplier base from used-book sellers to anyone who had titles for sale, it ultimately brought to purchasers a worldwide inventory database that, as of 2008, comprised 110 million books and 13,500 booksellers (in the broad sense of the term), which generated 4 million searches on the site each day and annual sales of $170 million.

The AbeBooks model is radical in yet another way. By combining new and used titles, it frees book purchasers from a dependence on the passing product launch, the marketing and publicity efforts of publishers, and the stocking decisions of booksellers. Easy availability is no longer confined to a short period following the release of a title, but is extended indefinitely. AbeBooks is to the long tail what J.K. Rowling is to bestsellers.

Amazon bought AbeBooks in the summer of 2008, and it will be interesting to chart the company's future direction and development. Certainly Amazon is not standing still; it is now using BookSurge[387] and CreateSpace to make printing part of its services, to save one element of the shipping costs. Of course, it is doing so by charging a premium for printing that requires the originating publisher to share the resulting savings. Indeed, Amazon is insisting that publishers print through BookSurge or CreateSpace if they wish their product to be listed as an Amazon book. BookSurge is in competition with Trafford and Ingram's Lightning Source[388] as well as Lulu.

The logic of POD mandates proximity to demand. Within a few years, bonded print-on-demand services capable of producing books will be familiar storefront occupants. Such services will be able to access print files, or, conversely, publishers' computers will be able to access in-store printers, in response to a consumer request. Staples is actually not far from providing such a service. The only limitation on the market will be competition from e-books, which we will consider next.

Business applications 5: E-books

E-books take full advantage of databases, and they will benefit from new media as software facilitates the integration of other media with print; the delays are ease of integration and size of image and sound files. Perhaps the current e-book format, text files adapted for e-book reading devices, will go the way of silent pictures. Perhaps not! No doubt there will be instances, and educational publishers will probably find them first, where audio and visual images will increase readers' understanding of content. When we have a better understanding of brain function, e-book developers may deploy soundtracks and background visuals to further enhance appreciation or understanding. But there will be limits. For example, it is difficult to believe that reading philosophy would be enhanced by much in the way of pictures and sounds.

Today's e-books are the ground floor. They are sites for creative experimentation. It is easy to imagine a user talking into an e-book reader to record his or her thoughts, with the recording tagged to the relevant passage and available to be played back or displayed as text. New media come into their own in a digital environment, and the richer the digital environment, the greater the opportunity for new media to shine. As noted, the e-book is a new medium itself. But beyond the e-book lies the possibility of expressing considered thought and human creativity in a richer media environment. Mike Matas presents an impressive early example (see link in endnote).[389]

As mentioned in the discussion of Kobo in Chapter 6, digital display and consumption allow publishers to understand far more about readers and reading than they have ever done before. Kobo will not be the only device or company collecting such information; educational publishers may very well make a business of selling interactive profiles of students' use of learning materials to educators. Amazon is already using generic user data for upsell promos to purchasers, informing shoppers that others who looked at the title they are examining have been interested in a family of other titles. Combinations of these data, together with brain imaging, may yet reveal findings — for example, that music of a historical period enhances the reading experience of a novel based in the same period. Interesting as such data might be, the possibility of governments or the courts gaining access to information on one's reading habits is a very real threat to privacy.

Business applications 6: Apps

Apps are multiplying at an astounding rate, and their expanding numbers are an indication of their profound significance. The basic reason for their expansion and significance is that they can successfully monetize online multimedia. The independent creative developer benefits, the distributor/mediator benefits, and users benefit. That said, there is more to a successful app than meets the eye. Happily, understanding the variables that determine success is growing easier. And, to some extent, they derive from the three technology trajectories introduced in this chapter. Apps are retrieved from a database, many include images and sounds as well as text, and many also depend for their success on interactivity. There is a legal element as well.

The term "apps" comes from two other terms, application software and applets. Generally the term apps is used to talk about software that is smaller and lighter than, say, Adobe suite software, Microsoft Office or Open Office, all of which are examples of application software. While software implies functionality, the main feature of an app can be content, as in an e-book. An e-book is an app in that it contains programming that allows it to be viewed and marked up in certain ways on certain devices. In the case of the e-book, the functionality can be fairly standard to the e-book form, circa 2011. As time goes by, books may be sold on a foundation of unique functionality (that will then likely to be copied) as well as unique content.

One of the strengths of apps is that they allow for a controlled interface; they are not always universally and openly accessible. They can be developed for universal usage or in conjunction with proprietary systems such as the Kindle, or Apple, RIM, BlackBerry, or Android products.

Apps meant for general consumption benefitted greatly from the demonstrated success of Apple's model of distribution of music files. Apple's model arose in response to two things: a music industry tied to the distribution of physical packages of products (CDs or albums of twelve or so songs), and largely illegal file sharing from sites such as Napster and then BitTorrent (the company). The well-known story of Apple's ninety-nine-cent single-song sales, which brought in masses of micropayments, does not need repeating. But the mechanisms behind that success are worthy of attention. Most important is that the Apple brand had a sufficient market presence to allow it to serve as an effective distributor as it became the largest world retailer of music. In turn, that helped Apple gain market share for its various devices and surge past Microsoft as the largest computing-based entity in the world. This expansion increased its effectiveness as a distributor. Equally important, the system Apple set up allowed for sales of virtual, rather than physical, products, and rights management. Apple took its share, sales were tracked, and the creators and producers received their share. And consumers were neither gouged for content they did not want, nor overcharged for what they did want. In addition, consumption was made convenient. On a whim, for a buck, users could download a song of choice from a vast database (catalogue) of content.

Apps operate on essentially the same sales model. Once devices and operating systems are developed, independent developers — like the musicians who produce MP3s — come up with ideas for apps containing useful content or functionality, undertake the programming, and submit them to the distributor of the device or operating system. After testing and agreement on pricing, the app is made available and income is shared according to a contracted formula. Success depends on the uptake of many users who are seeking to extend the functionality of whatever devices they own or use.

So important have apps become that the uptake of devices such as tablets has become dependent on the number of apps independent developers create for the device. Operating systems are likewise dependent on app-developer uptake. So when Google brought forward its Android operating system, it was able to establish the system in the marketplace in competition with Apple by giving it away to cell phone companies. In August 2011, HP walked away from its tablet and RIM was struggling with its Playbook for lack of developer, and hence consumer, uptake. The dynamics of uptake are coming to interact with demographic variables to determine market growth and acceptance. For example, it would make good sense to develop an app targeted at thirty-year-olds for the iPhone first, because of the iPhone's extensive penetration of that age group. Developing apps for sixty-year-olds might mean tailoring them to a number of different devices and operating systems, because a wide variety of devices are used by that age group.

Beyond simple e-books, there is a great deal of app development in educational and guidebook publishing. Most recently, as mentioned in the section on interactivity, companies such as Pearson have been successful in creating online labs. The companies claim that the labs enhance learning. An increase in this activity is being motivated by pressure from the education sector to include educational copying as fair dealing within the Copyright Act and, beyond that, to rule as fair dealing the copying of classroom sets. As copying without compensation increases, the business model for educational apps that provide online access to content and effective learning tools for a period of time becomes better and better, compared to selling a book to a student or an institution where it can be copied, legally or not, and passed on or sold to other students.

Overview and conclusion

As database technology provides the means for publishers to organize and vastly increase their efficiencies of production, dissemination and use of information and content, the capacity to do much more with the content they have on hand increases exponentially. As new media of expression expand and enrich opportunities for human beings to represent reality and thereby create meaning, the capacity to attain new understanding emerges, both through consumption of richer representations and by increased access to the tools of production that change our abilities to see, understand, and appreciate. In turn, databases made up of enriched multimedia representations are increasing the opportunities for interaction-mediated publishing that connects creators with their readers in a much more intimate manner and, hence, changes their voices. Developments following these three technology trajectories are not proceeding separately. They, too, are developing interactively.

Established members of the publishing and allied industries are seizing opportunities to use technical innovation that will extend, redefine, and realign their traditional roles. New entrants are also bringing technological expertise to publishing, with little allegiance to traditional forms and practices. Both groups are inventing more agile creation, production, and retailing practices; they are also envisioning non-publishing extensions of publishing activities. New channels of distribution are emerging alongside the potential for instant availability of any title. In all of this change, there is considerable positive volatility and invention that will let publishers position themselves advantageously to secure their future.

The implications of technological change for Canadian book publishing are mostly second-order implications, not direct results of technology. This is partly because, for the most part, although they are not bleeding-edge developers, Canadian publishers are relatively early adopters of technological change. Even so, they are running into a fair amount of difficulty making good decisions in the context of both technological change and the wide variety of other factors with which they must contend. Given the need to consider the implications of technology in interaction with the other variables discussed throughout this book, this discussion is most appropriately undertaken in the following chapter.

CHAPTER 8

Changing realities and the future of Canadian book publishing

Introduction: At the birth of modern Canadian book publishing ... 293

The political environment and the contemporary political economy of Canadian cultural production 295

The political, legal, and policy environment 301

    Law   302
    Creative Commons   305
    Policy and support programs   306

The economic environment 315

    A service model of publishing   318
    A service model for scholarly monographs   319
    A service model in educational publishing   324
    A service model in the trade sector   325
    Barriers to the acceptance of a service model   327

The cultural environment: A changing Canadian market 327

The educational environment 330

The ever-evolving technological (and organizational) environment 332

At the rebirth of modern Canadian book publishing:
    Overview and conclusion   333

Introduction: At the birth of modern Canadian book publishing . . .

The birthing years of modern Canadian (print) book publishing were profoundly influenced by a rather intense set of preceding events and realities: for example, a devastating world war, the proliferation of nuclear warheads capable of destroying virtually all life on the planet, and a sometimes violent assertion of civil rights in Europe, Africa, and North America. The period itself brought a nearly live-televised, U.S.-initiated, undeclared and unwinnable imperial war; a spate of kidnappings and assassinations of public figures; politically motivated riots by students in the streets of Paris; and the kidnapping and murder of an Italian prime minister.

Underlying portions of the civil mayhem was an intense idealism expressed both violently and peaceably, admirably captured in the movie *Across the Universe*. Baby boomers evinced a desire to remake the world in the image of the social ideals on which they had been weaned. Talk about violent overthrows, in some cases followed by actions, was juxtaposed with sit-ins, bed-ins, and love-ins. In Canada, Prime Minister Pierre Trudeau — with his young wife, and with his signature rose in his lapel — came to be identified more as a person on the side of flower power than as an official of the state. Nonetheless he proved willing to use the army to crush the nationalist dreams of discontented Quebecers, who themselves employed bombs, kidnapping, and murder (sometimes encouraged by undercover RCMP) in pursuit of their political goals.

Other substantial changes were afoot in Canada during this period. The final steps were taken in the transformation of the economy from one in which the dominant activity was agriculture to one based on industrial manufacture and resource extraction. A massive population boom came of age. This generation benefitted from growing wealth and a liberal education that, partly as a result of the writings and actions of the Bloomsbury Group, led them to sweep aside the attitudes of the previous generation — the quasi-Victorianism, and the conservatism brought on by world conflict, the privations of pioneer childhoods, and the Great Depression. Canada was shedding the last vestiges of colonial rule by creating its own unique socio-cultural milieu and the institutional foundations for full independence and self-government. Nations, races, cultures, aboriginal groups, and women were asserting their distinct identities.

From the 1960s through to the mid-1980s, the environment was not generally seen as being at risk. Rachel Carson and her many followers warned of the destruction of nature, but in the mainstream public consciousness the environment remained a paradise, pure and unspoiled. The back-to-the-land movement of the time was focused on turning on (with drugs), tuning in (to the potential for a peaceable and caring worldwide human community), and dropping out (of the rat race — i.e., not devoting one's life to one's employer), and Henry David Thoreau was a contemporary hero. Criticizing the environmental impact of printing a surplus of books to ensure access by all potential book purchasers would have been viewed as a colossal misplacement of priorities.

As well, both creators and users of all forms of media did not understand the conceptual foundations and the broad scope of copyright. In the 1980s, many assumed that possession of a material artifact gave the owner all rights to exploit it as he or she wished. Few thought it reasonable to compensate authors for the sales lost when their books were available for borrowing from libraries. Photocopying was a new technology and, overall, moral rights went virtually unrecognized. Canada was considering whether it should sign the Berne Convention and the Universal Copyright Convention, and position papers on copyright toyed with the idea of copyright being exhausted upon publication. The Copyright Act contained no provision for authors' copyright collectives.

During these years, a wide variety of authorial voices emerged and, through poetry, drama, children's books, short stories, novels, and non-fiction, they defined a social constitution, in a sense. The non-fiction encompassed the entire spectrum of existence — the personal, political, social, cultural, educational, historic, linguistic, geographic, artistic, architectural, communicational, and technological milieux. Readers seized on these books as affirmations of identity, and their support lifted Canadian book publishing to a somewhat exalted status as the medium of record for the era. Ideas, not gizmos, ruled social consciousness.

However, the times have changed. This concluding chapter assesses the near- and medium-term viability of Canada's book publishing industry, given its intrinsic characteristics and the relevant external environments within which book publishing in general exists in a changing world.

The political environment and the contemporary political economy of Canadian cultural production

The establishment and evolution of the Canadian book publishing industry, from a small clutch of agency-based firms in the late 1960s to an active industry of many more engaged primarily in originating titles, would have never been possible without a sympathetic political environment. That environment evolved steadily from the 1970s forward, built on a preceding discontinuous history of assertive nationalism led by determined visionaries such as William Lyon Mackenzie, Egerton Ryerson, Lorne Pierce, and Jack McClelland. As the twenty-first century unrolls, the foundations are being laid for further development of book writing and publishing, alongside other sectors of the creative economy. True, it may be that the value of cultural production will take some time to be fully recognized by successive governments, but in the long term, the emerging understanding of the role of creative activity will hold firm, because a political environment sympathetic to cultural production appears to be inherent in a nascent worldwide paradigm shift.

In Canada, this shift was publicly and clearly articulated in an April/May 2008 report with background studies commissioned by the Department of Canadian Heritage entitled *Valuing Culture: Measuring and Understanding Canada's Creative Economy*.[390,391] The report was compiled and written by the Conference Board of Canada, an economic and policy think tank that contributes to, and takes its lead from, economic thinking of the day.

Valuing Culture fully confirms the framework and analysis articulated first in the Ontario government report *The Business of Culture*;

again in the SAGIT report on cultural industries, *New Strategies for Culture and Trade*; also in the call for a New International Instrument for Cultural Diversity (NIICD) and in the ratification of the Convention on Cultural Diversity (CCD); and finally, in the development of the Supply Chain Initiative (all discussed in Chapter 4). In its definition of the cultural sector, the report adheres to the categorization system of national statistics-gathering agencies, including Statistics Canada,[392] which describe it as a sector composed of "written media [including newspapers], the film industry, broadcasting, sound recording and music publishing, performing arts, visual arts, crafts, architecture, photography, design, advertising, museums, art galleries, archives, libraries, and culture education."[393] Also included within the Statistics Canada categorization is the whole of the creative chain, including creation, production, manufacturing, distribution, and support activities (such as management, promotion, and preservation) — in short, everything involved in the "transmission of an aesthetic or intellectual concept . . . intended to elicit an emotional or cognitive response, and contain intellectual property rights."[394] The categorization additionally includes non-core services, for example, engineering, "where only part of the creative chain is in scope for culture," as in architecture.[395]

Using this categorization, the Conference Board estimated that "the real value-added output by cultural sector industries totalled $46 billion in 2007,"[396] approximately 3.8 percent of the total Canadian gross domestic product (GDP). The economic footprint of the culture sector is much larger when all of its direct, indirect, and induced effects are included. From this view, the full contribution of the sector as of 2007 was $84.6 billion, about 7.4 per cent of real GDP, and 1.1 million jobs.[397,398] To arrive at this calculation, the report employs a productivity multiplier of 1.84,[399] reflective of a "methodology [that] is meant to allow us to assess . . . the extent to which a wide range of culture industries flow through, and affect, our economy."[400] However, the above figures do not take volunteerism into account: if they did, the real value as a percentage of GDP would grow measurably further.[401]

In crafting its report, the Conference Board positioned the cultural and creative industries as measurable contributors to both the economy and society. At least two of the background papers suggest that "participation in various forms of arts and culture contributes to society in many tangible and intangible ways."[402] The report also cites the United Nations Development Programme's statement that "culture provides the social basis that allows for stimulating creativity, innovation, human progress and well-being. In this sense, culture can be seen as a driving force for human development in respect of economic growth, and also as a means of leading a more fulfilling intellectual, emotional, moral and spiritual life."[403]

In its examination of inputs, *Valuing Culture* identifies seven key drivers of the creative economy: consumption dynamics, innovation, technology, talent, diversity, social capital and collaboration, and capital investment. In its assessment of outputs, it notes that "the creative economy is itself a driver of competitiveness and growth in the broader economy."[404] This viewpoint is in stark contrast to the notion that the cultural industries are weak and in need of government subsidies. The contrast can be explained by the fact that the creative economy perspective focuses on macro-economics, while the weak cultural industry perspective focuses on micro-economics. We will return to macro- verses micro-economics shortly.

In short, *Valuing Culture* brings forward three significant facts: First, the creative economy accounts for a substantial contribution to Canada's GDP (7.4 percent or $84.6 million). Second, it generates 1.1 million jobs — as many as agriculture, forestry, fishing, mining, oil, gas, and utilities combined. Third, its contribution to economic growth is 6.6 percent.[405]

The Canadian cultural community was elated to see such clear statements of its contributions; however, *Valuing Culture* is not a document without peer. Also notable is a United Nations document of the same year called the *Creative Economy Report, 2008*.[406] That document notes:

> The concept of the "creative economy" is gaining ground in contemporary thinking about economic development. It entails a shift from the conventional models towards a multidisciplinary model dealing with the interface between economics, culture and

technology and centred on the predominance of services and creative content. (p.3, 4)

It continues:

> For some people, the "creative economy" is a holistic concept dealing with complex interactions between culture, economics and technology in the contemporary globalized world that is dominated by symbols, texts, sounds and images. (p.4)

To give a sense of the size and growth of the trade in creative goods and services at the international level, the report notes that between 2000 and 2005 this trade increased annually by 8.7 percent, reaching a value of $424 billion — 3.4 percent of total world trade.

In 2008, the same year that *Valuing Culture* and the *Creative Economy Report* were published, the U.K. government produced a creative economy strategy document called *Creative Britain: New talents for a new economy*.[407] The report is a comprehensive document recognizing the growth of the creative sector and preparing Britain to lead in this sector. It underlines providing the "best possible support structures" (p. 6) reflective of "the Government's fundamental belief in the role of public funding to stimulate creativity and sharpen Britain's creative edge" (p. 6). It calls for education programs to help unlock and then develop creative talent by teaching skills, and through apprenticeships, talent pathways, and innovative learning centres. It commits government to putting creativity at the heart of the economy by, among other things, calling on Britain's Technology Strategy Board to stimulate R&D in the creative sector and to encourage technology transfer to creative industries. The report identifies as a priority assisting the growth of creative business by encouraging projects that demonstrate creative excellence and commercial potential and attract investment to the sector. It also commits to fostering and protecting intellectual property and supporting creative clusters. The report puts forward the notion that Britain can become "the world's creative hub" by promoting cultural exchange and an annual "World Creative Business Conference." Finally, the report commits to keeping the strategy up to date.

Two years later, the United Nations produced a follow-up report to its initial *Creative Economy Report*. The 2010 report included ten key messages about creative economies: they aid diversification; boost South-South trade; benefit from a multidisciplinary creative nexus; require adapted intellectual property rights regimes; cut across the arts, business, and connectivity sectors; allow for participant partnership and ownership; are best built on local educational, cultural, social, and environmental foundations; respond to an eagerness for participation in memorable events often connected to brands; and are often underdeveloped in at least one area in each country.[408]

Complementary reports have also appeared from Nova Scotia, Newfoundland, Quebec, and Ontario, as well as Europe,[409] where discussions of the creative economy led to 2009 being declared the European Year of Creativity and Innovation. The main objective of the European year was "to raise awareness of the importance of creativity and innovation for personal, social and economic development; to disseminate good practices; to stimulate education and research, and to promote policy debate on related issues."[410]

Such documents and attendant discussions point to a major shift in the politics of the creative, cultural, or copyright industries (select the name of choice, depending on the context) in Canada. The creative sector is becoming, and already has become in Ontario and Quebec at the very least, a sector to be managed as a contributor to both economy and nation. In this new view, the businesses of the creative sector are economic engines, rather than consumers of discretionary government income.

Perhaps the greatest contribution of the concept of "the creative economy" is that it promotes a view of cultural production as parallel to other economic sectors, and implies that, like other sectors, creative economies require infrastructure investment. Such investment can take a wide variety of forms — cultural, industrial, and structural — many of which have been discussed throughout this book. They begin with grants to creators and production subsidies and extend into business support, ownership regulations, content quotas, tax credits, distribution laws, and so forth. In the final analysis, all such mechanisms are infrastructure investments that parallel similar support for other industries — for example, the oil industry's tax deferments, geological education, and technology research and development. The net effect of infrastructure investment in the creative economy is a sizable direct, indirect, and induced contribution to GDP, job creation, and, beyond these, to society through the generation of social capital and, hence, further spin-off economic growth — all with the added value of minimal negative impact on the environment. As the Conference Board put it, while Canadian government spending on the creative industries in 2007 was estimated to have reached $7.9 billion, consumer spending reached $18.9 billion and exports reached $5.0 billion, a return on investment of 300 percent.[411]

For many, these figures and this new political approach represent a welcome reversal of perspective: a liability has become an asset. The trick is to understand how to exploit the asset. To explain by analogy: Just as sports franchises have evolved from community subsidized operations to major money spinners for both owners and players, based on public investment, we are beginning to see how cultural productions can contribute to, rather than draw from, the public purse. In film and the performing arts, this transformation has been led by such festivals as the Toronto International Film Festival and organizations such as the Cirque du Soleil, and by the continual emergence of multimillionaire popular music and acting celebrities that only a generation ago were barely scratching out a living. Author and book events have just begun to be exploited as earning entities. Prevailing notions of authors as single-mindedly pursuing their craft, without complementary professional identities or organizational liaisons, may be inappropriate and outmoded in an information age. Similarly, as publishers establish themselves in niches and see vertical opportunities, they may begin to align themselves more closely with specific communities and associated organizations that, in tandem with digital communication, address irritating distribution and retailing inefficiencies that undermine profit possibilities.

Talk of indirect and induced effects in discussions of the creative economy make apparent how impoverished our insight is into such phenomena as the impact of early participation in the making and publishing of books. Perhaps such activities lead to a sense of enfranchisement, or to participation in community building or additional forms of meaning making and communication. We have little idea of the impact on young people of the acclaim given to successful literary role models from unknown towns and only slightly better known cities across Canada. Many come into early adulthood with dreams of becoming creative writers, journalists, or editors working for trade publishers or consumer magazines, or of participating in established online efforts or starting their own. What is the social gain of encouraging those who wish to pursue such ambitions?

We also know little about the universe of book spin-offs. The obvious transformation of books into movies and documentaries, we do understand. But what about the community formation engendered by radio discussions such as Canada Reads, book clubs, university courses, and professional education — not to mention the upstart businesses, tourism, and vast numbers of other activities that derive from books?

Given this emerging creative economic framework of understanding, it is useful to conclude this study with an assessment of the state of the various publishing-essential environments laid out in Chapter 1 in the context of the creative economy.

The political, legal, and policy environment

As the UN Creative Economy Reports of 2008 and 2010 make explicit, the creative economy concept realigns the political environment to call attention to the positive macro-economic realities of the creative sector, particularly book publishing and other cultural industries. It does so by noting their contribution to, for example, GDP, job creation, and social well-being. Now in development are measures that will allow us to more fully appreciate the contribution of the creative sector, even if today's governments have yet to recognize the wealth being generated before their very eyes. Complementing the political environment are laws, policies, and support programs that address micro-economic realities with far-reaching consequences.

Law

The strengths of Canada's current Copyright Act (1997) with respect to book publishing are its inclusion of a distribution right, its conformity to international standards, its inclusion of both property and moral rights, and, some might say, its limitations on the duration of copyright. Beyond those elements, it is clearly a mixed blessing.

The proposals of Bill C-32, brought forward in June 2010, and repeated in Bill C-11, brought forward in late 2011, did a better job of drawing a line between normal personal use of purchased copyrighted products and commercial piracy. Given the digital existence of any work — a film, music, an image, a book — and the ease of making a "perfect" copy of it, it is folly to expect consumers to follow a set of rules for copying that prevent them from using technology for their own convenience once they have purchased a copyrighted product. Such a misalignment of ownership and technological capacity threatens an orderly market, if only because it creates criminality where there should be none. On the other hand, effective anti-piracy laws that address unauthorized commercial exploitation are both acceptable and have tradition. Without an orderly marketplace in which creators and publishers can earn revenues from the sale of copies, creation and the value added by publishers will wither and die, or creative investment will become a handmaiden to sponsors and advertisers who, in turn, will determine the nature of what gets created and how it is disseminated in pursuit of business interests. To hand over the creativity in society to sponsors and advertisers would weaken considerably the broad demand dynamics of a direct public marketplace for creative goods. Especially in the case of books, which have traditionally stood as a medium without indirect commercial subsidy, this would be tragic.

There is further complexity to designing an appropriate legal environment. A prodigious amount of content of great value is freely distributed via the Internet, far exceeding that from which income is earned by means of controlled access. This creates an environment in which the pursuit of creators' property rights is often the exception rather than the rule. Scholarship is an apposite example of copyrighted content that would best be freely distributed. The freer the access to and attributed use of scholarship and research, the greater is its value to society, and the greater is the reputation of the scholar and the accompanying indirect rewards. Health information is much the same. Governments attempt to minimize healthcare costs by freely disseminating healthcare information, while commercial healthcare providers, such as clinics and pharmaceutical companies, provide free advertorial information to stimulate consumption of goods and services.

Bills C-32 and C-11 attempted to define two means of drawing the boundary separating free content from content for which producers want revenue. One is explicit notification of copyright on Internet materials by those who seek compensation for use. The second is technological protection measures (TPM), otherwise known as DRM (digital rights management). The difficulty with the proposed TPM regime was the illegality of breaking TPM protection, because that effectively trumps users' fair dealing rights. By signing a World Intellectual Property Organization (WIPO) treaty some years back, Canada is committed to TPMs. Canada is also a signatory to the Berne Convention, which includes provision of fair dealing. The challenge is to identify a good balance. As noted in an earlier chapter, including education as fair dealing and also granting publishers unassailable TPMs may set up negotiations that amount to a duel with cannons rather than pistols.

Bills C-32 and C-11 also attempted to distinguish between reasonable personal use and systematic piracy (i.e., the large-scale copying or dissemination of content with no recompense to the copyright holder) and created problems in doing so that the courts would have to work out. It is difficult indeed to distinguish between educational activities that fall within fair dealing, such as passing uses and private study uses, from those that do not, such as systematic copying and distribution of both print and digital educational materials. When Bill C-32 died on the order paper, the creator community breathed a cautious sign of relief, having been told by all parties in the House of Commons that the inclusion of education in fair dealing was a done deal. As of March 2012, with the bill being studied by committee, Bill C-11 contains little to constrain the very broad notion of copying for educational purposes. However, it is possible that technical amendments will find their way into the bill as it moves forward to passage.

Perhaps the greatest disappointment in the controversy surrounding copyright reform was the lack of interest from both creators and users in seeing the fundamental balance between the social interest and fair treatment of creators and publishers respected. Such a balance is an absolute necessity if Canada is to expand the creative sector. The main cry of authors was outrage at "expropriation" of their intellectual property.[412] Meanwhile, the Council of Ministers of Education made representation to the parliamentary committee reviewing the bill that they sought inclusion of multiple copies for classroom use as part of the education exemption.[413]

The contentiousness surrounding copyright will eventually be resolved. It is plainly counterproductive for copyright holders to stand in the way of a full use of intellectual property. Yet it is also counterproductive for society to deny creators and copyright holders a fair benefit from their intellectual and financial investments. As Michael Heller illustrates in *The Gridlock Economy*, there is a growing consciousness that the greater good is being sacrificed at the altar of patents and copyright.[414] Giuseppina D'Agostina counsels users and rights holders to bargain amongst themselves, arguing that it is clearly counterproductive and likely socially dysfunctional for law to stand in the way of what technology readily offers, especially if the means can be found to allow users to take full advantage of technology with appropriate compensation going to rights holders.[415] The decisions of a fair number of universities in the summer of 2011 to cease dealing with Access Copyright is certainly not the best way to pursue the common good.

Copyright issues are further exacerbated by the imbalance of power and the resulting public tension between creators and publishers, particularly in the music industry and in parts of the publishing industry (i.e., newspapers and magazines), where class action lawsuits have been mounted so that creators could collect proper recompense.[416] The excessive exploitation of creators by large corporate publishers has given rise to creators' well-publicized attempts to gain direct access to audiences. The actions of the rock group Radiohead are relatively well known. When the group completed its seventh album, it made the recording available for download and allowed listeners to decide what they would pay for it. Less well known was the apparent systematic disregard of creators' rights in the Canadian music industry, where albums were released, permissions were never pursued, and billions in royalties were never paid by music producers to artists.[417] If sectors of the industry, by underpayment or non-payment of royalties, show a lack of respect for their partner creators, it can hardly be expected that members of the public will respect the laws that the industry itself is flouting.

As part of a general attempt to reclaim ownership of intellectual property from commercial publishers for the public sector, public research funding agencies are increasingly insisting that the results of research be made publicly available within a short time after their initial publication. The National Institutes of Health (NIH) in the United States and the Canadian Institutes for Health Research (CIHR) are both advising authors that they, the authors, hold copyright on submitted manuscripts, even if a publisher claims the rights to the published version.[418] This stance clears the way for open access to research results.

The relationships between trade book publishers and their authors appear to be less fractious than those between some periodical publishers and freelance writers or between musicians and music producers, perhaps because there is a better balance in what each side brings to the table. It is true that large book publishers have a certain amount of marketplace power, specifically a marketing and distribution infrastructure superimposed on editorial and layout expertise. But in most situations it is the author and the quality of the writing that sell a book, not, as in the case of magazines and newspapers, the title of the publication. The exception here is Harlequin.

Creative Commons

The problems with copyright described above gave rise to an attempt, led by U.S. intellectual property lawyer Lawrence Lessig, to come up with a formal alternative, which became the Creative Commons. As the organization's website explains, Creative Commons is "a nonprofit corporation dedicated to making it easier for people to share and build upon the work of others, consistent with the rules of copyright."[419] The licences it developed allow authors and publishers to signal to users that they are free to use content in certain ways without needing to clear copyright. Most appropriate to book publishing is the Attribution Non-commercial No Derivatives (CC BY-NC-ND) licence. As noted on the Creative Commons website:

> This licence is the most restrictive of our six main licences, allowing redistribution. This licence is often called the "free advertising" licence because it allows others to download your works and share them with others as long as they mention you and link back to you, but they can't change them [your works] in any way or use them [your works] commercially.[420] [My clarifications added]

The other main licences allow various levels of alteration to the original work, specifying whether these can be done for commercial purposes or not. In all cases, the original creator must receive attribution.

An overview of the legal environment suggests that there is little doubt, especially given recent technological developments, that the legal structures surrounding book publishing are in flux. Creators' and publishers' rights are not being dismissed, but the value of free information and the impact of technology require new practices that take into account creators' rights, users' rights, social benefit, and technology. Presumably the copyright industries have sufficient strength to protect their own interests. If they do not, Canadian society stands to lose.

Policy and support programs

As copyright law continues to evolve, the combined support programs of the Canada Council, the federal government through the Department of Canadian Heritage and the Department of Foreign Affairs, and provincial arts councils and governments in British Columbia, Ontario, Quebec, Saskatchewan, and, more recently, Alberta appear to be sufficient, barring any substantial funding cuts, to allow the Canadian-owned industry to tick along in its current state without major financial crises.[421] The foreign-owned sector is also maintaining its course of publishing Canadian authors and even contributing to their development, at least for the time being.[422] As explained in Chapter 5, government programs are not the sole supports of book publishing. Awards also assist (though some are financed by municipal governments), along with the media and other cultural partners.

While there are no major or pressing issues with current policies and the support now being provided, there are certainly items of concern to the industry. Adjusting to new technologies tops the list, and a number of studies and initiatives are underway to ease this transition. The policy of providing funding to book publishers is always at some risk, and many saw the re-election of the Harper Conservatives in 2008 as a clear and present threat to existing programs. The general financial crisis of the same year, which forced the government to adopt a policy of stimulus spending, was seen as a saviour. In 2009, funding was renewed for BPIDP and the program was renamed the Canada Book Fund. For years there has been no threat to this successful program. The next step will be to survive the inevitable spending cuts that the government will introduce in the coming years. In British Columbia, for instance, in the wake of vast spending on the 2010 Olympics, arts funding was cut substantially. After much determined lobbying, the arts community won back that funding, and the tax credit program for book publishers was renewed in 2011. In late July, the government of British Columbia notified B.C. book publishers that the tax credit would be calculated on their CBF grant, excluding funds they might receive for export assistance, but it was silent on the extension of the program.[423] Later, it extended the program. In 2010 the Harper Conservatives initiated a review of foreign ownership policy in book publishing. The review was suspended with the 2011 election, but industry members expected it to re-emerge after the election, with particular attention being paid to distribution and retailing. That has yet to happen, but in January 2012, McClelland & Stewart was sold to Random House without any public consultation. Many saw the sale as setting a precedent for other firms to claim the right to seek foreign buyers and, hence, as nullifying the Canadian ownership regulations at the heart of book publishing. A short account of the event and its context is worth considering.

Avie Bennett joined the Board of Directors of McClelland & Stewart in 1984 because he had made a small investment in M&S.[424] That investment helped Jack McClelland with his banks and provided, along with other funds, further working capital. The company continued its publishing program with Jack McClelland at the helm until, one day, he called Bennett to say that he needed more funds. McClelland sounded so dejected that Bennett found himself asking if he wanted to sell the company. It was only on asking the question that Bennett realized that he was willing to buy it.[425] McClelland replied that he had had enough, and so, in 1986, Bennett took over the storied McClelland & Stewart. Very soon after he acquired the company, he installed future Governor General Adrienne Clarkson as publisher, and in September 1988, he appointed resident editor Douglas Gibson (whom he had hired in 1986 to start his own editorial imprint within M&S) to replace her.

Under Bennett's control, M&S saw fourteen years of substantial publishing that reflected the continuation of a valued brand. While Gibson, as publisher, is generally credited with the quality of M&S's publishing program in the years of Bennett's ownership (1986–2000), Bennett extended the value and prestige of the firm by bringing in such high-profile and celebrity authors as Pierre Trudeau, Jean Béliveau, Rohinton Mistry, Karen Kain, and John Crosbie. He also purchased *The Canadian Encyclopedia* from Mel Hurtig, arranged for its translation into French, and saw to its publication on CD-ROM; purchased Tundra Books from May Cutler and oversaw its development; and oversaw the reorganization and development of the New Canadian Library series, creating of it a vibrant cultural asset. In addition there was Gibson's work with such authors as Alice Munro, Robertson Davies, and Alistair MacLeod, and M&S also continued to publish well-regarded and best-selling fiction titles under Ellen Seligman. The praiseworthy results were in evidence in the form of an almost continuous presence on bestseller lists in fiction and non-fiction, but also in the press's roster of Giller Prize and Governor General's Award winners. M&S published the Giller winners in 1994, 1995, 1998, and 2000, and dominated the shortlist during the same time with seven other shortlisted titles.

During this period, the firm was supported by federal and provincial grants totalling around $1 million annually. Yet M&S remained unprofitable, and Bennett dipped into his personal wealth to come forward with significant subsidies each year, which some believe roughly matched the public grants to the firm. In 2000, when Bennett was seventy-two and concerned about succession, he decided to take action. Constrained by federal ownership regulations, and operating in very difficult times, with massive returns coming in from Chapters, Bennett considered other firms and individuals that might have the resources and inclination to take over M&S. Perceiving none, but having respect for John Neale, who was then CEO of Random House and who had started his career with M&S, Bennett developed a plan that would keep M&S in Canadian hands but effect a sale that might see the firm achieve profitability. He donated 75 percent of the company to the University of Toronto, personally financially guaranteeing its value, and sold, for an undisclosed sum, 25 percent to Random House Canada, the maximum percentage he could sell to a foreign firm at the time.[426] Bennett also hived off *The Canadian Encyclopedia*, which went to Historica as a separate profitable operation.

The deal saw Random House providing services to M&S in accounting, computing, distribution, and ultimately, marketing and sales, in an effort to achieve increased economies of scale on "non-competitive" functions not central to the core elements of publishing. The line of influence was drawn — in part because of ownership and control regulations — at editorial, inclusive of acquisitions, where M&S was left to be independent.[427] Also reflective of regulations, the operations of M&S were to be overseen by a seven-member board, with Avie Bennett as the initial Chair. Five members were appointed by the University of Toronto and two by Random House. As well, the agreement specified that U of T could not be tapped for any financial losses.[428] Were losses to occur, they would be registered as loans receivable by Random House.

Nationalist cries were clearly audible at the time, but the private deal was done, with predictions by those same voices of an eventual takeover by Random House.

Following the sale, Douglas Gibson remained publisher and CEO and continued to run M&S as an independent Canadian publishing house. In 2004, Gibson was pushed out of the publisher's chair and encouraged to devote his entire time to his "Douglas Gibson Books" imprint, established in recognition of his editing acumen and his illustrious stable of authors, who might have followed him to another house. Doug Pepper, a Canadian then working for Random House in New York, was brought in as publisher, and although the level of losses decreased on a year-to-year basis during his tenure, partly because the firm was downsizing, M&S was unable to address its outstanding debt.[429] During this period, the number of Giller awards and nominations also declined and M&S faded as a beacon of the best Canadian books, especially national non-fiction titles.

Staff shrinkage within M&S led to some integration of additional non-editorial functions with Random House. The separate M&S sales force disappeared, M&S publicists began working for both Random House and M&S authors, and M&S's subsidiary rights department was closed, with responsibilities for rights sales of both Random House and M&S titles eventually being placed in the hands of the Canadian-owned Cooke Agency.[430] At the time of the 2000 sale, staff numbered fifty; just before the late 2011 sale, exclusive M&S staff numbered fifteen. But framing this as a reduction from fifty to fifteen overstates the shrinkage. Brad Martin noted: "[A]t the time of takeover ... there were many more people working on [M&S's] books who were on the Random House payroll. We consciously moved people off the M&S payroll to reduce costs."[431] At the same time, of course, that meant staff members' effort was spread over both M&S and other Random House titles. Martin noted further that with regard to the M&S publishing editorial team, "the numbers on the [editorial] publishing side are relatively constant over the past 12 years; we have just changed skill sets."[432]

The first two questions that many in the industry asked about the sale were, first: given Canadian ownership policy, how could Investment Canada have allowed the company to turn over U of T's 75 percent ownership of M&S to Random House, and second: how, especially, could it have allowed a sale in which U of T received no funds from Random House?[433] On the first point, given the review of ownership policy begun by the Harper government and aimed at loosening that policy, it would be hard to believe that, at the political level, there would have been significant resistance. That said, Investment Canada officials would have had to operate within current guidelines, including the requirement that the deal be of net benefit to Canadians. Regarding the second point, according to Brad Martin, in spite of the fact that U of T received no money, Random House paid for the acquisition by swallowing the press's debt — significantly more than $10 million — in addition to other millions in write-offs that Random House had already taken on with the company.[434] These costs were in addition to the sum that Random House paid for its 25 percent share.

In persuading Investment Canada to allow the takeover, Random House had a number of advantages on its side, including its 25 percent ownership and a loan of between $10 and $20 million. Random House Canada is also a solvent, profit-making, trade book publishing company with a substantial Canadian publishing program. Random House had the knowledge that "M&S could have filed for bankruptcy protection, but its shareholders chose not to allow that to happen."[435] Random House did express willingness to Investment Canada that it would carry the M&S brand forward, including the continued publishing of Canadian authors.[436] It also argued that it planned to integrate both M&S and Tundra — a separate children's book imprint of M&S — as imprints into its overall operations and invest in the growth of both brands in a global context.[437] Random House's short-term growth and profitability targets for their acquired company are both ambitious and impressive. Random House also had evidence that there was no Canadian-owned trade publisher with the appetite to acquire M&S and pay off its debt to Random House. It could have pointed to the net benefits it had generated since Bertelsmann assumed control of Random House Canada, including its support of Canadian authors and editors and support of Canadian publishing education programs, such as the Masters' and Summer Workshops at Simon Fraser University. It did agree to take on additional "net benefits," including an annual M&S lecture series, the continuation of the publishing of poetry, continuation of the Journey Prize, and participation in various Canadian cultural activities.[438]

Whatever the terms of the agreement, Investment Canada was persuaded to grant the sale. Time will tell whether, in the absence of grants in the order of $1 million annually,[439] the integration of the company into Random House and the planned investment in these two acquired imprints in an international context will turn them into profitable operations. It will certainly be a challenge, but in the immediate aftermath of the approval, Brad Martin firmly believed growth and profitability were achievable.[440] Time will also tell whether Random House's treatment of the imprints will continue to provide opportunities for Canadian authors and whether their Canadian character will be maintained.

Was the sale of the whole company to Random House inevitable? Was the takeover inevitable, as some of the voices favouring protection of Canadian ownership predicted? In interviews, both Bennett and Martin insisted that the takeover resulted from marketplace challenges. Martin resisted the notion that the company became appreciably smaller, noting that title production was maintained (in the approximate range of sixty-five to eighty-five or more titles per year). Bennett talked of an unlucky period of time for M&S. Bennett also expressed his satisfaction with the services provided by Random House between 2000 and 2012 — apparently, very fairly, based on a percentage of sales. Neither spoke of the competitive disadvantage faced by M&S in pursuing the niche, national bestseller fiction and non-fiction market, where its competitors were three large international conglomerates — Random House (including Doubleday and Knopf), HarperCollins, and Penguin. But the question was not asked. Given business realities that included Random House's minority status and the barrier erected against its control of the publishing/editorial function, it was up to the chosen publishing/editorial team of M&S to earn its continuing independence. It did not.

The stark reality of marketplace challenges is this: Over the long term, 1972 to 2012, in spite of all its renown, and all its awards and accomplishments, Canada's foremost English-language trade book publisher of leading fiction and non-fiction primarily for the Canadian market proved to be financially unviable. Even its ability to take advantage of the economies of scale and the wisdom of Canada's largest trade book publisher on all but acquisitions and editorial preparation could not rescue its viability. This unviability, although it is but a single if notable instance, speaks directly to the continuing need for a supportive framework for culturally significant, non-conglomerate, Canadian-owned book publishing, most importantly for the development of Canadian writers and the publication of non-fiction works of national interest. After all, evidently, bestselling titles still only account for approximately 25 percent of sales.[441]

So in late 2011, with debt between $10 and $20 million, "an amount that exceeded top line sales,"[442] the Board of Directors decided to sell and Random House stepped forward as a willing buyer. Why 2011? According to both Bennett and Martin, the time had come; the accumulating debt and the possibility of the company ever being able to pay it back had reached a tipping point.

No matter what the financial statements said, at the time of its sale, M&S was not a worthless enterprise.[443] In fact, it had considerable value beyond its brand and grants access. Its backlist contained a significant portion of Canadian literary and intellectual history, work that will continue to generate revenue for many years to come. A Canadian buyer could readily have purchased this library and used the sales of backlist titles to continue a notable publishing program. As well, there was undoubted value in signed contracts with authors, as demonstrated by Random House's plans to continue on more or less the same publishing course. However, according to Martin, no Canadian buyer was willing to pay the amount Random House wanted to abandon its interest in the company,[444] presumably without swallowing a loss.

The sale of M&S has its own importance — Canada has lost ownership of more "family silver," although not its control by Canadian managers. But the sale also has significance for the long-term prognosis of the domestically owned book publishing industry, provided that the industry won't be defeated by technologically based restructuring. Avie Bennett is a good and generous Canadian. With eyes open, he took on an important, but money-losing, Canadian cultural institution and threw both his energy and resources into developing, focusing, and growing the company. In 2000, he went looking for another "good

Canadian" as a successor. When none were forthcoming, he placed M&S on hold with U of T and the foreign-owned Random House Canada. Twelve years later, the ownership fell to Random House alone.

What about other sizable Canadian-owned firms? Anna Porter was able to persuade the Canadian-owned distributing agent H.B. Fenn to take operational control of Key Porter. But then Fenn went into bankruptcy (not from the burden of the takeover) and Key Porter effectively vanished, with a resulting loss, mostly to Fenn and some to Anna Porter. The largest Canadian trade book publisher left standing at the time of this writing was Douglas & McIntyre. In 2007 Scott McIntyre was able to persuade his reading club friend Mark Scott (who in 2012 was also a director of the market-*uber-alles* think tank the Fraser Institute) to take some of his considerable earnings and put them to good use trying to usher D&M into the digital age. McIntyre and his associate Rob Sanders (publisher of the Greystone imprint) stayed on, but in a March 2012 move (after the M&S sale) that seemed to anticipate a second company sale, the positions of COO and Publisher were handed to Jesse Finkelstein and Trena White, two relatively young, outstanding graduates of Simon Fraser University's Master of Publishing program. McIntyre and Sanders stayed on as mentors and to assist with strategic direction,[445] and thus, with the long-term owner/operators acting as advisors and capable young managers in place, the firm was in a position to be sold. After the sale of M&S, one could imagine Simon & Schuster persuading Investment Canada to allow it to convert its distributor status into that of a publisher by acquiring a Canadian company. Of course, HarperCollins (its distributor), Penguin, perhaps Wiley or, again, Random House could readily fold it into their operations.

However, an equally viable and policy-respecting alternative would be to make Canadian ownership non-negotiable, as it was for a number of years. The creation of a process to allow Canadian cultural properties to restructure, perhaps using some of the same methods that firms use to protect themselves from bankruptcy, could help with succession. Within that environment, if needed, financial and publishing experts could be brought in to work with the company to allow staff or other interested Canadian investors to purchase the company, in whole or in part, thereby preserving much of its publishing program as well as its domestic ownership. The big do not necessarily have to get bigger, especially when size compromises cultural value. From a domestic ownership perspective, it would be quite wonderful if, as this was being written, the staff members of D&M Inc. were putting the capital in place to take over all or part of the firm. They could readily be assisted by a facilitative restructuring environment.

The interaction of technology and policy is bound to twist and turn in the near future. For instance, e-book distribution might be deemed to fall within the provisions of the Broadcasting Act, with the result that the broadcasting regulator, the Canadian Radio-television and Telecommunications Commission (CTRC), may become involved. With an Amazon distribution service moving into Canada, Chapters/Indigo's sale of its Kobo e-book reader, and the proliferation of other devices such as the iPad, it is likely that the terrain will become more complex before it is re-simplified.

The economic environment

What does the currently evolving economic environment have in store for book publishers? From a macro-economic perspective, which emphasizes the role of the creative economy, job creation, and social capital, and perhaps extends to the worldwide recognition of a small cadre of Canadian authors, the economics of book publishing work extremely well. From a micro-economic perspective, which looks at all authors and titles, the operational and market realities of book publishing do not work well at all. They fail to generate a reasonable profit for authors *or* publishers for about 80 percent of published titles. How does this apparent paradox come about?

The micro-economics of book publishing don't work because there are too many books in the Canadian marketplace, a good number of which are run-on copies of mediocre titles from foreign markets. As noted in Chapter 5, Statistics Canada figures suggest that the average sales of a Canadian-originated title are $14,000–$20,000.[446] This level of per-title sales is fundamentally unsustainable as a business model. It may make sense as a cultural participation policy, or even a talent development policy, but the desire of human beings to convey information and create meaning, or to be part of that process, exceeds the interest sufficient numbers of readers have in reading what can readily be made available. As Shelley, Wordsworth, and Keats each noted, making meaning by writing is inherently satisfying.[447] That others may read what one has written, talk about it, and praise it, multiplies the satisfaction, but does not drive the initial determination.[448] When low-priced imports are coupled with the drive of Canadians to write, the result will always be that there are too many books.

The organization of book production further complicates matters, insofar as it is essentially feudal. With or without publishers' approvals, authors toil away without a wage to complete books for a small share of the return — often, for unestablished authors, as little as 5 percent of retail price. This represents about 10 percent of what the publisher receives. So, on average, Canadian-originated titles earn their authors $1,400 – $2,000. This feudal structure survives because both publishers and authors dream of winning big — which, in the book environment, means either earning pots of money or being lavished with critical acclaim or major national and international awards. As the headlines tell us, winning is possible and frequent. The problem is that it is highly improbable. The psychological problem for a writer is that to admit this improbability seems to mean accepting that he or she has insufficient talent.

There are organizational alternatives to the financial status quo. For publishers, corporate concentration is one. Corporate concentration refocuses decision-making on the likelihood of profit, rather than the intrinsic value of the book; it tends to produce economies of scale that increase percentage earnings; it also gives resulting firms sufficient marketplace power to undermine market entrants that threaten their profits. But it undermines cultural value. In concrete terms, the difficulty, as the U.S. film industry illustrates so well, is that corporate concentration narrows product variety. Canadian-owned publishers have been unwavering in their resistance to corporate concentration, arguing, quite rightly, that an industry that is heterogeneous in terms of size, location, and genre orientation serves the cultural and literary interests of Canada far better than an industry dominated by a few large firms.

The micro-economics of book publishing also don't work because, within the industry, such goals as human creativity and cultural value overpower profit seeking and, hence, undermine risk management. (By the way, prize-winning author/psychologist Daniel Kahneman has written a wonderful treatise on decision making and risk taking that provides considerable insight into the behaviour of book publishers.[449]) Risk management is further undermined by the history of book publishing. In the past, rulers' fear of regime-disrupting ideas resulted in book publishers being licensed, based on their trustworthiness. The net result has been that publishers, and, for the most part, publishers alone, spend funds on the development of new titles each season. In doing so, they combine venture capitalism with manufacturing, thereby courting risk rather than minimizing it. In general, publishers diminish risk to some extent by keeping their backlist in print and spreading costs over a number and mixture of titles, authors, and genres,[450] combining sure-selling genres, such as hockey books and cookbooks, with lower-selling new fiction. Based on the notion that marketing has a relative rather than an absolute impact,[451] publishers also budget for a standard set of (tiered) marketing, promotion, and publicity activities that bring a certain level of attention to a title. Canadian trade book publishers have also managed to decrease the overall risk in the marketplace by about 25 to 30 percent — roughly the average investment governments and other donors make in title origination. But the failure of the vast majority of titles to recover their costs strongly suggests that book publishers and authors are preoccupied by thoughts other than managing risk and keeping an eye on a reasonable return on their investments of energy and effort.

Is book publishing the victim of what economists call market failure, and does this justify government intervention? The question cannot yet be answered. Market failure usually refers to the value of a good or service that emerges in due course, compared to the value received relatively immediately by the creator. The classic examples used are paintings that fetch small prices during an artist's lifetime, but millions after the artist's death. To classify books as victims of market failure, economists would want evidence that, in general, titles generate considerably more revenue much later in the authors' lives, or after their deaths, than they did in the years following their initial release. While this is true for a considerable number of titles, it is not true for the vast majority. If books were to be classed as victims of market failure, a case would need to be made, using measures capable of bringing forward evidence, that in terms of social capital dynamics, the contributions of the vast majority of titles were being inadequately rewarded.

The ubiquity of technology is also having substantial impact on the economics of publishing. There is considerable talk of the economies of "free" as in free information.[452,453] While such talk is dressed up as new, and in book publishing it is somewhat novel, the counter-example is that missionaries rarely sold Bibles, but rather gave them away. "Free" means nothing more than indirect payment. By "free," people generally mean open electronic access. This sort of "free" works for blockbuster authors because, like selling religion, it amounts to advertising for events or other considerations, or, an enticement to purchase the printed book. Free books, in this case short printed narratives that capture the psychographic of a target audience, even work for BMW.[454] But in general, only some titles, in some cases, benefit from freely accessible electronic versions. In the end, no matter how willing the author is to make the work openly accessible, the intellectual work and financial resources required by publishing must be compensated.

To encapsulate this discussion of the economic environment, publishers' motivations, like authors', are similar to those of soldiers going over the top: pursuit of the greater good in the face of a hail of potentially death-delivering sales figures. Publishers use business methods for the higher end of disseminating ideas and being players in the great game of nurturing creativity and creating culture. Since they see their job as generating social capital (although they don't regularly use such a term), the risk they incur is a necessary means to a higher end. A high risk is one thing; no opportunity for financial reward is quite another.

A service model of publishing

A host of firms followed in the footsteps of the fee-for-service model that Trafford (now part of Author Solutions, which also owns AuthorHouse) reinvigorated. Once Trafford demonstrated that considerable demand existed for a fee-for-service model of publishing, that fee-for-service publishing changes the economics of a firm, and that it can earn publishers, and even some authors, normal profits, others followed. Trafford's list of authors also demonstrated that legitimate writers were interested in such a service, and when these authors used Trafford, they generally made more money than they would with a traditional publisher. It is also the case that if the same authors were to use a printer, rather than a fee-for-service publisher, they would make even more money. Trafford's publishing services model is not inexorably tied to print-on-demand (POD) publishing, although, in itself, POD reduces risk because printing is not undertaken in anticipation of a market. The most important point, first made by Trafford and then by other similarly organized firms, is that, like doctors, dentists, and bankers (and, historically, vanity publishers), publishers could offer a professional service based on fees set at a level that would generate profit.

Publishers appear unable to embrace a publishing services model because of hubris and history: hubris, in the sense that they have no wish to step down from being a member of an august industry that chooses what the public will read and, hence, what ideas and creativity will circulate in book form; history, because they believe that in order to maintain such a position, they must continue the industry tradition of being the sole investors in making manuscripts public. But it appears to be a simple matter to preserve the august position while separating it from the obligation to invest. Publishers could evaluate and select manuscripts (for a fee) and then require the manuscript's owner to seek investors to bring the book to market.

A service model for scholarly monographs

Scholarly monographs provide a good example of an unacknowledged fee-for-service model that has effectively been in place, at least in Canada, for years. Since the 1970s, many British and North American universities have reduced, and then stopped providing, financial support to their presses, arguing that they should sustain themselves through marketplace earnings. While Oxford, Cambridge, Harvard, Princeton, Yale, MIT, and other top-tier university presses were able to do so, partly based on the value of what they publish and partly based on brand, many smaller and less prestigious presses were not. As a result, valuable scholarly and research monographs languish in the filing cabinets and on the computers of researchers. Those that are published, if they are in English, often have sales of 300 to 500 institutional buyers worldwide, and many achieve fewer than fifty private purchases by scholars. In languages other than English, universities have not been so quick to abandon their presses, and scholarly publishing is quite explicitly subsidized by a variety of means.

University presses often insist that contribution to knowledge determines which titles they publish.[455] Yet because presses work with a revenue-generating framework populated by sales and access to grants, considerations including the number, and to some extent the types, of titles that get published; the topic areas that the press considers; and the extent of marketing effort (given expected sales) all constrain choice, and often affect the choice and the perceived value of the work by other researchers. Moreover, the administration and dynamics of peer review — for instance, the fact that reviewers are chosen from a very limited pool of scholars whose interests, research, teaching, and administrative programs afford them the time to devote to this rather onerous task — also affects the choice process. Subsequent to acquisition, the demand resulting from the marketing efforts of the hundreds of university presses — mostly carried out in a six-to-twelve-month period surrounding the release of a new title, assisted by the literature search activities of scholars and students, and by the title being listed in the press's catalogue — purportedly tap the limits of worldwide demand and, to some extent, contribute significantly to the perceived value of the title. (The level of British scholarship in Indian university libraries is a case in point. It is affected by trade patterns as well as scholarly value.) In a larger frame of commercial trading, various studies demonstrate that commercially published scholarly journals often have higher circulations and command higher prices than the not-for-profit journals that are ranked higher in research value by researchers.[456,457] Studies documenting the restricted circulation of science research published in developing countries and in languages other than English also speak to an imperfect knowledge dissemination system.[458,459,460] The shortcomings introduced by such dynamics do not call for abandoning the sale of research monographs and journals, or peer review, but rather for their reconfiguration.

To adapt the words of BioMed Central, now a private sector publisher of open-access science, technology, and medicine journals, publication could and should be treated as the last phase of the research process.[461] Such a publication system would ensure a much more complete record of knowledge, including, perhaps, a valuable record of negative results that would save researchers from treading down well-worn but now invisible pathways to dead ends. The system would include peer review, although, it would be configured more as an investment of resources to reach an appropriate knowledge and presentation standard than as a simple gatekeeping mechanism. This would be a positive development, reflective of the reality of, for example, this book; many, if not most, of the key documents cited here are valuable and authoritative publications that arose from such a process, rather than from blind, arm's-length peer review. The Massey Commission and Ontario Royal Commission reports are leading examples. Research articles and monographs could equally travel down a parallel path of quality control that might or might not include a final arm's-length judgement. As peer review is currently practised, it is overly restrictive, if only because it is constrained by budgets uncoupled from research effort. The continuous launching of new journals and the considerable energy being devoted to establishing online humanities monographs reflect, in part, the report-as-last-phase model and attest to the belief of scholars and librarians that many valuable research articles and monographs are going unpublished.

The persuasiveness of this last-phase model comes also from the existence of two markets in scholarly monograph publishing.[462] One is author demand for publishing services; the second is reader demand, mainly defined by library purchases as a stand-in for that demand. On the author-demand side of the equation is the support system to ensure publishing opportunities are in place. In Canada, that system includes university press eligibility for BPIDP (now CBF) funding on easier market-performance terms than those applicable to trade publishers. The average grant appears to be approximately $2,000 per title. As well, Canadian university presses are the most frequent drawers on funds from the Aid to Scholarly Publishing program (ASPP) in the amount of $8,000 per title. Canadian university presses are also eligible for Canada Council grants in the neighbourhood of $3,000 per title. In addition to these grants, scholarly presses generally ask authors to pay for indexing and sometimes other editing costs, and it is not uncommon for the presses to ask for additional subventions from authors' research funds or from their universities for any unusual costs or to subsidize titles where the press expects low sales. Quite commonly, this last source can bring in an additional $5,000 or more for a single monograph. In total, such support amounts to $18,000 per title.

At that rate, this support system is so extensive that, in effect, scholarly presses are fee-for-service publishing, where the fees are paid by the academic/research community and its agencies. The parallel of advertisers contributing most of the funding needed to run newspapers, in exchange for potentially gaining the attention of readers, is notable. The market may make a crucial contribution, just as readers' subscription fees do for newspapers, but it is a relatively minor source of revenue in comparison with the system subsidies. The rough math on the contribution that sales make towards covering costs is easy to do. The number of copies sold, multiplied by the price per copy, factoring in the discount to the retailer or wholesaler, minus the costs of printing, warehousing, fulfillment and shipping costs, and scant royalties for the author, suggest that $6,000 in net sales revenue per title would not be unusual (about 25 percent of the cost of production).

Given these financial and dual-market realities, rather than focus on acceptance and rejection, it could be better for university presses to focus on the potential value of a manuscript, identify the nature and extent of the resources needed to turn it into a valuable document, inform the author of the cost, and leave it to him or her to assemble the necessary funds. The purpose would not be to avoid peer review, but to focus on meeting demand for publishing services. Elsevier professor Anita De Waard makes a plausible case for reducing the advance-censorship role of peer review, by claiming that, in reality, the findings of scientific papers are transformed from claims to facts on the basis of citation, which is now trackable and measurable.⁴⁶³ With authors responsible for assembling the full support needed from research grants and institutional grants, over and above what is available from BPIDP (now CBF), ASPP, and the Canada Council, such a system would acknowledge the fee-for-service nature of university press operations, and recognize that peer review, editorial development, layout, and production could be paid-for core services. This system would not detract from the reputation of the press, because the press would still be in a position to accept or refuse manuscripts and make constructive use of peer review. Were market earnings to be replaced with up-front fees, a broader range of scholarly monographs could be made available in electronic form to both public and commercial databases. Researchers could sign up for various RSS feeds to receive alerts of new titles in their chosen areas of interest. Citation and usage analysis could be set up to evaluate published manuscripts, and provision could be made for commentary to be added by readers. Print copies could be sold on a print-on-demand, cost-recovery basis. Arguably, such a fee-for-service model combined with electronic publishing and marketing would be a far more effective strategy for knowledge dissemination.

Commercial publishers, as well as established university presses outside Canada and less established university presses within Canada, such as Athabasca University Press, are beginning to consider a system that would freely disseminate basic knowledge while allowing them to earn revenue through value-added services built around the basic monograph. For example, Bloomsbury, the publisher of J.K. Rowling's Harry Potter series, plans to make all books in its Bloomsbury Academic division openly accessible on the web. Led for a time by Frances Pinter, formerly with the Soros Foundation, where she was very much involved in promoting open-access journal publishing, Bloomsbury Academic will allow users to read, download, and print the electronic versions. Pinter intends to add features to the basic electronic manuscript to generate revenue, and she also plans to sell printed and bound copies to those who wish to purchase them, using a POD system. Finally, she is looking to acquire texts that will encourage the formation of communities of contributors and lend an attractive, ever-growing coherence to her list. In turn, this could lead to access to research funds that would help sustain the publishing program, to the development of speaker services and conference events, and in turn, to social impact.[464]

The benefits of service-based, open-access, electronic publishing for both researchers and scholarly presses would be immense, essentially because the model minimizes costs and risk and maximizes access. Among the efforts in progress to help minimize production costs is Open Monograph Press (OMP), which will take what Open Journal Systems has done in the scholarly journal world and apply the model to scholarly monographs.[465] OMP is manuscript management software, now in beta format, that greatly eases the acquisition and administrative origination of a title.[466] In the United States, steps are being taken to bring libraries and scholars closer together outside a market model to ensure the dissemination of research findings.[467] And on February 29, 2012, OAPEN announced the formation of the Directory of Open Access Books (DOAB), "a searchable index to peer-reviewed monographs and edited volumes published under an Open Access business model, with links to the full texts of the publications at the publisher's website or repository."[468] Given the knowledge value but financial marginality of scholarly monographs, developments that parallel journal publishing would seem a sensible method to recoup the investments society and scholars make to support research. Such efforts would lead to the creation of digest sites that would specialize in ranking, and hence marketing, titles in their specific subject areas.

A service model in educational publishing

The custom publishing done by Pearson and other educational publishers, such as McGraw-Hill Ryerson and Wiley-Blackwell, is, in effect, service publishing for teaching professors, except that the fees are collected from students, and they are collected after, rather than prior to, production. This model might equally be termed subscription-based publishing, because the size of the initial market can be closely estimated. Custom publisher reps identify professors who have manuscripts or compilations, generally known as course packs. Once the reps determine class size and calculate the cost of permissions, editing, and production, they inform the professors, who agree that the book will be required reading, which means students in the class must purchase it. The publisher then goes ahead and produces the book. Run-on sales beyond the professors' own classes bring benefit to both the professors and the company, and students should also benefit from a lower price. The economies of custom publishing are such that one company, Pearson, can create reasonably priced custom anthologies of public-domain material for classes with enrolment as low as twenty students.[469]

A service model in the trade sector

Bringing a service model to trade publishing requires consideration of a host of details. But consider this initial outline. Some years ago the Ontario Arts Council created two programs for writers: the Writers' Reserve and the Writers' Works in Progress. The Writers' Reserve, designed "to assist professional writers in the creation of new work ... is administered by third-party recommenders from the literary community." These recommenders are "Ontario-based, Canadian-owned book and literary magazine publishers chosen by the OAC."[470] Writers' Reserve grants amount to seed funding to help an author develop a manuscript that the publisher feels would be worth publishing. This program is backed up by the Writers' Works in Progress fund, which allows "Ontario-based professional writers [to] apply for support for the continuation of new work in poetry or prose, including graphic novels," thereby helping them "complete book-length works of literary merit."[471] These two programs make a public investment in the author to pay part of the costs of manuscript development, hence allowing publishers to attract writers (in lieu of an advance against royalties), obtain good manuscripts, and augment authors' earnings from royalties.

The Canada Council for the Arts could convert its publishing support programs to encourage a service-based model as well, a move that might encourage a variety of provinces to move in the same direction, given that many provinces run programs that are parallel to those of the Canada Council. Currently, the Canada Council funds a percentage of the standard deficit publishers incur for books of each genre funded. While emerging publishers receive title grants for specific publications, established publishers receive block grants in the amount of their expected deficits, based on their programs during the previous year and their projected number and type of titles for the coming year. There is also room in the evaluation process for peer-judged professional excellence.

These days, a deficit-inspired schedule starts in exactly the wrong place. Given the growing acceptance of the macro-economic value of the creative economy, the Canada Council could instead establish a professional fee schedule for the standard costs of developing and publishing a title and fund a percentage of such fees, with the requirement that the funding be used to pay publishing professionals according to the schedule used for calculating the value of the support. Such a scheme would require the publisher to pay the set fee — say $50 an hour for design or $55 for substantive editing — in order to access Council funding. The provinces, at least the majority whose funding programs emulate the Canada Council's model, could follow suit. In turn, Canadian Heritage's Canada Book Fund (CBF) could couch its support in terms of the professional fees payable for the origination of titles. If the CBF also insisted on a higher minimum profit level, perhaps around 7 percent, such funding could place firms in a position of greater responsibility for the costs of their long-term survival, for professional development, and for investment in technology. Such a system would be simple to understand and justify; it would also greatly simplify the provision of assistance, reduce government administrative costs, and put publishers in charge of investing in their own development. This approach also addresses succession, as 7 percent profit could make a firm worth buying, and it addresses the culture of entitlement to funds for industry development of every kind.

A service-based publishing model offers a number of other advantages. It opens up opportunities for third-party investors, as well as authors and publishers, to contribute to development expenses. A publisher could negotiate with an author or sponsor to cover any category of costs — manufacturing, marketing (including sales and publicity), and distribution. The inevitable hierarchy of prestige that already exists among publishers would continue, as certain publishers, editors, layout professionals, marketers, and other publishing professionals would be able to charge higher fees than others. Fee-based services would also bring publishing in line with other service professions and would complement the manner in which other cultural agencies, such as the Canadian Television Fund, operate, where grants to producers cover development and production costs, and other revenues derive from sales. Such a policy would accord with the process by which many other industries are assisted by government, where certain categories of costs are considered eligible for calculation of various forms of assistance.

Barriers to the acceptance of a service model

If a service model has so much to recommend it, why have publishers not adopted it? Besides the effects of hubris and history, mentioned earlier, two factors stand in the way of the widespread adoption of service-based publishing. One is the lack of characteristics clearly and immediately discernible to the consumer that differentiate professional products and services from those of non-professionals. The second is the over-supply of labour: there are far more people who wish to pursue a career in publishing than the market can support. Both these factors could be addressed if the industry wished to do so by, for instance, increasing the requirements for training, assessment, and credentials, and by unionization. The economics of talent-based industries do not have to mean the exploitation of the many who are not celebrities.

The cultural environment: A changing Canadian market

As an analytical framework, economics can encompass all human activity, but culture is an equally all-encompassing perspective on human affairs. The cultural environment is composed of institutional, ideational, and attitudinal influences that nurture the creative activities of its citizens. To review the relevant cultural background largely detailed in this book: A cultural environment friendly to book publishing grew gradually in Canada, in counterpoint to the nation's political and economic colonial shadow identity with respect to the Great Powers (the United Kingdom, the United States, and, to a lesser extent, France). Building on the determined will of a number of historical figures acting at particular junctures, including John Lovell, Egerton Ryerson, Lorne Pierce, and Jack McClelland, the Massey Commission laid the foundations for the evolution of modern Canadian culture. The Ontario Royal Commission on Book Publishing took up the task of marrying book publishing to the cultural goals inherent in the Massey Commission and laid out a plan to patriate control over both trade and educational book publishing. Against the background of these inspirational documents, Canada entered a period of policy and support-program development that eventually allowed book publishers and authors to provide the Canadian reading public with a substantial stream of trade books written from a Canadian perspective and often written about Canadian subjects.

As a result of the general success in Canadian cultural development — including the emergence of a Canadian film-making community; a lively music scene built on minimum Canadian content regulations on radio; and magazine, television, radio, and newspaper industries dominated by Canadian ownership — Canadian cultural production is now standing on its own. The self-confidence resulting from that success means that fretting over the survival of a distinct culture, let alone an independent country — the very essence of the good fight that Canadian book publishers embraced in the 1970s and 1980s — is yesterday's struggle. In many circles it is taken for granted that there is a continuing need to ensure appropriate policies are in place in support of Canadian expression and interaction.

The development of community-building and social networking technologies, the enthusiastic uptake of communication technology in general, and the global realities within which we now live are encouraging young Canadians to reach out from a solid foundation of confident Canadianness into international arenas. Many are interested and engaged in creating Canadian games for world markets, playing or listening to Canadian music that will tour the world, creating or viewing Canadian film and television that will receive international accolades and be sold into a wide variety of export markets, and writing or reading Canadian books that will win prestigious international prizes. These developments have created a culture of realistic expectations that will determine the nature and extent of government support policies for the creative industries. And just as governments will respond to expectations inherent in the cultural environment, book publishers, given the growth and dynamism of new media, will find their future in world markets and in evolving formats and creativity bounded not by print on paper but by technology, inventiveness, and the imagination. In short, the future of Canadian book publishing lies in its ability to engage a media-evolving culture that unapologetically embraces both the Canadian and the global.

In this context, consider the mixed blessings of "net benefits" and their interplay with ownership policy in the case of three respected editors. Louise Dennys was a principal in two Canadian-owned companies, both of which went into bankruptcy, before she found a home at Random House Canada, where she has brought forward a number of Canadian authors now known worldwide.[472] Phyllis Bruce has done similar exceptional work from within HarperCollins Canada, as did Cynthia Good when she was president and publisher of Penguin Canada. By providing homes for all three of these talented editors, mainly in response to federal policy allowing foreign ownership in return for "net benefits," foreign-owned, Canadian-based publishers have enhanced the possibilities for Canadian creators' participation on the world stage.

Within a cultural context, the contrast between the changing orientations of young Canadians, who are moving towards global realities, and the specialization of Canadian publishers and the authors they publish, which deal largely with domestic realities, is important to note. The future of a significant number of Canadian-oriented titles is unknown in our globalizing world. Will demand re-emerge as this generation gets older? Will Canadian authors and publishers leave national realities behind and address questions of interest to readers around the world? Will Canadian authors and publishers develop new treatments of Canadian and world realities where the two intermingle powerfully and alluringly? Will other publishers follow the lead of ECW and create books on Canadian popular culture?

The answers to these questions are not yet discernible, although the emergence of a local/global social dynamic may encourage a parallel literary dialectic. Indeed, we cannot yet know whether the market for books — that is, long form texts, with or without sound and images — will contract, expand, or continue at its current size. As our universe of ideas unfolds, it is increasingly apparent that writers and publishers have lost some of their pre-eminence. Other media, such as music, film, and the Internet, are increasingly influential. From a national perspective, as *The Business of Culture* and *Valuing Culture* make so very clear, if governments fail to invest in creative industries to sustain a noteworthy presence for Canadian creators and publishers in the marketplace, there are inexhaustible flows of foreign cultural products that will swamp the Canadian cultural scene, remove creative Canadians from the markets of their native land, and eliminate their opportunities to participate in world markets.

The educational environment

Historically, educational publishing has served as a foundation for the development of literacy, authorship, and trade book publishing. For good or ill, the trade and education sectors have been separated in Canada and elsewhere in the developed world, and educational publishing has been largely lost as an incubator of trade publishing.

Educational publishing is typified by publishers' extensive investment in adding various kinds of value such as the articulation of objectives, the sequencing of content in pursuit of those objectives, the introduction, development, expansion, and testing of vocabulary and concepts, and, more recently, the custom design of feedback based on diagnosis of responses. Reflecting this investment, educational publishers, rather than authors and editors, are often the holders of copyright. Educational publishing, like reference publishing, is almost wholly the province of large, specialty publishers, although there are significant exceptions in the form of even larger companies spanning education and trade, and small, niche-oriented domestic educational publishers.

The above said, the current content of classroom materials in Canadian schools and universities differs substantially from that of the 1970s and 1980s. The very public evolution of both Canadian culture in general and trade publishing specifically has spilled over into the curricula of schools and post-secondary institutions, where Canadian specificities have become studied educational realities, even if those specificities are not all-pervasive. There are still debates about including contemporary writing in the curriculum, Canadian or otherwise. Adapted science books still rely too often on examples provided in the original foreign texts. But today's classrooms are staffed mainly by Canadians with a knowledge of Canadian realities, and, in the context of multiculturalism, there is a greater recognition of schools as socializing institutions. The administrators and faculty of post-secondary institutions also recognize their role in preparing students to be Canadian citizens, not to dream of being citizens of the world with no understanding of their home country. To a much greater extent than in the period from 1960 to 1980, the education system as a whole contributes to the socialization of Canadian students as Canadian citizens in a global world challenged by both local and global realities for which we all bear responsibility. Given that contribution, it would be reasonable to believe that Canadian readers will continue to be engaged by Canadian fiction and non-fiction.

The element of the educational environment that has perhaps the greatest impact on publishers is the view that because education is a social good, learning materials should be free to the educational system, and educators should be permitted to make whatever number of copies they require. Such a stance is untenable in that it deprives publishers of legitimate income. Sections of the education community have been reluctant to agree to proposed tariffs by Access Copyright that have been based on low-cost, fair estimates of the "unauthorized" copying that takes place within educational institutions. The continuous resistance of the education community represents a continuing assault on creators and publishers and undermines respect for copyright in society as a whole.

The ever-evolving technological (and organizational) environment

The emergence of a market for digital versions of books, built on a foundation of ICT-enabled databases and the presence of functional, battery-operated reading devices — Kindles, Sony Readers, iPhones, iPads, and Kobos — is changing the nature of books and the book industry fundamentally. Such changes include, but are not restricted to, the production of digital and print-on-paper texts, their electronic or material distribution, their retailing, the possibility and status of digital returns, technologically differentiated print title management, the dominance of book publishers as the entry point to the market as opposed to self-publishers and service publishers, the marketing of both online and printed books, the role of libraries, the social practices of reading and lending books, environmental impact, and the potential for interactivity. And this is just the beginning. Opportunities presented by new media will serve as catalysts to realities yet to be imagined.

The long-term impact of self-publishing is probably overrated. At the beginning of many new technologies, the leap forward made possible by new production tools appears to overshadow the need for professionals. But just as the "grammar and spelling" of word-processing programs does not create admirable stylists, so software programs are unlikely to eclipse the talents and training of graphic artists — which is not to say that industrial production never renders craft obsolete. That said, the presence of three times the number of new titles being published outside, as opposed to inside, the traditional book publishing industry (U.S. figures) is changing the market, if not for printed books, then certainly for e-books. Where this new production vibrancy will take book publishers and society in general is difficult to say.

Interactivity, on the other hand, may be underrated. The advent of liquid books such as Paul Shoebridge's and Michael Simons' *Welcome to Pine Point*,[473] in which various optional sub-narratives are available within an overall narrative, suggests that there may be further opportunities lurking beneath our dreamt-of possibilities for emerging book forms.

Certainly, the enthusiasm of Canadians for technology and the embrace of technological development by governments and business are creating new opportunities for book publishers. The digitization projects of the Ontario Media Development Corporation and the Canadian Knowledge Research Network, as well as that of Google, are encouraging participation in digital markets, as is the struggle for market share among Amazon, Sony, Apple, Kobo and Indigo, Barnes & Noble, and others in the market, or soon to enter it. Already, funds are flowing from digital sales to publishers and authors, and professional development workshops are helping both authors and publishers understand digital markets.

At the rebirth of modern Canadian book publishing: Overview and conclusion

This book has brought forward a consideration of the development of Canadian book publishing, with particular attention to public policy, that has taken cognizance of the nature of the book, Canadian book publishing history, and contemporary realities and trends. Before Canada was Canada, book publishing began as both printing and bookselling. Book publishing then entered a long and discontinuous gestation, in which Canadian authorship and publishing in book form rose to the fore only occasionally. Modern book publishing was born in an era focused on the articulation of culture and the assertion of a national identity, as a collaboration between industry and government on the foundation of a shared understanding of the importance of book publishing to the evolution of Canada and Canadian culture. Canadian book publishing also reflected the ever-present desire of Canadian writers to express themselves and of Canadian readers to access those expressions. But times have changed. The twenty-first-century priorities of Canada and Canadians differ substantially from those of the 1970s and 1980s. Canada exists in an integrating world of nations where a plurality of economic and cultural power centres are emerging, led by China and followed by India, Russia, and Brazil. In this context, cultural plurality is becoming more apparent. And technological opportunity has also introduced new possibilities in even greater plurality.

For Canada, given the value of the creative economy and the determination of Canadians to participate in cultural expression, alongside the desire of many policy makers and governments to foster participation in cultural production, the advantages of full participation in the creative economy are many. For individual book publishers and their employees, however, as for authors, the financial realities remain difficult. A financial challenge exists at the level of the firm: the manner in which book publishing is carried out and publishing services are provided are uneconomic. These dynamics can change, but the will to recognize and address inefficiencies in the organization of work in book publishing is not yet apparent. How can this change?

Technological development is playing a major transformative role, in the organization of the industry and in the embrace of electronic formats. Immediate opportunities, such as digitizing works, and challenges, such as new competitors like O'Reilly and publishing services firms, are arising from technology, but the longer-term technological and cultural opportunity and challenge for book publishers is to reinvent, in a modern socio-technological context, the attributes that have carried printed books through nearly 600 years in the Western world — the extended exploration, the intimacy of the reader with a single creative voice, the generation of ideas in readers' minds, the social interaction and discussion of meaning that books call out.

If publishers don't embrace new media opportunities while preserving the powerful attributes that give books their prominence, and if they don't organize production, distribution, and financial rationalization in response to technological capacity, competing industries will reduce the time people spend with "books." Put another way, if publishers don't embrace evolving opportunities in every sphere of book publishing, the already substantial gap between the contributions to limited economic growth made by the printing and publishing industries and the more robust contributions made by other industries of the creative sector, identified in *Valuing Culture*, may increase.

The opportunities and challenges for book publishers were demonstrated dramatically at a November 2011 seminar with staff of the ACP's eBOUND project, which is aimed at assisting ACP members to create electronic editions of print book. As a demonstration of what is now technologically possible, one of the panel members presented an adaptation of *Peter Rabbit* for the iPad, with enriched and interactive media. Almost immediately, the customary discussion about cost and the feasibility of earning back initial investment arose, with the publishers in the room appearing to conclude that such projects were financially unfeasible in the Canadian trade market. This seems wrongheaded, and it comes from a misunderstanding of how events are likely to unfold.

To illustrate: In small markets, it is the author who, most often, invests the most time and energy and takes on the greatest risk, motivated by the desire to create, to say something excellently, or to provide original insight. In small market publishing it will therefore most likely be authors, not publishers, who will underwrite the creation of digital products, because publishers won't be able to afford to do so at an early stage in the development of a publishable work. Moreover, it is likely, in Canada, that arts councils, including the Canada Council, will encourage authors to take up the creative possibilities that are emerging from new technologies.

In large market publishing — the multinational mass market, professional, and educational publishing — two models are appearing. Educational publishers are developing interactive technologies of learning, and are hiring authors on a contractual basis to fill in the content. On the other hand, new technology-driven publishers, such as Push Pop Press, now owned by Facebook, are partnering with powerful authors such as Al Gore to create multimedia properties.[474]

The small-market-oriented book publishers who were speaking at the ACP seminar saw themselves as the creators of digital products because they envisioned digital products as conversion editions — as electronic files, mainly of text, that approximate the same reading experience as the book. Such a vision falls short, leading them to dismiss the invention of an iPad image of a windmill that turns when the user blows on the device as a gimmick. Such functionality is hardly a gimmick for the intended audience. It is a marvelous, even magical, feature developed by Push Pop Press[475] that will delight children, as well as adults with less allegiance to text alone. With the availability of production and communication technology (developed by Apple and others), access to the Internet (courtesy of Tim Berners-Lee), and a variety of places to test the market such as Facebook (developed by Mark Zuckerberg) and YouTube (developed by Steve Chen, Chad Hurley, and Jawed Karim), a whole new world of possibilities is emerging on the horizon.

The printed book is a marvellous socio-technical ensemble that has been with us, if one includes incunabula, for centuries.[476] But for some time the printed book has been obsolete, in Marshall McLuhan's sense: it is congruent with society and, as a medium, is not having a transformative effect or influencing our current worldview.[477] Currently, media-based social transformations are driven by blogs, podcasts, wikis, social networking, websites, and mash-ups.[478] New organizations and forms of social interaction are arising around them. These emerging media also carry a whole discourse of engagement that is taking shape outside the realm of the book, some of which is trivial, but much of which will have major consequences, in part because these new media forms reconstruct the social interaction associated with their use — for example, crowd-sourcing the development of a media product.

Along with other creative industries, book publishing remains critical to the cultural project that is Canada. Moreover, writing, publishing, and their associated government support structures are an important part of the distinctive contribution Canada is making to the world. Arguably, in embracing multiculturalism, diversity, and heterogeneity, Canadian book publishing remains critical to the cultural project that is civilization in its totality. That said, as the information media expand, book publishers may find that they will have to compete more vigorously for the services of capable and creative employees in editorial, design, production, and marketing. For the same reasons, competent authors may become scarcer, and publishers may be required to offer greater financial reward upfront. Such information-economy-derived dynamics may be healthy, in that they may better align book publishers and their employees with business and financial norms. But the transition is likely to be painful.

The Business of Culture made the case that Toronto could and should set its sights on becoming a world-notable centre of cultural production. It has done so. A much more powerful and interesting goal, in keeping with the principles that have made Canadian book publishing a cultural success, would be to have Canada set its sights on becoming a notable country of cultural production, an objective that embraces the creative economy and national and international cultural diversity. The Convention on Cultural Diversity elaborates this orientation by laying the conceptual foundations for Canada and other countries to assume such a role within a diversity-respecting and interactive globalized world, and *Valuing Culture* describes how both Canadian society and the Canadian economy would benefit from such a development.

The infrastructure that makes it feasible for Canada to increase its share of international and domestic markets derives from the foundations identified in this book. Out of the cultural confidence thus achieved have come complementary, still-evolving, political, legal, and policy environments. These environments now exist nationally and internationally in the affirmation of the principle of the promotion and protection of cultural diversity, formalized by the voluntary signing of the Convention on Cultural Diversity by over ninety nations. They have been further entrenched by the attention being paid at the UN and around the world to the creative economy. On the basis of the principles and policies of both the CCD and creative economy initiatives, cultural markets will gradually restructure, which may lead to the injection of a multiplicity of diverse local expressions into global entertainment trade flows. As Canadians and others are exposed to greater diversity, invidious comparisons of local to global products may be diminished by the plurality of locals — the productions of other smaller nations — to create a much more receptive cultural environment for a multiplicity of local expressions. In the uniqueness that is Canada, it is easy to imagine Canadian multicultural values and the entrepreneurial spirit of Canadian book publishing joining with Canada's outward-looking education system to affirm our cosmopolitanism, our dual allegiance to Canada and to the larger world, further strengthening the celebration of diversity.

Table 5.1, extended
Book publishing revenue 2006 and 2004 (in thousands of dollars)

	Survey portion	Eng-lang surveyed	Share of of sales French & English	Cdn-controlled firms	Share of total foreign- and Cdn-controlled	Ontario	% Ont
Total Operating Revenue							
2006	**$2,131,451**						
Survey portion	**$2,013,792**	$1,636,349	**81%**	$1,130,906	56%	$1,296,803	79%
Sales in Canada	**$1,501,159**	$1,196,524		$768,356	51%	$1,025,700	86%
Own titles	$921,573	$680,465	74%	$590,705	64%	$550,467	81%
Educational	$445,382	$321,227	72%	$242,774	55%	$315,864	98%
Trade incl. children's	$379,616	$297,940	78%	$260,127	69%	$197,329	66%
Schol., ref, prof.,tech	$96,574	$61,297	63%	$87,804	91%	$37,275	61%
As exclusive agents	$579,586	$516,059	89%	$177,651	31%	$475,232	92%
Educational	$218,286	$205,935	94%	$83,918	38%	$205,920	100%
Trade incl. children's	$331,955	$290,412	87%	$85,366	26%	$249,705	86%
Schol., ref, prof., tech	$29,345	$19,710	67%	$8,367	29%	$19,607	99%
Combined Own and Agency titles	$1,501,159	$886,400	59%	$768,356	51%	$1,025,699	116%
Educational	$663,668	$611,639	92%	$326,692	49%	$521,784	85%
Trade incl. children's	**$711,571**	**$588,352**	83%	**$345,493**	49%	$453,302	77%
Schol., ref, prof., tech	$125,919	$81,007	64%	$96,171	76%	$56,882	70%
Exports & other foreign sales	$244,699	$227,225	93%	$231,125	94%	$174,380	77%
Other revenue (grants, rights)	$267,934	$212,600	79%	$131,425	49%	$96,723	45%
Total Operating Revenue							
2004	$2,055,559						
Survey portion	$1,958,986	$1,597,151	82%	$1,128,010	58%	$1,271,558	80%
Sales in Canada	$1,454,531	$1,191,419	82%	$732,441	50%	$1,021,230	86%
Own titles	$888,701	$671,438	76%	$570,794	64%	$549,609	82%

	Survey portion	Eng-lang surveyed	Share of of sales French & English	Cdn-controlled firms	Share of total foreign- and Cdn-controlled	Ontario	% Ont
Educational	$405,212	$296,600	73%	$223,482	55%	$288,444	97%
Trade incl. children's	$353,887	$279,263	79%	$250,958	71%	$217,803	78%
Schol., ref, prof., tech	$129,602	$95,575	74%	$96,354	74%	$43,362	45%
As exclusive agents	$565,829	$519,981	92%	$161,647	29%	$471,622	91%
Educational	$235,603	$223,461	95%	$84,733	36%	$222,063	99%
Trade incl. children's	$304,673	$275,807	91%	$70,014	23%	$229,780	83%
Schol., ref, prof., tech	$25,553	$20,714	81%	$6,900	27%	$19,779	95%
Combined Own and Agency titles	$1,454,530	$1,191,419	82%	$732,441	50%	$1,021,231	86%
Educational	$640,815	$520,061	81%	$308,215	48%	$510,507	98%
Trade incl. children's	$658,560	$555,070	84%	$320,972	49%	$447,583	81%
Schol., ref, prof., tech	$155,155	$116,289	75%	$103,254	67%	$63,141	54%
Exports & other foreign sales	$234,428	$218,962	93%	$230,930	99%	$174,542	80%
Other revenue (grants, rights)	$270,027	$186,771	69%	$164,639	61%	$75,785	41%

Source: Statistics Canada Catalogue no. 87F0004X

Policy History

Primary focus of document

Ontario Royal Commission recommendations, 1971
> Established the notion of cultural benefit and need for domestic predominance over economic contribution

Government statements and actions 1972-1980
> Established recognition of Canadian book publishing and of its need for subsidies and business support

ACP statements and recommended actions, 1980
> Asked for the financial support required to maintain a heterogeneous group of entrepreneurial, risk-taking, widely competent owner/managers engaged in cultural production

ACP statements and recommended actions, 1984
> Identified key needs of book publishing business: access to capital, distribution, ownership, taxation

Federal government statements, commissioned reports, and actions, 1985-1989
> Industrial commitment: Clear resolve by federal government to help publishers become viable businesses, producing cultural titles via structural intervention and industrial support

ACP statements and recommended actions, 1989
> Industrial assistance: Determination to identify means of making publishing financially viable while maintaining hetereogeneity of firms pursuing culturally significant works

Federal government statements and actions, July 1990; commissioned Audley report, 1990
> Structural analysis: Attempt to institute structural mechanisms, i.e., other than direct financial assistance, to support industry

Federal government statements, actions, and commissioned reports, 1992-1993
> Evaluation Report: Primacy of culture undermines business development; business performance should be required for grant access

Ontario government, 1994 (preceded by review document, 1989)
> Review and strategy documents: Main recommendation: Establish Toronto as a leading, technologically advanced cultural production centre, active in international and domestic entertainment and information markets

ACP statements and recommended actions, 1995
> Restatement of industry financial and operational needs: market-based challenges, primacy of cultural value over profits and firm growth

Federal government 1995–1998, Commissioned Report 2000
> Government flipflop: Apparent determination by some in government to renounce continuing government support for book publishing conflicts with equal determination by others to strengthen support

Federal government 1999–2004; UNESCO 2001, 2005
> Realigned development: Strategy emphasizing market participation, but focused on legitimacy of measures to strengthen national cultural space

Ownership and market share

Ontario Royal Commission recommendations, 1971
> Restrict ownership to Canadians
> Encourage development of scholarly publishing

Government statements and actions 1972–1980
> 1974 – Foreign Investment Review Act restricted new ownership of book publishing firms to increase access of Canadian firms to foreign titles

ACP statements and recommended actions, 1984
> Recommended avoidance of ownership concentration
> Emphasized cultural orientation of small companies

Federal government statements, commissioned reports, and actions, 1985–1989
> 1985 – Baie-Comeau Agreement forced divestiture to Canadians on the sale of a foreign-owned company. Disallowance of new foreign ownership
> Goal of 50% ownership of all new companies within two years

ACP statements and recommended actions, 1989
> Called for strengthening of restrictions on foreign ownership
> Called for an increase in market share by Canadian firms

Federal government statements and actions, July 1990; commissioned Audley report, 1990
> Audley: Recommended strengthening Baie-Comeau by increasing Canadian ownership (with financial measures outlined under Financial, below)

Ontario government, 1994 (preceded by review document, 1989)
> 1989 – Ontario paper observes that market share of Canadian publishers is shrinking, especially for Canadian-originated titles
> 1994 – *The Business of Culture* Recommendation (BoC Rec): Increase domestic and international market share of Canadian-owned, Ontario-based cultural producers

ACP statements and recommended actions 1995
> Continuing need for foreign investment regulation

Financial

Ontario Royal Commission recommendations, 1971
> Established that a need existed for working capital from government-guaranteed loans to compensate for abnormal industry structure
> Proposed grants based on number of titles published

Government statements and actions 1972–1980
> 1972 – $300,000 set aside for Canadian Book Publishing Board. Cultural support: $1.9 million to Canada Council to help develop block grants for Canadian book publishing; also, marketing/industrial support export assistance via AECB
> 1974 – Cultural support: Ontario Arts Council funding increased from $50,000 to $340,000
> 1977 – Cultural support: Funds provided for book purchase program
> 1979 – Industrial support: Book Publishing Development Program established $2.5 million

ACP statements and recommended actions, 1980
> Diagnosis: Cost-based lack of competitiveness with foreign sector; lack of access to lower cost production, i.e., importing and distribution; lack of access to subsidiary rights market; lack of distribution right; inadequate distribution mechanisms, unfriendly banking system, lack of accesss to sufficient working capital
> Industry-identified needs: Increased grants and development of industrial support measures (beyond title subsidies) such as loans and market development, to be administered by a government-funded, arm's-length agency, i.e. Canadian Publishing Board
> Industry-identified needs: Structural support such as regulatory measures and foreign investment restrictions

ACP statements and recommended actions, 1984
> Called for means to attract venture capital for buy-outs of foreign-owned publishers by Canadians
> Recommended increasing access to financial assistance and innovation funds based on sales levels
> Called for a national loan guarantee program (modelled on Ontario's)

Federal government statements, commissioned reports, and actions, 1985–1989
> 1986 – Industrial support: BPIDP established, $75 million to publishers over 3 years. Fairly minimal financial performance criteria set for BPIDP eligibility
> 1987 – *Vital Links* promoted support to achieve self-financing for larger diversifying publishers, and cultural subsidies for smaller publishers for author development
> 1989 – BPIDP provided an additional $110 million to publishers
> 1989 – Postal Assistance program wound down

ACP statements and recommended actions, 1989
- Recommended creation of income tax credit for title development; increased funding for Canada Council and SSHRC (cultural and scholarly grants); increased flexibility of federal loan program; removal of GST from books
- Admitted structural weakness in comparison to foreign-owned firms

Federal government statements and actions, July 1990; commissioned Audley report, 1990
- 1990 – Cultural Industries Development Fund created
- Audley: Recommended tax credits for title origination; support for distribution infrastructure; and increases in loan access for title origination.
- Audley: Noted constraints of high interest rates, lack of access to handling imports, implementation of GST

Federal government statements, actions, and commissioned reports, 1992–1993
- 1992 – Fox-Jones: Data indicate that financial health of industry remains unchanged despite loans, financial assistance, ownership restrictions
- 1992 – Fox-Jones: Primacy of culture undermines normal market conditions and rationalization for profitability
- 1993 – BPIDP increased emphasis on business performance for eligibility, but this was subsequently relaxed
- Continued support for exports, cultural titles and postal assistance

Ontario government, 1994 (preceded by review document, 1989)
- 1989 – Ontario paper: Shrinking market share of culturally significant books requires increased financing to compensate
- 1994 – BoC Rec: Create a stable, equitable operating environment for Canadian cultural industries, and an industry/government partnership to pursue growth of sector

ACP statements and recommended actions 1995
- Business challenges defined as: cost-of-sales disadvantage of 9.2% compared to foreign-owned firms; investment capital requirements greater than revenues; grants generate 7 times earned revenues; focus of industry is cultural development, not profit or firm growth
- GST revenues outstrip grants to industry
- Called for: production investment with, if necessary, returns on investment to government for successful titles; loan access; deficit funding for cultural titles; national action plan to strengthen book publishing

Federal government 1995 to 1998, Commissioned Report 2000
- 1995 – Announced cuts of 55% to BPIDP (actual cuts = 36%)
- 1998 – Restoration of BPIDP funding

> 1998 – No tax credit offered in restoration of BPIDP

> 2000 – Donner and Lazar: Diagnosis: poor profits and inadequate capitalization; prices externally determined, no competitive economies of scale, lack of access to investment capital, weak business relationship with bookstore chain Chapters

> 2000 – Donner and Lazar: Proposed measure: improve competitive position by consolidation to achieve economies of scale, combined with refundable tax credit

Federal government 1999–2004; UNESCO 2001, 2005

> Support for cultural industries re-legitimized by the value of cultural diversity in cultural production and the interaction of communities

Market Design

Ontario Royal Commission recommendations, 1971

> Goal: Create structures to compensate for a domestic market of insufficent size to warrant a domestic book publishing industry

> Goal: Ensure flow of Canadian ideas and literary creativity

> Goal: Establish Canadian developed school textbooks, scholarly publishing, review and criticism

> Means: Marketplace intervention in aid of cultural goals

> Goal: Secure market opportunities for Canadian authors

> Means: Establish Book Publishing Board to oversee development of the industry. 1972 Initial funding $300,000

> Goal: Ensure authors receive income due. Suggested means: Royalties insurance

> Goal: Eliminate public sector competition. Means: Contract out government publishing to Canadian publishers

Government statements and actions 1972–1980

> Diagnosis: Import-based pricing constraints

ACP statements and recommended actions, 1980

> Need: Address competitive disadvantage of Canadian firms

Federal government statements, commissioned reports, and actions, 1985–1989

> 1987 – *Vital Links* cast global products as undesirable; criticized dominance of foreign subsidiaries and foreign-owned distributors, dominance of large firms, suppression of rights-based market, parallel importation by institutions and libraries (buying around Canadian rights holders), lack of capital access, lack of success in educational markets

> 1987 – Free Trade Agreement (FTA) signed, with culture off the table

> 1989 – Peat Marwick: "hard to imagine" Canadian firms surviving on trade publishing for Canadian readers

ACP statements and recommended actions, 1989

> Emphasized industry commitment to maintaining primacy of cultural goals using business practices.

> Called for focus on Canadian books for Canadian readers, published profitably by Canadian firms; also, for support for publishers' chosen fields, in order to promote cultural and regional diversity

> Diagnosed financial problems as based in the industry's domination by imports, which lead to high comparative manufacturing costs and inadequately remunerative pricing; in the industry's participation mainly in the high-risk, low-return trade sector; and in its limited access to capital

> Recommended increased access to education and library markets

Federal government statements and actions, July 1990; commissioned Audley report, 1990

> Audley diagnosed financial problems as: Dominance of imports, making market inaccessible to Canadian firms; barriers to entry to educational markets; higher production costs for Canadian firms resulting from primary activity being title origination (80% of new Canadian-authored titles are published by Canadian publishers)

> Audley's recommendations: See Ownership and Financial categories.

Ontario government, 1994 (preceded by review document, 1989)

> 1994 – BoC Rec: Reorient policy towards support for market participation, growth of firms, and increased market share or cultural expression

> 1994 – BoC Rec: Create a Toronto-based cultural industries production centre; establish arm's-length Board to oversee development (became Ontario Media Development Corporation); emphasize international and domestic participation in entertainment and information markets

ACP statements and recommended actions, 1995

> Emphasized success of Canadian writers and demand for Canadian titles by Canadian readers, noting that Canadian firms published 85% new titles, and that 71% of new titles were Canadian authored, 80% of which were published by Canadian firms. Also noted that export sales were at 20%

Federal government 1999–2004; UNESCO 2001, 2005

> 1999 – New Strategies for Culture and Trade (SAGIT Report) aimed at fortifying exemption of culture in trade agreements

> 1999 – SAGIT called for acceptance of cultural support policies to create market space for domestic products

> 2001 – UNESCO: Universal Declaration on Cultural Diversity adopted

> 2005 – UNESCO: Convention on the Protection and Promotion of the Diversity of Cultural Expressions (Convention on Cultural Diversity, CCD) adopted

> 2005 – CCD requires that nations take into account the relevant provisions of CCD when interpreting other treaties to which they are parties, or when entering into other international obligations

> Post-2000 – Notion of social capital begins to attain legitimacy

Marketing and promotion

Ontario Royal Commission recommendations, 1971

> Fund reprints of out-of-print Canadian titles
> Fund literary awards
> Create display of Canadian books in London UK
> Fund a review journal for Canadian books
> Increase awareness of Canadian books among educators, librarians, and media
> Train educators in awareness of cultural role
> Reduce balkanization of provincial education markets
> Increase contact among educators, authors, and publishers
> Increase awareness of nonfiction titles useful for education
> Encourage collaboration in title development between education ministry officials and Canadian publishers
> Develop culturally sensitive evaluation materials for students (OISE)

Government statements and actions, 1972–1980

> 1975 – Cooperative promotion and distribution program established

ACP statements and recommended actions, 1980

> Goal: Organize related groups, such as writers, booksellers, librarians, teachers, school trustees, students, parents, Quebec publishers, and other cultural industries, in support of book publishers' major aims

ACP statements and recommended actions, 1984

> Recommended public procurement through retail, in return for Canadian title stocks; incentives for aggressive sales campaigns; funding for travelling to book fairs; a tax rebate for booksellers who stock Canadian titles
> Called for increased Canada Council tour and promotion money for all supported books
> Expressed alarm regarding centralized buying by retail chains

Federal government statements and actions, July 1990; commissioned Audley report, 1990

> Federal government attempted to open provincially controlled educational markets to Canadian publishers

Ontario government, 1994 (preceded by review document, 1989)

> Addressed in 31 recommendations and 40 tactics in *The Business of Culture* (1994)

Training

Ontario Royal Commission recommendations, 1971

> Provide business training for publishers

> Provide business consulting services for publishers

Ontario government, 1994 (preceded by review document, 1989)

> Addressed in 31 recommendations and 40 tactics in *The Business of Culture* (1994)

Technology

ACP statements and recommended actions, 1984

> Identified need for electronic data interchange for ordering and distribution

Ontario government, 1994 (preceded by review document, 1989)

> 1994 – BoC Rec: Create an industry/government framework for taking lead in development of multimedia products

ACP statements and recommended actions 1995

> Made note of experimentation with CD-ROM versions of encyclopedias

Data monitoring

Ontario Royal Commission recommendations, 1971

> Establish a general survey of industry (established by StatCan 1973)

Ontario government, 1994 (preceded by review document, 1989)

> Addressed in 31 recommendations and 40 tactics in *The Business of Culture* (1994)

Federal government 1999–2004; UNESCO 2001, 2005

> 2002 – BookNet Canada established; begins to collect and then publish sales data

> 2004 – Department of Canadian Heritage undertakes annual survey of performance of client firms

Exports

Ontario Royal Commission recommendations, 1971

> Provide sales assistance for exports (AECB established 1972)

Government statements and actions 1972–1980

> 1972 – AECB established

ACP statements and recommended actions, 1984

> Commended AECB assistance

Ontario government, 1994 (preceded by review document, 1989)

> 1994 – Primary focus on international participation and necessary technology and infrastructure

Distribution

Ontario Royal Commission recommendations, 1971

> Establish distribution support (PDAP established, 1994; Supply Chain Initiative 2001)

Government statements and actions 1972–1980

> 1975 – Cooperative promotion and distribution program established

ACP statements and recommended actions, 1980

> Called for support for electronic data interchange

ACP statements and recommended actions, 1984

> Import-based market accepted

> Called for mandated 2% annual growth of Canadian-owned distributors

ACP statements and recommended actions, 1989

> Called for enhanced distribution and availability of Canadian titles, including support for exports

Ontario government, 1994 (preceded by review document, 1989)

> Addressed in 31 recommendations and 40 tactics in *The Business of Culture* (1994)

Copyright

Ontario Royal Commission recommendations, 1971

> Strengthen ability of Canadian publishers to acquire rights

> Protect Canadian editions

> Compensate publishers for multiple use of library books; PLR was established in 1986. Creators' rights collective established 1988.

ACP statements and recommended actions, 1984

> Identified need for distribution right

ACP statements and recommended actions, 1989

> Called for establishment of distribution right within Copyright Act, to prevent parallel importation

Federal government statements and actions, July 1990; commissioned Audley report, 1990

> Initial moves made to forbid parallel importation

> Audley: Recommended that parallel importation via distribution right be disallowed in Copyright Act

Ontario government, 1994 (preceded by review document, 1989)

> Addressed in 31 recommendations and 40 tactics in *The Business of Culture* (1994)

ACP statements and recommended actions 1995

> Highlighted the need for protection of creators' rights, and for a distribution right

Federal government 1995–1998, Commissioned Report 2000

> 1996 – Distribution right established

Research

ACP statements and recommended actions, 1984

> Recommended profiling purchasers of Canadian titles

> Recommended identifying regional challenges and genre participation barriers

NOTES

ULTRA LIBRIS

1. See Susan Juby, "Editor to Author: Some Personal Reflections on Getting Published," in *Publishing Studies: Book Publishing 1*, ed. Rowland Lorimer, John Maxwell, and Jillian Shoichet (Vancouver: CCSP Press, 2005), 16–42, for a discussion of authors' attitudes to publishers.
2. Carla Hesse, "Books in Time," in *The Future of the Book*, ed. Geoffrey Nunberg (Los Angeles: University of California Press, 1988), 21–36.
3. Statistics Canada, Service Industries Division, *Book Publishers*, 2008, cat. no. 87F0004X (Ottawa: 2010).
4. It should be noted that while Simon & Schuster is excluded from the Statistics Canada survey, the survey includes exactly the same importing and distribution activity carried out by Random House Canada and Penguin Canada, although probably not HarperCollins, because the Canadian publishing company and the importing and distribution arm are two separate legal entities.
5. Knowledge of the industry indicates that agency sales by Canadian firms are concentrated among a limited number of Canadian-controlled firms, such as Raincoast, H.B. Fenn, and Firefly.
6. Statistics Canada, *Book Publishers*, 2008, cat. no. 87F0004X (Ottawa: 2010).
7. Turner-Riggs, *The Book Retail Sector in Canada* (Ottawa: Department of Canadian Heritage, 2007).
8. In March 2010, Amazon began a campaign to establish a Canadian base of operations. It currently processes Canadian orders in the United States and subcontracts delivery to an arm of Canada Post.
9. The figure was calculated by doubling the wholesale value of the domestic industry to obtain its retail value ($3 billion), adding on 20 percent to cover non-surveyed book distribution ($600 million), and multiplying that by 5 percent (the current GST rate).
10. Turku School of Economics and Business Administration and Rightscom, *Publishing Market Watch: Final Report* (Brussels: DG Enterprise of the European Commission, 2005). Available

online at ec.europa.eu/information_society/media_taskforce/doc/pmw_20050127.pdf.
11 See, for example, M. Zifcak, "Australia without Retail Price Maintenance," *Logos*, 2 (1991): 204-8.
12 Australian Council for the Arts, *Arts funding guide, 2009*, 24. www.australiacouncil.gov.au/grants/fundingguide.
13 Australian Council for the Arts, "Literature — July 2009" (assessment meeting report), www.australiacouncil.gov.au/grants/assessment-reports/literature2.
14 A number of studies of Harlequin have been conducted, including Paul Grescoe's *The Merchants of Venus: Inside Harlequin and the Empire of Romance* (Vancouver: Raincoast Books, 1996).

CHAPTER 1

15 UNESCO defines a book as follows: "Non-periodic printed publication of at least 49 pages exclusive of the cover pages ... made available to the public."
16 In Canada, because of its small market, publishers may seek to sell rights in another English-speaking market at the time they acquire the manuscript, and may continue pursuing such a sale up to the point of printing. If successful, they can combine their own initial print run with that of the foreign partner, especially if the partner is based in the United States, and thereby achieve lower per-unit costs through economies of scale in the print run.
17 Shyla Seller, "Implications of Authorship: The Author/Editor Relationship from Proposal to Manuscript" (Master of Publishing project report, Simon Fraser University, 2005).
18 I am grateful to David Kent for this story of the launch. The video is available on YouTube: www.youtube.com/watch?v=WVT2MPNCqgM.
19 Roy MacSkimming, *The Perilous Trade: Book Publishing in Canada 1946–2006* (Toronto: McClelland & Stewart, 2007), 122.
20 Ibid., 129ff.
21 Jack David, email to author, October 7, 2008.
22 David's moves turned out to be prescient in light of major grant cutting in 1995 and 1996 (detailed in Chapter 3) and the slide of the Canadian dollar, which was worth less than USD$0.75 by 1995 and declined to USD$0.68 by 1998, values that provided ECW with a 30 percent bonus to export earnings.

23 Information from University of Delaware Library, Special Collections Department, "Seventy Years at the Hogarth Press: The Press of Virginia and Leonard Woolf," www.lib.udel.edu/ud/spec/exhibits/hogarth/.

24 The group took its name from the fact that it met in the evening around the time of the full moon; in the days before electric lights, it was easier to find one's way home at night by the light of the moon.

25 Jenny Uglow, *The Lunar Men: Five Friends whose Curiosity Changed the World* (New York: Farrar, Straus and Giroux, 2002).

26 Carla Hesse, "Enlightenment Epistemology and the Laws of Authorship in Revolutionary France, 1777–1793," in *Law and the Order of Culture*, ed. Robert Post (Berkeley: University of California Press, 1991), content.cdlib.org.

27 Robert Darnton, *The Literary Underground of the Old Regime* (Cambridge, MA: Harvard University Press, 1982), 193–94.

28 Robert Darnton, *The Forbidden Best-Sellers of Pre-revolutionary France* (New York: Norton, 1995).

29 Roger Pearson, "Introduction," *Candide and other Stories* (Toronto: Oxford University Press, 1990), vii.

30 Ibid.

31 Paul Grescoe, *The Merchants of Venus* (Vancouver: Raincoast Books, 1996).

32 Here and throughout the book, by "generative," I mean the power of a text to allow readers to understand a word or concept in one context and apply it to a new context, and thereby generate new understanding. This usage of "generativity" is similar to "extensibility," from computing science, which is the x in XML and refers to the capacity of a system to be extended further by new users based on the established rules of the system.

33 Jonathan Rose, *The Intellectual Life of the British Working Classes* (New Haven, CT: Yale University Press, 2001).

34 Paul Tiessen, "From Literary Modernism to the Tantramar Marshes: Anticipating McLuhan in British and Canadian Media Theory and Practice," *Canadian Journal of Communication* 18, no. 4 (1993), 451–68.

35 Marshall McLuhan, *The Mechanical Bride: Folklore of Industrial Man* (New York: Vanguard Press, 1951); *The Gutenberg Galaxy: The Making of Typographic Man* (Toronto: University of Toronto Press, 1962); and *Understanding Media: The Extensions of Man* (New York: McGraw Hill, New American Library, 1964).

36 James Lorimer and Susan Shaw, *Book Reading in Canada* (Toronto: Association of Canadian Publishers, 1981); Nancy Duxbury, *The Reading and Purchasing Public: The English-Canadian Trade Book Market in Canada*

(Toronto: Association of Canadian Publishers, 1995); and Rowland Lorimer and Lindsay Lynch, *The Latest Canadian National Reading Study, 2005: Publishers' Analysis* (Vancouver: Canadian Centre for Studies in Publishing, 2007).

37 Janice Radway, *Reading the Romance: Women, Patriarchy and Popular Literature* (Chapel Hill: University of North Carolina Press, 1984).

38 Some trace this attack to an anti-book book, *The Incoherence of the Philosophers* by Abu Hamid Muhammad al-Ghazali (trans. Michael Marmura; Provo, UT: Brigham Young University Press, 1997). al-Ghazali, a faith-inspired Sufi mystic, questioned philosophy and reason and, in doing so, gave rise to a faith-over-reason social movement that is still being played out. One wonders if this backlash was, in part, caused by limited literacy, or perhaps a discourse that was beyond the reach of the many.

39 Lev Vygotsky, *Thought and Language* (Cambridge, MA: MIT Press, 1968).

40 Jerome Bruner et. al., eds., *Studies in Cognitive Growth: A Collaboration at the Center for Cognitive Studies* (New York: Wiley, 1966).

41 David R. Olson and Nancy Torrance, eds., *The Making of Literate Societies* (Malden, MA: Blackwell, 2001).

42 R.H. Tawney, *Religion and the Rise of Capitalism* (London: John Murray, 1926).

43 Gavin Menzies, *1434: The Year a Magnificent Chinese Fleet Sailed to Italy and Ignited the Renaissance* (New York: William Morrow [HarperCollins] 2008).

44 Harold Innis, *The Fur Trade in Canada: An Introduction to Canadian Economic History* (Toronto: University of Toronto Press, 1930), and *The Cod Fisheries: The History of an International Economy* (Toronto: University of Toronto Press, 1942).

45 Donald Creighton, *The Commercial Empire of the St. Lawrence* (Toronto: Ryerson Press, 1937).

46 United Nations, *Creative Economy: Report 2010, Creative Economy: A Feasible Development Option* (United Nations, UNCTAD, 2010. xxiii).

47 Paul Litt, *The Muses, the Masses, and the Massey Commission* (Toronto: University of Toronto Press, 1992).

48 Interviews with publishers, October and November 2008.

49 Statistics Canada, *Culture, Tourism and the Centre for Education Statistics*, cat. no. 87F0004X. See also Statistics Canada, "Book Publishing Industry," *The Daily*, July 10, 2008.

50 Rowland Lorimer, "A Canadian Social Studies for Canada," *The History Teacher* 16, no. 4 (1981), 45–55.

51 Rowland Lorimer, "Your Canadian Reader," *Lighthouse* 2, no. 3 (1978), 6–16.

CHAPTER 2

52 George L. Parker, *Beginnings of the Book Trade* (Toronto: University of Toronto Press, 1985), 13.
53 A description of some of the details of this activity is provided by Mary Lu MacDonald, "Subscription Publishing," in *History of the Book in Canada*, vol. 1, *Beginnings to 1840*, eds. Patricia L. Fleming, Gallichan Gilles, and Yvan Lamonde (Toronto: University of Toronto Press, 2004), 78–80.
54 Parker, *Beginnings of the Book Trade*, 53.
55 Ibid., 106.
56 Parker, *Beginnings of the Book Trade*, 54, citing Catharine Parr Traill's letters.
57 Ibid., 166
58 H.P. Gundy, "The Development of the Book Trade in Canada," in *Background Papers*, Ontario, Royal Commission on Book Publishing. (Toronto: Queen's Printer for Ontario, 1972), 1–38.
59 Ibid., 166.
60 Satu Repo, "From Pilgrim's Progress to Sesame Street: 125 years of Colonial Readers," in *The Politics of the Canadian Public School*, ed. George Martell (Toronto: James Lorimer, 1974), 120,121.
61 Gundy, "The Development of the Book Trade in Canada," 18. The publishing house was subsequently owned by the United Church of Canada and then sold to McGraw-Hill in 1970 to form the company McGraw-Hill Ryerson. This sale was a key event in establishing an independent Canadian-owned book publishing industry.
62 Parker, *Beginnings of the Book Trade*, 117.
63 Strictly speaking, the first of these might more accurately be called a resistance, since the Canadian government had yet to establish legitimate control over Red River.
64 A more detailed account of various dynamics of early book publishing in Canada has recently appeared in the form of the three-volume History of the Book project. Patricia L. Fleming, Gallichan Gilles, and Yvan Lamonde, eds., *History of the Book in Canada*, vol. 1, *Beginnings to 1840* (Toronto: University of Toronto Press, 2004); Yvan Lamonde, Patricia L. Fleming, Fiona A. Black, eds., *History of the Book in Canada*, vol. 2, *1840 to 1918* (Toronto: University of Toronto Press, 2005); Carole Gerson

and Jacques Michon, *History of the Book in Canada*, vol. 3, *1918 to 1980* (Toronto: University of Toronto Press, 2007).
65   Parker, *Beginnings of the Book Trade*, 185.
66   Ibid., 193.
67   Ibid., 193.
68   Gundy, "The Development of the Book Trade in Canada," 18, 19.
69   Ibid., 21.
70   Ibid.
71   Ibid.
72   The spread of books was helped considerably by Andrew Carnegie's matching grants to local libraries.
73   Gundy, "The Development of the Book Trade in Canada," 22.
74   Ibid., 25.
75   Ibid., 26–28.
76   Ibid., 29.
77   Urvashi Butalia, "English Textbook: Indian Publisher," *Media, Culture and Society*, 15 (1993): 217–32.
78   Philip Altbach and Edith Hishino, *International Book Publishing: An Encyclopedia* (New York: Garland, 1995). See section on Africa, 366–423.
79   There were notable nonparticipants in this dynamic of change, particularly in Asia and the USSR.
80   Canada, *Royal Commission on National Development in the Arts, Letters and Sciences* (Ottawa: King's Printer, 1952).
81   Consistent with what he was calling for in the evolution of Canada, Massey ascended to the vice-regal throne as the first Canadian-born Governor General of the country in 1952, one year after delivering his report. That he was appointed so soon after the report was delivered suggests that it was warmly received by the elites of Canada, both Liberal (under Louis St. Laurent) and Conservative (under John Diefenbaker). In 1958, Diefenbaker extended Massey's vice-regal appointment for one year.
82   Canada, *Royal Commission on National Development in the Arts, Letters and Sciences*, 377.
83   Ibid., 363.
84   Hilda Neatby, *So Little for the Mind* (Toronto: Clarke Irwin, 1954).
85   Rowland Lorimer and Patrick Keeney, "Defining the Curriculum: The Role of the Multinational Textbook in Canada," in *Language, Authority and Criticism: Readings on the School Textbook*, ed. Suzanne De Castell, Allan Luke, and Carmen Luke (London/Philadelphia: Falmer Press, 1988), 171–83.

86 Rowland Lorimer, *The Nation in the Schools: Wanted, a Canadian Education* (Toronto: OISE Press, 1984).
87 Gundy, "The Development of the Book Trade in Canada," 34.

CHAPTER 3

88 Roy MacSkimming, *The Perilous Trade* (Toronto: McClelland & Stewart, 2003), 172.
89 Ontario, Royal Commission on Book Publishing, *Canadian Publishers and Canadian Publishing* (Toronto: Queen's Printer, 1972).
90 These events were captured in the first and third interim reports of the commission. The commission's final report (Ontario, Royal Commission on Book Publishing, *Canadian Publishers and Canadian Publishing*) was submitted August 20, 1971, and was followed by a volume of very useful background papers — Ontario, Royal Commission on Book Publishing, *Background Papers* (Toronto: Queen's Printer, 1972) — in December 1972.
91 This generally accepted figure was originally developed by John Huenefeld, a U.S. book industry consultant, in whose newsletter, the *Huenefeld-PubWest Survey of Financial Operations* (2001): 125, it first appeared. See also John Huenefeld, *The Huenefeld Guide to Book Publishing* (Bedford, MA: The Huenefeld Company, 2001). As Talon Books editor and publisher Karl Siegler points out (in an email to the author, October 7, 2008), that amount would probably be higher now, given that booksellers are taking longer to pay the bills.
92 Ontario, *Canadian Publishers and Canadian Publishing*, 221.
93 Ibid., 237.
94 Ibid., 240–84.
95 "No matter how bad it is," some of them seemed to mutter under their breath.
96 Canada, Department of Canadian Heritage, *Publishing Measures* (Ottawa: DCH, 2004, 2005, 2006).
97 The majority of such studies have been commissioned by the Association of Canadian Publishers or Canadian Publishers Council. All are available to members, some to the public.
98 Peter Lougheed's Alberta government invested $8.3 million in the creation of learning materials with Canadian content, which were underused in schools in Alberta and across Canada.
99 Rowland Lorimer and Patrick Keeney, "Defining the curriculum: The role of the multinational textbook in Canada," in *Language, Authority and*

Criticism: Readings on the School Textbook, ed. Suzanne De Castell, Allan Luke, and Carmen Luke (London/Philadelphia: Falmer Press, 1988), 171–83.

100 As of July 2010, the Department of Canadian Heritage's Canadian Studies program still had a website, www.pch.gc.ca/pgm/pec-csp/info-eng.cfm. Program cancelled; site no longer available.

101 See, for example, Rowland Lorimer, *The Nation in the Schools: Wanted: a Canadian Education* (Toronto: OISE Press, 1984).

102 The fact that copyright rests with the publisher, rather than with the editors or authors, is a rough indication of their relative importance.

103 Robyn Matthew, "Entering a New Market: Oxford University Press Canada's Foray into French-as-a-Second-Language Publishing" (Master of Publishing project report, Simon Fraser University, 2007).

104 It is worth noting that educational publishers, especially K–12 publishers, generally pay authors a flat rate for their services rather than paying them royalties. This business arrangement reflects the substantial investment the publisher makes over and above that of paying the authors.

105 See the IDRC website at www.idrc.ca/en/ev-66174-201-1-DO_TOPIC.html.

106 Gérard Pelletier, *Federal aid for publishing* (text of speech presented February 11, 1972 in Montreal), *Quill & Quire* (March 1972), 12.

107 Department of Communications (DOC), news release, June 18, 1986, 3.

108 Nancy Duxbury, *The Economic, Political, and Social Contexts of English-Language Book Title Production in Canada, 1973–1996* (Ph.D. thesis, Simon Fraser University, 2000), 405.

109 Given this structure, were a publishing firm to have lower costs than average, or were it to be more successful in the market than average, it would still receive a subsidy based on the average deficit and hence be able to recover a greater percentage of its costs, even to the point of being able to make a profit (after grants).

110 Patricia Aldana, *Canadian Publishing: An Industrial Strategy for its Preservation and Development in the Eighties* (Toronto: Association of Canadian Publishers, 1980).

111 en.wikipedia.org/wiki/History_of_General_Motors.

112 Association of Canadian Publishers, *A Mid-Decade Assessment* (Toronto: ACP, 1985).

113 Roger Barnes and Rowland Lorimer, *Book Publishing: The Act* (Toronto: ACP, 1996).

114 Roger Barnes and Rowland Lorimer, *Book Purchasing in Canada: 1997 Survey*, commissioned report (Toronto: Canadian Publishers Council, 1998).
115 See the Canadian Publishers Council website, www.pubcouncil.ca/membership.php.
116 Innovative projects were intended to encourage publishers to explore alternative methods and markets, with the belief that doing so might make them more profitable. At this time, federal policy was still driven by the notion that government was a temporary partner of book publishers.
117 The irony of naming the restrictive-of-free-trade Baie-Comeau Agreement after the birthplace of Prime Minister Brian Mulroney was certainly appreciated by industry members, if not the prime minister himself. Mulroney was seen as a champion of free trade and of transforming the Foreign Investment Review Agency into a welcome mat for foreign investors — the reverse of its initial role, which was to keep foreign investment at bay; the Baie-Comeau Agreement, named for the place where the agreement was reached, worked to do exactly the opposite for the book publishing industry. Baie-Comeau was initially administered by Investment Canada, whose motto, coined by the prime minister, was "Canada is Open for Business." Eventually the contradiction became so problematic that administration of Baie-Comeau was handed over to the Department of Canadian Heritage.
118 In 1992, Communications Minister Perrin Beatty announced a climbdown from the "forced divestiture" measures of the Baie-Comeau Agreement. He announced that indirect acquisitions of Canadian-based book publishers that were already foreign owned could take place, if foreign investors would make commitments that were "of net benefit to Canada and to the Canadian-controlled sector." Investment Canada was to assess these undertakings on their merits. The benefits were to fall within one or more of the following categories:
   1. a commitment to the development of Canadian authors, such as undertaking joint publishing ventures with Canadian-controlled publishers;
   2. a commitment to the infrastructure of the Canadian book distribution system;
   3. access for Canadian-controlled publishers to the company's domestic or international marketing and distribution infrastructure; or
   4. a commitment to education and research in Canadian publishing.

119 Canada, Ministry of Communications, Information Services, "Masse Announces New Book Publishing Policy Development Program and Additional Funding for the Canada Council," press release, June 18, 1986, 14.

120 See Chapter 5 for a discussion of the current dynamics and structure of the program.

121 Roy MacSkimming, "Baie-Comeau Gets Sealed," *Quill & Quire* 58, no. 10 (1992), 16.

122 The links referred to in the title are the links internal to each sector that make book publishing or feature filmmaking viable: for example, ensuring distribution and retail display for Canadian productions.

123 Canada, *Vital Links: Canadian Cultural Industries* (Ottawa: Department of Communications, 1987), 17.

124 Ibid., 31.

125 Association of Canadian Publishers, *Book Publishing and Canadian Culture: A National Strategy for the 1990s* (Toronto: ACP, 1991).

126 Peat Marwick Consulting Group and Bill Roberts, *English-Language Book Publishing and Distribution in Canada: Issues and Trends, Final Report* (Ottawa: Department of Communications, October 1989).

127 Ibid., IV 17–18.

128 Paul Audley and Associates, *Book Publishing Policy: A Review of Background Information and Policy Options* (Ottawa: Department of Communications, 1990).

129 Ibid., 148.

130 Ibid., 149.

131 Many industry members and others have argued that the presence of the GST on books stabilizes funding for the book publishing industry, in that federal support programs are completely paid for by the GST alone (of all taxes), a fact that the industry can use in negotiating support with government.

132 Fox Jones, *Evaluation of the Book Publishing Industry Development Program: Economic Study* and *Financial and Market Impact Study* (Ottawa: Department of Communications, 1992).

133 Evidently, the bureaucrat who was the architect of these cuts was Peter Nicholson. Gloria Galloway, "Q&A. Peter Nicholson: The man Paul Martin chose to craft the cost-cutting 1995 budget reflects with Gloria Galloway on Ottawa's challenge this time around," *Globe and Mail*, March 1, 2010, A4.

134 Manning became leader of the official opposition in 1997 and remained in that position until March 2000.

135 Association of Canadian Publishers, *New Directions: Rethinking Public Policy for Canadian Books* (Toronto: ACP, 1995), 3.
136 ACP, *Setting Priorities for Federal Book Publishing Policy* (Toronto: ACP, 1997).
137 Ibid., 2.
138 Roy MacSkimming, *Making Policy for Canadian Publishing* (Toronto: ACP, 2002).

CHAPTER 4

139 John English, *Citizen of the World: The Life of Pierre Elliott Trudeau*, vol. 1, *1919–1968* (Toronto: Knopf, 2006), and *Just Watch Me: The Life of Pierre Trudeau*, vol. 2, *1968–2000* (Toronto: Knopf, 2009); Max Nemmi and Monique Nemmi, *Young Trudeau: Son of Quebec, Father of Canada*, trans. William Johnson (Toronto: McClelland & Stewart, 2006).
140 To be strictly accurate, those who were first engaged by Trudeau were born just ahead of the baby boom. As such, they had seen society being transformed just behind them and better understood the ongoing changes than their younger brothers and sisters.
141 Howard White, *Books and Water* (Video portrait of Harbour Publishing, 1992).
142 See, for example, James Laxer and Robert Laxer, *The Liberal Idea of Canada: Pierre Trudeau and the Question of Canada's Survival* (Toronto: James Lorimer, 1977); Kari Levitt, *Silent Surrender: The Multinational Corporation in Canada* (Toronto: Macmillan, 1970); Robin Mathews, *The Struggle for Canadian Universities: A Dossier* (Toronto: New Press, 1969); Mel Hurtig, *The Betrayal of Canada* (Toronto: Stoddart, 1991).
143 ACP, *New Directions: Rethinking Public Policy for Canadian Books*.
144 Ontario, *The Business of Culture: The Report of the Advisory Committee on a Cultural Industries Sectoral Strategy (ACCISS)* (Toronto: Queen's Printer, 1994).
145 Ontario Ministry of Culture and Communications, Cultural Industries and Agencies Branch, *Canadian Book Publishing in Ontario* (Toronto: Author, 1989).
146 Ibid.
147 Ontario, *The Business of Culture*.
148 In this program, investors were given tax credits for investing in film. The trouble was not with the intent but with the way in which deals were structured. The deal makers made lots of money because they were paid up front for their efforts, but many of the films were junk, and few investors ever realized any long-term profit. Book publishing and sound recording have attracted less grandiose schemes.

149 Ontario, *The Business of Culture*, 11.
150 Ibid., 15.
151 Ibid., 16.
152 Mike Harris's "common sense" Tories took over from the NDP. Among other things, they cancelled the publishers' long-standing loan guarantee program established by the Ontario Royal Commission.
153 This discussion began in the Association of Canadian Publishers' early 1990s assessment of the industry and its needs, *Book Publishing and Canadian Culture: A National Strategy for the 1990s* (Toronto: ACP, 1991), and was re-stated in the 1995 ACP paper *New Directions: Rethinking Public Policy for Canadian Books*, discussed in Chapter 3.
154 Arthur Donner and Fred Lazar, *The Competitive Challenges Facing Book Publishers in Canada* (Ottawa: Department of Canadian Heritage, Cultural Industries Branch, 2000).
155 Ibid., 6.
156 Arthur Donner and Fred Lazar, *The Canadian Book Publishing Industry: Competitive Challenges and the Need for Restructuring* (Ottawa: Department of Canadian Heritage, Cultural Industries Branch, 2000).
157 ACP, *Book Publishing and Canadian Culture: A National Strategy for the 1990s*.
158 It is interesting to note that, of the four largest Canadian-owned publishing firms that may have benefitted from consolidation, only one remains active and independent in 2010. General Publishing declared bankruptcy in 2002; Key Porter was sold to distributor H.B. Fenn in 2004; 75 percent of McClelland & Stewart was donated to the University of Toronto in 2000, with 25 percent sold to Random House Canada, a move that resulted in a full sale to Random House for zero dollars in 2012 (see Chapter 8 for a review of this sale). Only Douglas & McIntyre remains an untied, independent publisher, purchased in 2007 by outside investors new to book publishing.
159 Most recently, Lazar has turned his attention to the airline industry, commenting on the value of the Boeing Dreamliner to Air Canada. www.theglobeandmail.com/globe-investor/for-air-canada-787-a-game-changer/article1962224/ Accessed March 30, 2011.
160 ACP, *Book Publishing and Canadian Culture: A National Strategy for the 1990s*, 1.
161 Foreign Affairs and International Trade Canada, *New Strategies for Culture and Trade: Canadian Culture in a Global World*, February 1999, www.international.gc.ca/trade-agreements-accords-commerciaux/fo/canculture.aspx?lang=en.
162 Ibid.

163 UNESCO, *Many Voices: One World: Report of the International Commission on Communication Problems* (MacBride Commission) (Paris: Unipub, 1980).
164 Peter Grant and Chris Wood, *Blockbusters and Trade Wars* (Vancouver: Douglas & McIntyre, 2004).
165 Richard E. Caves, *Creative Industries: Contracts between Art and Commerce* (Cambridge, MA: Harvard University Press, 2000).
166 Grant and Wood, *Blockbusters and Trade Wars*, 56. Grant and Wood provide an important and succinct definition of "public good" as a technical term used by economists to describe a good "whose production cost is independent of the number of people who consume it."
167 See Colin Hoskins and Rolf Mirus, "Reasons for the U.S. dominance of the international trade in television programmes," *Media, Culture and Society*, 10 (1988), 499–515; and Colin Hoskins, Stuart McFadyen, and Adam Finn, "Cultural Industries from an Economic/Business Research Perspective," *Canadian Journal of Communication*, 25, 41 (2000), 127–44; www.cjc-online.ca/index.php/journal/article/view/1146/1065.
168 Grant and Wood, *Blockbusters and Trade Wars*, 315.
169 See, for example, the papers noted, all from *Canadian Journal of Communication*, 27, 2 (2002), and others in the same volume. Greg Baeker, "Sharpening the Lens: Recent Research on Cultural Policy, Cultural Diversity and Social Cohesion," 179–96; John Hannigan, "Culture, Globalization and Social Cohesion: Towards a Deterritorialized Global Fluids model," 277–88; Jane Jenson, "Identifying the Links: Social Cohesion and Culture," 141–52.
170 World Trade Organization, "Canada — Certain Measures Concerning Periodicals," Dispute DS31, 1996, 1997.
171 This is understandable because, ultimately, all manner of producers of branded products can claim distinctiveness and, hence, a need for protection.
172 Grant and Wood, *Blockbusters and Trade Wars*, 267, 366.
173 Ibid., 268. While Grant and Wood identify and discuss the content variable, they do not discuss the producer to retailer market.
174 An analysis of the merchandising perspectives of independent bookstores can be found in Rowland Lorimer and Roger Barnes, *Merchandising in Independent Bookstores* (Toronto: Association of Canadian Publishers, 1998).
175 Grant and Wood, *Blockbusters and Trade Wars*, 268.
176 Rowland Lorimer and Roger Barnes, "Book Reading, Purchasing, Marketing, and Title Production," in *Book Publishing* 1, ed. Rowland Lorimer, John Maxwell, and Jillian Shoichet (Vancouver: CCSP Press,

2005), 220–56.
177 Grant and Wood, *Blockbusters and Trade Wars*, 269.
178 E. Parker and A. Furnham, "Does Sex Sell? The Effect of Sexual Programme Content on the Recall of Sexual and Non-Sexual Advertisements," *Applied Cognitive Psychology* (2007): 1217–28.
179 Grant and Wood, *Blockbusters and Trade Wars*, Chapters 7 to 14, 139–326.
180 Section 3 of the Broadcasting Act presents a full set of reasons for public ownership. See *The Broadcasting Act,* SC 1991, c. 11, laws-lois.justice.gc.ca/eng/B-9.01/.
181 *Investment Canada Act, RSC,* 1985, c. 28 (1st Supp.), ss. 20–24, laws-lois.justice.gc.ca/eng/I-21.8.
182 It is interesting that Costco has managed to circumvent the intent of this Investment Canada legislation. As noted in the introduction to this book, Costco now accounts for a greater market share than all independent bookstores. Although foreign corporations are not allowed to set up book businesses, including bookstores, Costco was able to get around this rule because its business is not primarily bookselling.
183 Grant and Wood, *Blockbusters and Trade Wars*, 315.
184 The text of the declaration is available at unesdoc.unesco.org/images/0012/001271/127160m.pdf or through the UNESCO portal at portal.unesco.org (under "Resources" and then "Declarations").
185 The text of the convention is also available through the UNESCO portal under "Resources" and then "Conventions."
186 Garry Neil, "The Convention as a response to the cultural challenges of economic globalisation," in *UNESCO's Convention on the Protection and Promotion of the Diversity of Cultural Expressions: Making It Work*, ed. Nina Obuljen and Joost Smiers (Zagreb: Institute for International Relations, 2006), 39–70, www.culturelink.org/publics/joint/diversity01/Obuljen_Unesco_Diversity.pdf.
187 Pierre Bourdieu, "The Forms of Capital," in *Handbook of Theory and Research for the Sociology of Education*, ed. John G. Richardson, 241–58 (New York: Greenwood Press, 1986), www.marxists.org/reference/subject/philosophy/works/fr/bourdieu-forms-capital.htm.
188 OECD, *The Well-Being of Nations: The Role of Human and Social Capital* (Paris: OECD Centre for Educational Research and Innovation, 2001), 41.
189 Robert Putnam, *Making Democracy Work: Civic Traditions in Modern Italy* (Princeton: Princeton University Press, 1993).
190 Robert Putnam, "Bowling Alone: America's Declining Social Capital," *Journal of Democracy* 6, 1 (1995), 65–78.

191 John Helliwell, *Globalization and Well-Being* (Vancouver: UBC Press, 2002).

192 Canada, Standing Committee on Cultural Heritage, *The Challenge of Change: A Consideration of the Canadian Book Industry* (Ottawa: Government of Canada, 2000), www2.parl.gc.ca/HousePublications/Publication.aspx?DocId=1031737&Language=E&Mode=1&Parl=36&Ses=2.

193 Ibid., 2.

194 Ibid., 4.

195 As far as I can determine, the provenance of this opportunity can be traced to the initiative of Allan Clarke, at the time head of book publishing support programs in the Department of Canadian Heritage.

196 Heather MacLean, "The Supply Chain Initiative: The Inception and Implementation of a New Funding Initiative for the Department of Canadian Heritage" (Master of Publishing program project report, Simon Fraser University, 2009).

197 Marc Laberge, "The Canadian Book Industry Supply Chain Initiative Business Plan: Prepared for the Canadian Book Industry Supply Chain Initiative Steering Committee" (Ottawa: Department of Canadian Heritage, 2002), 4.

198 Both figures are to be found in Divine Whittman-Hart, "Canadian Book Industry: Transition to the New Economy," slide presentation to the Association of Canadian Publishers and Canadian Publishers Council, 2001.

CHAPTER 5

199 In an article entitled "Why's everybody always picking on us?" (*Globe and Mail*, February 14, 2009, F14), Cormorant publisher Marc Côté argues that schools and the media together create a cultural environment that suppresses demand for Canadian-authored books.

200 The figure of 20 percent was provided by Kevin Hanson, CEO of Simon & Schuster, a major distributor who would know as much as anyone. A 20 percent increase from the activities of foreign distributors translates into an 8 percent market share reduction for title-originating Canadian publishers. (120/50 = 41.7.)

201 Statistics Canada, Service Industries Division, *Book Publishers*, 2006, cat. no. 87F0004X (Ottawa: Statistics Canada, 2009), www.statcan.gc.ca/pub/87f0004x/87f0004x2008001-eng.pdf.

202 Ibid.

203 Canadian Heritage, *Publishing Measures* (Ottawa: Department of Canadian Heritage, 2009).

204 This estimate is arrived at by dividing total trade sales by the number of trade titles published plus the number of trade titles reprinted. The inclusion of the number of reprinted trade titles as a stand-in for all backlist trade titles estimates overall backlist trade sales at 26 percent of all trade sales. Excluding backlist trade sales altogether brings average per-title trade sales up to $19,258 — not a great deal more encouraging.

205 It is important to note the differences in the data systems used in 2004 and 2006. Using the data from Table 5.1 to create a parallel table for 2006 yields changes in sales by foreign-controlled firms of minus 4 percent, minus 5 percent, minus 14 percent and plus 20 percent for education; children's; other trade, all formats; and scholarly, reference, professional, and technical, respectively. These differences are far beyond the changes that might reasonably be expected over two years. They point to the impact of excluding the distributors and, to some degree, undermine the confidence one can have in Statistics Canada's monitoring of the book publishing industry.

206 *Book Importation Regulations*, SOR/99-324, are part of the *Copyright Act*, RSC 1985, c. C-42.

207 The mechanism for establishing this right is to forward a listing to a catalogue of books in print published by Bowker, which is almost universally used by booksellers.

208 Schedule C, "An Act imposing Duties of Customs, with the Tariff of Duties payable under it," which appears to date back to 1867, required customs officers to check each title crossing the border against a list of titles for which Canadian publishers held Canadian rights. While today this could be computerized, even when the import regulations were appended to the Copyright Act, this system of checking would have been a chore.

209 Manjunath Pendakur, *Canadian Dreams and American Control: The Political Economy of the Canadian Film Industry* (Detroit: Wayne State University Press, 1990).

210 Canadian Heritage, "The Canadian Film Industry and Investment Canada," text of fact sheet issued by Communications Canada, November 6, 2008, www.pch.gc.ca/invest/film-eng.cfm.

211 All figures come from *Public Lending Right Commission Annual Report, 2010–2011* which can be accessed through the Commission's website, www.plr-dpp.ca.

212 The PLR survey to determine the foundation for payments to authors no longer samples university and college libraries. This exclusion is based on the judgement of the PLR Commission that these are not public libraries, even though the vast majority exist in public institutions and are accessible to members of the public.
213 John Ibbitson, "Canada risks being shut out of Pacific trade pact, New Zealand Prime Minister warns," *Globe and Mail*, April 15, 2010.
214 BookNet Canada website, www.booknetcanada.ca.
215 Divine Whittman-Hart, "Canadian Book Industry: Transition to the New Economy," Slide Presentation to the Association of Canadian Publishers and Canadian Publishers Council, 2001.
216 BookManager website, www.bookmanager.ca.
217 Department of Canadian Heritage, "Book Publishing Industry Development Program," www.pch.gc.ca/pgm/padie-bpidp/index-eng.cfm.
218 Department of Canadian Heritage, "Canada Book Fund (formerly Book Publishing Industry Development Program)," March 9, 2010, www.pch.gc.ca/eng/1268182505843/1268255450528.
219 Ibid.
220 Canadian Heritage website, www.pch.gc.ca.
221 Department of Canadian Heritage, "Support for Organizations and Associations," www.pch.gc.ca/pgm/flc-cbf/soa/guide-eng.cfm.
222 Ibid.
223 Department of Canadian Heritage, "Livres Canada Books: Foreign Rights Marketing Assistance Program," www.livrescanadabooks.com/en/funding/frmap/.
224 Department of Canadian Heritage, "Livres Canada Books: Export Marketing Assistance Program," www.livrescanadabooks.com/en/funding/emap/. Site no longer available.
225 Association for the Export of Canadian Books website, www.aecb.org.
226 Department of Canadian Heritage, "Livres Canada Books: Access Market Intelligence," www.livrescanadabooks.com/en/market_intelligence/.
227 Department of Canadian Heritage, Book Policy and Programs Division, *Printed Matters* (Ottawa: DCH, 2005).
228 See Department of Canadian Heritage, "Intersections: Reading the cultural landscape," www.pch.gc.ca/pc-ch/org/sectr/ac-ca/pblctns/anl-rpt/2007-2008/index-eng.cfm.
229 Canada Council for the Arts, "Book Publishing Support: Block Grants," www.canadacouncil.ca/grants/writing/ap127723094273982142.htm.

230 Ibid.
231 This calculation was based on the total block grant funding given to University of British Columbia Press, McGill-Queens University Press, and the University of Toronto Press (drawn from www.canadacouncil.ca/grants/recipients/) divided by the number of titles submitted (email correspondence from Canada Council program officer). The number is approximate because the calculation is somewhat complex.
232 Foreign Affairs and International Trade Canada website, www.international.gc.ca. The Harper government's cancellation of programs supporting cultural diplomacy and international travel for authors and other creators is, unfortunately, shortsighted and ideologically narrow. I expect that Canadians will find a way to reinstate such programs in pursuit of a respect for creativity and plurality within Canada and the advantages of cultural representation and interaction at the international level.
233 Ibid.
234 See Department of Canadian Heritage, "Canada Book Fund, Business development support," www.pch.gc.ca/pgm/flc-cbf/sae-sfp/guide/103-eng.cfm and "Canada Book Fund, 2010–2011 — Application Guide," www.pch.gc.ca/pgm/flc-cbf/soa/guide-eng.cfm.
235 British Columbia, Ministry of Finance, "British Columbia Book Publishing Tax Credit," *Tax Bulletin* CIT 008 (2009), www.sbr.gov.bc.ca/documents_library/bulletins/cit_008.pdf.
236 Ibid.
237 British Columbia Arts Council website, www.bcartscouncil.ca.
238 Ontario Media Development Corporation, "Ontario Book Publishing Tax Credit," www.omdc.on.ca/Page3397.aspx. Note that in 2000, consultants Donner and Lazar recommended that the federal government implement a 30 percent tax credit. Arthur Donner and Fred Lazar, *The Canadian Book Publishing Industry: Competitive Challenges and the Need for Restructuring* (Ottawa: Department of Canadian Heritage, Cultural Industries Branch, 2000). Had Donner and Lazar's suggestion been instituted, Ontario book publishers would have been receiving a 60 percent tax credit for their Ontario spending that, combined with all other forms of support, could have reduced title investment costs to nearly zero. Little wonder the measure was rejected.
239 Ontario Arts Council website, www.arts.on.ca.
240 Castledale Inc. and Nordicity, *A Strategic Study for the Book Publishing Industry in Ontario* (Toronto: Ontario Media Development Corporation, 2008).

241 Ibid., 7.
242 Alberta Foundation for the Arts, *Moving Arts: 2006/2007 Review*, 16, www.affta.ab.ca/resources/AFA-06-07-Year-in-review.pdf.
243 Book Publishers Association of Alberta, *Alberta Book Publishers Operating Support Initiative*, 2009, 2. Cited in Kelsey Everton, "Alberta Bound: Support for Book Publishing in Alberta" (unpublished student paper, Master of Publishing program, Simon Fraser University, 2010), 13.
244 Government of Manitoba, Manitoba Culture, Heritage and Tourism, "Arts Branch: Services and Programs," www.gov.mb.ca/chc/arts/.
245 Newfoundland and Labrador, Department of Tourism, Culture and Recreation, "Newfoundland and Labrador Publishers Assistance Program," www.tcr.gov.nl.ca/tcr/artsculture/cedp/publishers_assistance_program.html; Newfoundland and Labrador, Department of Tourism, Culture and Recreation, "Publishers' Assistance Program Provides $200,000 Investment," news release, August 4, 2009, www.releases.gov.nl.ca/releases/2009/tcr/0804n02.htm.
246 I am aware of counter-examples, but they still do not constitute an industry trend. Lone Pine has achieved continuous expansion by repeating a successful funding formula. Dundurn is growing by means of acquisitions, as is Heritage House. The distributor H.B. Fenn purchased Key Porter and was running it as an imprint. Douglas & McIntyre's new owner, Mark Scott, has now acquired New Society Publishers after acquiring D&M and Greystone. But by far the majority of firms would not report having plans for growth because they lack the capital for growth. In the recent past, Jack Stoddart tended to acquire 10 percent of firms that he distributed, but that all fell apart with the demise of General Publishing and General Distribution Services.
247 *Vital Links*, 31.
248 As noted earlier, the cost of borrowing exceeds profit levels, leaving publishers in the position of not being able to borrow.
249 $(.25 \times \$30{,}000 \times 4) + (.21 \times \$30{,}000 \times 12) = \$30{,}000 + \$75{,}600$.
250 Peter Milroy, *Canadian Books in Review: A quantitative analysis of coverage given to books in the weekend review sections of selected Canadian newspapers* (Vancouver: CCSP Research Report #1, 1994).
251 Rowland Lorimer, "The future of English-language book publishing," in *Beyond Quebec: Taking Stock of Canada*, ed. Kenneth McRoberts (Montreal: McGill-Queen's University Press, 1995), 202–17.
252 Pat Cavill, *Collections: How and Why Public Libraries Select and Buy Their Canadian Books* (Toronto: Association of Canadian Publishers, 1998).

253 See, for example, Graham Orpwood, "Canadian Content in School Texts and the Changing Goals of Education," *Education Canada* 20, 1 (1980): 16–19.

254 Jean Baird, *Canadian Literature in High Schools: A Research Study* (Toronto: Writer's Trust, 2001), ii.

255 It appears that "moral fibre" refers to the ability of young people to resist temptation in its various forms: sex, drugs, certain linguistic forms, and perhaps even rock and roll. One reason why moral fibre might be weaker in Canadian literature is that the Canadian work is largely made up of literary trade titles, the "raw" artistic output of authors, whereas there is a well-developed, commercial, education-market-targeted juvenile literature in the U.S. that parallels textbooks in avoiding content that might be objectionable to just about any group.

256 Ibid., v.

257 Nicole Tomlinson, "Canadian books now mandatory in high schools," *Vancouver Sun*, July 5, 2008.

258 Rebecca Wigod, "New rules coming to get Canadian literature in schools," *Vancouver Sun*, July 26, 2008.

259 Rowland Lorimer and Stephen Osborne, *A Financial, Circulation and Publications Analysis of 16 Canadian Literary and 17 Canadian Arts Magazines* (Ottawa: Canada Council, 2000).

260 LSM Consulting, *Book Publishers: Training Gaps Analysis in Canada 2006* (Ottawa: Cultural Human Resources Council, 2006), www.culturalhrc.ca/research/CHRC_Book_Publisher_TGA-en.pdf.

261 Ibid., 3.

262 Ibid.

263 Canada Council for the Arts, "About the GGs," www.canadacouncil.ca/prizes/ggla/qz128686615675969592.htm.

CHAPTER 6

264 The figures reported here have been compiled from a wide variety of sources, including research undertaken for the British Columbia, Saskatchewan, and federal governments, as well as other published sources such as Thomas Woll's *Publishing for Profit* (Chicago: Chicago Review Press, 2006).

265 This may or may not be true. According to Halpape, "as of 2007, BookNet Canada Sales Data tracks the sales of about 675,000 unique titles available to the Canadian retail market. The German *Verzeichnis Lieferbarer Bücher VLB*, a research tool for the German book trade,

lists around 1,200,000 available titles." Jan Halpape, "Subsidies versus a Net Price System: A Comparison of the Canadian and German Book Industry Support Systems" (Master of Publishing student paper, Simon Fraser University, 2008). Halpape drew the BookNet Canada figure from Turner-Riggs, *The Book Retail Sector in Canada* (Vancouver: Turner-Riggs, 2007), 29, and the German figure from MVB Marketing-und Verlagsservice des Buchhandels GmbH, "Verzeichnis Lieferbarer Bücher (VLB)," www.vlb.de, accessed December 11, 2008. The claim is also doubtful because there is no money to be made importing designated titles in small numbers. In fact, some publisher/distributors, such as Raincoast Books, have been reviewing their agency operations and cutting back their representation of foreign publishers where price combined with sales makes that representation uneconomic. (Personal communication with confidential Raincoast source.)

266 Rowland Lorimer, *Build It and They Will Flow: Book Distribution in English Canada* (research report, Department of Canadian Heritage, Ottawa, 1997).

267 Just to give a sense of the size of the market, the member nations of the Commonwealth are

> Antigua and Barbuda, Australia, the Bahamas, Bangladesh, Barbados, Belize, Botswana, Brunei Darussalam, Cameroon, Canada, Cyprus, Dominica, Fiji Islands, the Gambia, Ghana, Grenada, Guyana, India, Jamaica, Kenya, Kiribati, Lesotho, Malawi, Malaysia, Maldives, Malta, Mauritius, Mozambique, Namibia, Nauru, New Zealand, Nigeria, Pakistan, Papua New Guinea, St. Kitts and Nevis, St. Lucia, St. Vincent and the Grenadines, Samoa, Seychelles, Sierra Leone, Singapore, Solomon Islands, South Africa, Sri Lanka, Swaziland, Tonga, Trinidad and Tobago, Tuvalu, Uganda, United Kingdom, United Republic of Tanzania, Vanuatu, Zambia.

268 Legal deposit figures not only include vanity and non-commercial titles but are also processing figures, as opposed to an accurate count of titles published in a year. Interestingly, as John Stegenga of Library and Archives Canada pointed out, there was a distinct dip in the number of titles processed from 2004 to 2007. Stegenga cautioned that the dip might be an anomaly of the process rather than a dip in publishing activity. In summary, in 2004 there were 22,972 titles published; in 2005 there were 20,670; in 2006, 18,214; and in 2007, 25,240. (Personal correspondence.)

269 Roger Barnes and Rowland Lorimer, *Book Purchasing: The Act* (Toronto:

ACP, 1996); Roger Barnes and Rowland Lorimer, *Book Purchasing in Canada 1997 Survey* (Toronto: Canadian Publishers Council, 1998).

270 Rowland Lorimer and Lindsay Lynch, *The Latest Canadian National Reading Study, 2005: Publishers' Analysis* (Vancouver: Canadian Centre for Studies in Publishing, 2007).

271 Rowland Lorimer and Roger Barnes, "Book Reading, Purchasing, Marketing, and Title Production," in *Book Publishing 1*, ed. Rowland Lorimer, John Maxwell, and Jillian Shoichet (Vancouver: CCSP Press, 2005), 220–56.

272 Angela Roberge, *Adolescents' Reading in the Lower Mainland (of Vancouver)*. (Honours project, Simon Fraser University, 1995.)

273 Créatec+, *Reading and Buying Books for Pleasure* (Ottawa: Department of Canadian Heritage, 2006).

274 Lorimer and Lynch, *Latest Canadian National Reading Study, 2005*.

275 Topic and type of book are combined into one category as a way of creating equivalent categories for non-fiction and fiction without doubling up on choice factors. That is to say, purchasers of novels are usually looking for a novel. To some degree, whether they buy a novel depends on the content, but form and content are inseparable. Separate categories of topic and type of book would probably result in less useful results.

276 Respectively, James Lorimer and Susan Shaw, *Book Reading in Canada* (Toronto: Association of Canadian Publishers, 1981); and Nancy Duxbury, *The Reading and Purchasing Public: The English-Canadian Trade Book Market in Canada* (Toronto: Association of Canadian Publishers, 1995).

277 Lorimer and Lynch, *Latest Canadian National Reading Study, 2005*, 19.

278 Interview with author, undated.

279 Rowland Lorimer and Roger Barnes, "Book Reading, Purchasing, Marketing, and Title Production," in *Book Publishing 1*, ed. Rowland Lorimer, John Maxwell, and Jillian Shoichet (Vancouver: CCSP Press, 2005), 220–56.

280 Barnes and Lorimer, *Book Purchasing: The Act*; Barnes and Lorimer, *Book Purchasing in Canada 1997 Survey*; Roger Barnes and Rowland Lorimer, *Merchandising in Independent Bookstores* (Toronto: Association of Canadian Publishers, 1998); Lorimer and Lynch, *Latest Canadian National Reading Study*.

281 Personal conversation with the author at the lunch celebration of the 25th anniversary of Douglas & McIntyre, 1995.

282 Roger Barnes and Rowland Lorimer, *Children's Book Purchasing* (Toronto: Canadian Publishers Council, 1999).
283 Ibid., 6.
284 Rowland Lorimer and Roger Barnes, "Book Reading, Purchasing, Marketing, and Title Production," in *Book Publishing 1*, ed. Rowland Lorimer, John Maxwell, and Jillian Shoichet (Vancouver: CCSP Press, 2005), 220–56.
285 Whereas guerrilla marketing is generally defined as unconventional marketing intended to get maximum results from minimal resources, it is often manifested through giveaways to high profile members of target markets, who may then give away minor promotional gear, such as T-shirts, to friends and acquaintances. In the book world, using guerilla marketing for certain authors, say Douglas Coupland, could work well.
286 The emphasis of Ms. McArthur's publishing program is on mainstream U.K. titles. It would appear that this is the reason her sales through Chapters/Indigo are high.
287 Conversation at ECW Press, March 25, 2010.
288 Erin Williams, *The Chapters Effect on British Columbia-Based Literary Publishers* (Master of Publishing project report, Simon Fraser University, 2006).
289 Chapters/Indigo: Our Company: Management. www.chapters.indigo.ca/our-company/management.
290 Competition Bureau, "Competition Bureau Files Application with Consent of Indigo and Chapters," April 20, 2002, www.cb-bc.gc.ca/eic/site/cb-bc.nsf/eng/00802.html.
291 James Adams, "Publishers' Plans Anger Bookstores," *Globe and Mail*, January 11, 2007, R1; James Adams, "Bookstores reach deal," *Globe and Mail*, January 13, 2007, R9.
292 The comment was made at the final project presentation of the 2007 Book Publishing Immersion Workshop, Simon Fraser University.
293 Caroline Skelton, "The Wal-Mart level of excellence," Quillblog/ *Quill & Quire*, September 2, 2005, www.quillandquire.com/blog/index.php/2005/09/02/the-wal-mart-level-of-excellence/.
294 Chris Anderson, *The Long Tail: Why the Future of Business is Selling Less of More* (New York: Hyperion, 2006).
295 Advertisement in *Globe and Mail*, June 12, 2010, F11.
296 Marina Strauss, "Nine Questions for Heather Reisman," *Globe and Mail*, May 27, 2010, www.theglobeandmail.com/report-on-business/rob-magazine/nine-questions-for-heather-reisman.
297 Review of the Revised Foreign Investment Policy in Book Publishing

and Distribution. www.pch.gc.ca/eng/1272486502392/1268255450528. Accessed March 2011.
298. Matt Hartley, "New chapter for Kobo as firm sold to Japan's Rakuten," *Financial Post. FP Tech Desk.* November 8, 2011. business.financialpost.com/2011/11/08/new-chapter-for-kobo-as-firm-sold-to-japans-rakuten/.
299. Marina Strauss, "Indigo aims to shake the dust out of its book bins," *Globe and Mail,* November 24, 2011, B3.
300. Marina Strauss, "Less Tolstoy, more toys," *Globe and Mail,* April 9, 2011, B6 and B7.
301. Publisher presentation at SFU's Summer Publishing workshops.
302. Marina Strauss, "Less Tolstoy, more toys." *Globe and Mail,* April 9, 2011, B6.
303. Howard Schultz and Joanne Gordon, *Onward: How Starbucks Fought for Its Life without Losing Its Soul* (Emmaus, PA: Rodale Books 2011).
304. Marketing Department, Indigo/Chapters, *Globe and Mail,* April 16, 2011, R15.
305. Rowland Lorimer and Roger Barnes, "Book Reading, Purchasing, Marketing, and Title Production," in *Book Publishing 1,* eds. Rowland Lorimer, John Maxwell, and Jillian Shoichet (Vancouver: CCSP Press, 2005), 220–56.
306. Undated interview with Harbour Publishing, spring 2011.
307. Indigo Books & Music Inc. 2011. *First Quarter Report* "EBITDA, defined as earnings before interest, taxes, depreciation and amortization decreased $19.1 million to a loss of $19.9 million for the 13-week period ended July 2, 2011, compared to a loss of $0.8 million for the 13-week period ended July 3 2010," (p. 5) www.chapters.indigo.ca/investor-relations/corporate-documents/. Accessed August 2011.
308. Stuart Woods, "Updated: H.B. Fenn initiates bankruptcy proceedings," *Quill & Quire,* February 3, 2011. www.quillandquire.com/blog/index.php/2011/02/03/h-b-fenn-initiates-bankruptcy-proceedings/.
309. Quill & Quire, *The Book Trade in Canada, 2008 Edition* (Toronto: Quill & Quire, 2008).
310. Richard Gu, *The Establishment of Bertelsmann in China* (Master of Publishing project report, Simon Fraser University, 2006). An excerpt from this report was published as Qianqiao Gu, "Bertelsmann in China: Low Profile, Patient Growth," *Logos* (2007), 173–81. Bertelsmann has since abandoned its operations in China.
311. McClelland & Stewart appears to continue on a long but not-so-winding road to oblivion as an independent, Canadian-owned firm. Since the heralded appointment of CEO Doug Pepper, under the 75 percent ownership of the University of Toronto and 25 percent

ownership of Random House Canada/Bertelsmann, the firm has distinguished itself by letting go of senior staff. By 2010 the firm was down to four editors and two publishers, which makes it smaller than Vancouver's D&M. The absence of a real owner/publisher whose veins run with books is apparent. This footnote was written before the 2012 sale of M&S to Random House.

312 The fourth firm, Simon & Schuster, has a significant market share in Canadian trade publishing, but as a result of past agreements it is a distributor in Canada rather than a publishing company. This means that the CEO, Kevin Hanson, arranges the publication of a few titles by Canadian authors out of the New York office, rather an ironic situation.

313 Before she stepped down from her position as president of Penguin Canada, Cynthia Good made this clear on a number of occasions.

314 Roy MacSkimming, *The Perilous Trade: Book Publishing in Canada, 1946–2006*, updated edition (Toronto: McClelland & Stewart, 2007), 406.

315 Earning out an advance means that the book sells enough copies that the royalties due from sales equal or exceed the advance given to the author.

316 "Remove this barrier to books: Readers, though bookstores, should be able to buy at a fair price, without a largely hidden subsidy to Canadian-based publishers," *Globe and Mail*, May 11, 2010, www.theglobeandmail.com/news/opinions/editorials/remove-this-barrier-to-books/article1563936/.

317 Susan Juby, "Editor to Author: Some Personal Reflections on Getting Published," in *Publishing Studies: Book Publishing 1*, ed. Rowland Lorimer, John Maxwell, and Jillian Shoichet (Vancouver: CCSP Press, 2005), 16–42.

318 James Adams, "$2.5-million voyage," *Vancouver Sun*, August 9, 2008, R1.

319 However, Canadian publishers are better at rights sales than they used to be, thanks to the programs provided by the Association for the Export of Canadian Books.

320 MacSkimming, *Perilous Trade*, 405, 406.

321 Ibid.

322 Williams, *Chapters Effect*.

323 Ibid., 68

324 Scott Anderson, "Heritage May Help GDS Clients Through Cash Crunch: But at Least One Publishing Executive Is Calling for a Longer-term Solution," *Quill & Quire* 67, no. 4 (April 2001), 4.

325 Williams, *Chapters Effect*, 69.

326 Rowland Lorimer, *Current Disruptions in Retail and Distribution and the*

State of British Columbia Book Publishers (Victoria: Government of British Columbia, 2001).
327 Rolf Maurer, "Brought to the Brink: It's time for all Canadians to rethink the role of literature in our lives," *Quill & Quire* 77, no. 7 (September 2011), 14.
328 Cross River Publishing Consultants, *PMA White Papers: Book Industry Returns: An Analysis of the Problem; Opportunities for Improvement; With a Focus on Independent Publishers* (Manhattan Beach, CA: Publishers Marketing Association, 2001).
329 Canada, 40th Parliament, 3rd Session, Legislative Committee on Bill C-32, Thursday March 24, 2011, Presentation by the Council of Ministers of Education. www2.parl.gc.ca/HousePublications/Publication.aspx?DocId=5071024&Language=E&Mode=1&Parl=40&Ses=3#Int-3823706.

CHAPTER 7

330 Alberto Manguel, *A History of Reading* (Toronto: Vintage, 1998).
331 Canadian Research Knowledge Network, "History," www.crkn.ca/about/history.
332 Mike Shatzkin, who writes on digital publishing, confirms that by 2008, only about one-third of all digital book sales were in e-book formats, as opposed to PDF files readable on a personal computer. "Digital publishing in the U.S.: Driving the industry to vertical niches?" *Logos* 19, 2 (2008): 56–60.
333 Authors Guild, "Authors Guild v. Google Settlement Resources Page," www.authorsguild.org/advocacy/articles/settlement-resources.html, provides information on the legal settlement.
334 Jenna Newman, "The Google books settlement: A private contract in the absence of adequate copyright law," *Scholarly and Research Communication*, 2011, 1(2), 36.
335 Iain Marlow, "Bid for Nortel patents marks Google's new push into mobile world," *Globe and Mail*, April 4, 2011. www.theglobeandmail.com/news/technology/tech-news/google-bids-for-nortel-patents/article1969788/.
336 United State District Court, Southern District Court of New York, The Author's Guild et al, plaintiffs against Google Inc., www.nysd.uscourts.gov/cases/show.php?db=special&id=115. Accessed March 31, 2011.
337 Harlequin's English subsidiary, Mills & Boon, reported e-books made

up 4 percent of sales by September 2009, the first anniversary of its making e-books available. Since that time, other sales figures show healthy growth. See, for example, Sean Poulter, "Sex Sells: Mills and Boon Boom Sparked by Downloaded Romances," March 28, 2010, posted on the Bookbee.net website, bookbee.posterous.com/sex-sells-mills-and-boon-boom-sparked-by-down.

338 Jay Yarrow, "Dan Brown's E-Book Sales Aren't So Killer After All," *Business Insider*, September 22, 2009, www.businessinsider.com/dan-browns-e-book-sales-arent-so-killer-after-all-2009-9.

339 "Android's Impact on eBooks, Reading," *Kindle Review*, October 27, 2009, ireaderreview.com/2009/10/27/androids-impact-on-ereading/.

340 Alison Flood, "Stieg Larsson becomes first author to sell 1m books," Guardian.co.uk, July 28, 2010, www.guardian.co.uk/books/2010/jul/28/stieg-larsson-1m-ebooks-amazon.

341 Xiao, Dongfa. Guest Presentation "Book Publishing in China (2011)," Simon Fraser University, March 30, 2011.

342 W. Bijker, "Do Not Despair: There Is Life after Constructivism," *Science, Technology and Human Values* 18 (1993), 113–38.

343 George Jones, "Kobo Vox and Pulse: 'How Is Kindle Going to Catch up with Us?'" *TabTimes*, October 28, 2011. tabtimes.com/news/media/2011/10/28/kobo-vox-and-pulse-how-kindle-going-catch-us.

344 Kimberly Budziak, "The influence of online reading communities on the publishing landscape" (Simon Fraser University, Master of Publishing program, student paper, December 2011).

345 The organization of the production and circulation of multiple copies of copied manuscripts prior to the printing press is described in Brian Patrick McGuire's *Jean Gerson and the Last Medieval Reformation* (University Park: Pennsylvania State University Press, 2005).

346 I have misplaced a, perhaps apocryphal, account that described Gutenberg being questioned on suspicion of theft because his cartful of Bibles represented so much wealth in his time that the authorities could not believe they were the property of one man.

347 It is noteworthy that in low-wage economies, such as that in current-day China, set-up costs for reprinting are not nearly as formidable as they are in high-wage economies, such as those of Europe and North America.

348 A film showing the operation of a linotype machine can be found at www.youtube.com/watch?v=pRYxOs1oCRY. Another showing a letterpress in use can be found at www.youtube.com/watch?v=yEoOowx6TQs&feature=fvw.

349 Heather Sanderson, "A PexOdyssey: Bibliographic Data Management and the Implementation of PexOd at the Dundurn Group" (Master of Publishing program report, Simon Fraser University, 2004).
350 See the Leanpub website at Leanpub.com.
351 Nicholas Carr, *The Big Switch: Rewiring the World, from Edison to Google* (New York: Norton, 2009).
352 Carla Hesse, "Books in Time," in *The Future of the Book*, ed. Geoffrey Nunberg (Los Angeles: University of California Press, 1988), 21–36.
353 Marshall McLuhan, *The Gutenberg Galaxy: The Making of Typographic Man* (Toronto: University of Toronto Press, 1962); Harold Innis, *The Bias of Communication* (Toronto: University of Toronto Press, 1951).
354 Roland Barthes, *Camera Lucida: Reflections on Photography*, trans: Richard Howard (New York: Farrar, Straus and Giroux, 1981).
355 Marsha Kinders, *Playing with Power in Movies, Television, and Video Games: From Muppet Babies to Teenage Mutant Ninja Turtles* (Los Angeles: University of California Press, 1991). Accessible via Google Books at books.google.com/books?id=raDNu1lThHQC&lpg=PP1&pg=PP1#v=onepage&q=transmedia %20intertexuality&f=false.
356 Vanessa Chan, Cynara Geissler, and Ann-Marie Metten, "Notes Towards a B.C.-OMDC: New Media Partnership Potential in British Columbia (Master of Publishing program student paper, Simon Fraser University, 2010).
357 CAREO Project, "Overview," www.ucalgary.ca/commons/careo/CAREOrepo.htm. Site no longer available.
358 "Earth at Night," http://apod.nasa.gov/apod/image/0011/earthlights2_dmsp_big.jpg.
359 Nova Scotia, www.novascotia.ca.
360 Simulating History website, home page, www.simulatinghistory.com; viHistory website, "How It Works," vihistory.ca/content/about/intro.php.
361 "Raymond G. Siemens: Profile," web.uvic.ca/~siemens/.
362 Elizabeth Eisenstein, *The Printing Press as an Agent of Change* (New York: Cambridge University Press, 1979).
363 Carla Hesse, "The rise of intellectual property, 700 BC–AD 2000: an idea in the balance," *Daedalus* (Spring 2002), 34, 35.
364 Cited in Hesse, "The rise of intellectual property," 34.
365 Christopher Aide, "A more comprehensive soul: Romantic conceptions of authorship and the copyright doctrine of moral right," *University of Toronto: Faculty of Law Review* 48, 2 (1990), 217–18.

366 Brian, "*Level 26: Dark Origins* by Anthony Zuiker & Duane Swierczynski reviewed," BSCReview, August 26, 2009, www.bscreview.com. Site no longer available.

367 Ellen M. George, "*Level 26* by Anthony Zuiker and Duane Swierczynski book review," AuthorsDen.com, September 28, 2009, www.authorsden.com/visit/viewarticle.asp?id=50960.

368 Michael Slade (pen name for lawyer Jay Clarke and various collaborators), respected author in the horror/crime genre, reports that one of his fans acts as a continuity consultant, highlighting inconsistencies between what the author proposes for an ongoing cast of characters and what he has already written about their pasts.

369 These insights are derived from one of many discussions of copyright that can be found on the web. In this case, it is on the website of Lift Studios (Lift Studios, *Copyright and the Web with Todd Sieling*, www.liftstudios.ca/research/lsb017/). They are also rooted in a research paper/editorial from the journal *Scholarly and Research Communication* (Rowland Lorimer, "Scholarly and Research Communication: A Journal and Some Founding Ideas," *Scholarly and Research Communication* 1, 1 [2010], www.src-online.ca).

370 Rowland Lorimer, *Magazines Alberta: Vibrancy, Growth, Interactive Community Leadership* (Calgary: Alberta Magazine Publishers Association, 2008).

371 See, for example, Daniel Levitin, *This Is Your Brain on Music* (Toronto: Penguin, 2007).

372 Shatzkin, "Digital Publishing in the U.S.: Driving the Industry to Vertical Niches?"

373 "About O'Reilly," O'Reilly Media website, www.oreilly.com/about.

374 Although I have been unable to carry out the research on this matter, it would seem that the computerization of offset presses would allow off-press preparation of files by means of a simulator, thus diminishing press set-up costs. At this point, the possibility remains speculative.

375 2008 figures, various manufacturer's websites.

376 See, for example, the HP website, h10088.www1.hp.com/cda/gap/display/main/index.jsp?zn=gap&cp=20000_4041_101__.

377 Edwards Brothers website, www.edwardsbrothers.com/.

378 Webcom website, www.webcomlink.com/tech/digital.htm.

379 Evidence of the varying organizational adjustments printers have made to maximize technological capacity comes from one book project where quotes were obtained from several printers for a short print run

of twenty-five copies (the book's specifications were 180 pages; 6 by 9; 24# white bond paper; 2-colour cover coated, 10 point, no bleeds; PDF for art). Even though all three printers were using digital printing, the first asked for $41.90 per copy for a print run of 10 and $14.00 for 50 (for twenty-five it would have been about $26.00); another asked for $18.00 per copy for twenty-five, and a third asked for $7.21 per copy plus $30.00 for one proof copy. The example comes from 2007. Price structures have changed dramatically in the face of competition.

380 Hugh McGuire, "Sifting through All These Books," O'Reilly Radar, June 14, 2010, radar.oreilly.com/2010/06/sifting-through-all-these-book.html.

381 Given this level of concentration and output, it would be surprising if these companies did not move gradually into a more traditional role by adding value and marketing effort to titles with market potential, thereby becoming a new wave of publishers.

382 A good example of such an organization is the PLAN Institute, which helps people with disabilities (www.planinstitute.ca). Other organizations, such as the Centre for Addiction and Mental Health (www.camh.net/), have their own in-house publisher and publishing operation.

383 Rowland Lorimer and Lindsay Lynch, *The Latest Canadian National Reading Study: Publishers' Analysis* (Vancouver: Canadian Centre for Studies in Publishing, 2007), 15, table 5.

384 Marc Slocum, "Amazon Growth Fuels Online's Book Market Share," O'Reilly Tools of Change for Publishing website, April 16, 2008, toc.oreilly.com/2008/04/amazon-growth-fuels-onlines-bo.html.

385 Wright Investment Services, "Wrightreports: Amazon.com, Inc.— Company Profile Snapshot," 2009, wrightreports.ecnext.com/coms2/reportdesc_COMPANY_023135106.

386 Chris Anderson, *The Long Tail: Why the Future of Business Is Selling Less of More* (New York: Hyperion, 2006), 23.

387 BookSurge website, www.booksurge.com.

388 Lightning Source website, www.lightningsource.com.

389 Mike Matas presents "A next-generation digital book" www.ted.com/talks/mike_matas.html.

CHAPTER 8

390 Conference Board of Canada, *Valuing Culture: Measuring and Understanding Canada's Creative Economy* (Ottawa: Author, 2008).

391 Conference Board of Canada, *Compendium of Research Papers: The International Forum on the Creative Economy* (Ottawa: Author, 2008), www.conferenceboard.ca/e-Library/Abstract.aspx?did=2701.

392 Statistics Canada, *Canadian framework for culture statistics*, cat.no. 81-595-MIE2004021 (Ottawa: Author, 2004), www5.statcan.gc.ca/bsolc/olc-cel/olc-cel?catno=81-595-MIE2004021&lang=eng.

393 Conference Board of Canada, *Valuing Culture*, 3.

394 Ibid., 4.

395 Ibid.

396 Ibid., iii.

397 Ibid., iv.

398 It is noteworthy that the written media account for 26.6 percent of the cultural sector as a whole, while the film industry comes in second at 14.8 percent, and broadcasting third at 9.2 percent. On the other hand, in the years leading up to 2007, publishing was the slowest-growing subsector. This growth trajectory may be turning around with the increase in electronic books and self-publishing.

399 Abeer Reeza and Charles Saunders, *Economic Contribution of the Canadian Magazine Industry* (Toronto: Informetrica Ltd., for Magazines Canada, 2006).

400 Conference Board of Canada, *Valuing Culture*, 30.

401 Presumably the measurements also exclude cultural materials for which there is no copyright claimed, e.g., newsletters and information pamphlets.

402 Conference Board of Canada, *Valuing Culture*, 7. The papers cited are Janet Ruiz, *A Literature Review of the Evidence Base for Culture, the Arts and Sport Policy* (Edinburgh: Scottish Executive Education Department, 2004) and François Matarasso, *Use or Ornament: The Social Impact of Participation in the Arts* (London: Comedia, 1997).

403 Conference Board of Canada, *Valuing Culture*, 6–7.

404 Ibid., 33.

405 The claim that culture contributes 7.4 percent to GDP gains credibility due to the fact that it can be traced through one of the background papers to an Industry Canada study of 2001. Specifically, Aaron Sawchuk and Graham Henderson ("Intellectual Property: Engine for Growth and the Backbone of the Creative Economy," in Conference Board of Canada, *Compendium of Research Papers*, 189–200) cite Sandra Charles, Gilles McDougall, and Julie Tran, *The Importance of the Intellectual Property Industries in the Canadian Economy* (Ottawa: Industry Canada, 2001).

According to these papers, in 2000 the "copyright industries" were "the third most important contributor to our economic growth," valued at 7.4 percent of GDP and with a growth rate at 6.6 percent (p. 190).

406 United Nations, *Creative Economy Report*, 2008. New York: United Nations. www.unctad.org/templates/webflyer.asp?docid=9750&intItemID=2068&lang=1.

407 U.K. Department of Culture Media and Sport, *Creative Britain: New Talent for a New Economy* (London: DCMS 2008).

408 United Nations, Creative Economy Report, 2010. New York: United Nations. www.unctad.org/Templates/WebFlyer.asp?intItemID=5763&lang=1.

409 See Dr. Žiga Turk, "Lisbon Strategy 2.1: Europe — The Most Creative Economy in the World." The site contains a presentation for a Pre-Presidency Conference, Ljubljana, December 3–4, 2007. The website of the European Office for Cultural Policy and Economy is also a useful site, www.european-creative-industries.eu.

410 European Year of Creativity and Innovation 2009 website, create2009.europa.eu/.

411 Conference Board of Canada, *Valuing Culture*, 29. The Conference Board notes the reasonable probability that these are underestimates of consumer spending and exports.

412 Canadian Conference for the Arts. C-32 — "An Act to modernize or to expropriate copyright?" www.ccarts.ca/en/events/Thinking Culture/. Accessed April 15, 2011.

413 Canada, 40th Parliament, 3rd Session, Legislative Committee on Bill C-32, Thursday March 24, 2011, Presentation by the Council of Ministers of Education. www2.parl.gc.ca/HousePublications/Publication.aspx?DocId=5071024&Language=E&Mode=1&Parl=40&Ses=3#Int-3823706.

414 Heller points out that scientists and other hard science researchers regularly work in direct contravention of copyright law: *The Gridlock Economy: How Too Much Ownership Wrecks Markets, Stops Innovation and Costs Lives* (New York: Basic Books, 2008), 66–68.

415 Giuseppina D'Agostina, "Healing Fair Dealing? A Comparative Copyright Analysis of Canadian Fair Dealing to U.K. Fair Dealing and U.S. Fair Use," CLPE Research Paper 28/2007, *McGill Law Review* 53, 2 (2008).

416 *Robertson v. Globe and Mail*; *Robertson v. Canwest Global, et al.*

417 *Baker v Sony, EMI, Universal Music, Warner, CMRRA, SODRAC*. Ontario Superior Court File number CV0800360651 00CP.

418 This is an interesting stance that does not appear to have been fully tested

in court. See Neil Thakur, National Institutes for Health (USA), "Funding Policies and Research Access — Round Table" (panel discussion, International Conference on Electronic Publishing (Elpub), Toronto, June 27, 2008).

419 Creative Commons, "About," creativecommons.org/about/.
420 Creative Commons, "About Licenses," creativecommons.org/licenses.
421 Alberta Foundation for the Arts website, www.affta.ab.ca/.
422 In May 2010, the Canadian Booksellers Association lobbied the government to remove distribution rights from the Copyright Act. On May 11, the *Globe and Mail* published an editorial supporting the idea (www.theglobeandmail.com/news/opinions/editorials/remove-this-barrier-to-books/article1563936/). If the distribution right disappeared, the Canadian publishing programs of the foreign-owned sector would also eventually disappear.
423 British Columbia, Corporate Income Tax, Tax Credits – "Book Publishing Tax Credit," August 2010. www.sbr.gov.bc.ca/business/income_taxes/corporation_income_tax/tax_credits/book_publishing.htm.
424 Telephone interview with Avie Bennett by author, March 7, 2012.
425 Telephone interview with Avie Bennett by author, January 17, 2012.
426 Brad Martin pointed out that in the early period following the 2000 sale, Random House was able to use its size to ensure more timely payments to M&S from Chapters during that extremely tumultuous period. Telephone interview with Avie Bennett and Brad Martin by author, March 16, 2012.
427 Bennett interview, March 7, 2012.
428 Robert Fulford, 2000. "Robert Fulford's column about the future of McClelland & Stewart," which appeared in *The National Post*, June 27, 2000. www.robertfulford.com/McClelland.html.
429 Telephone interview with Avie Bennett and Brad Martin, March 16, 2012 (Brad Martin speaking).
430 "The Cooke Agency: About Us." www.cookeagency.ca/about.htm.
431 Email from Brad Martin to the author, March 22, 2012.
432 Email from Brad Martin to the author, March 23, 2012.
433 CTV News, 2012. "Random House becomes sole owner of McClelland and Stewart. 'The university is receiving no compensation for this transaction,' U of T spokeswoman Laurie Stephens told The Canadian Press in an email." www.ctv.ca/CTVNews/Entertainment/20120111/random-house-mcclelland-stewart-120111/#ixzz1jwLa9P1l.
434 Telephone interview with Avie Bennett and Brad Martin, March 16, 2012.

435 Ibid., Brad Martin speaking.
436 Ibid.
437 Ibid.
438 Ibid.
439 M&S's 2010/2011 publishing program earned it $180,800 from the Canada Council Block Grant and Authors' Tours program. See Canada Council for the Arts, 2012. Searchable Grants Listing, www.canadacouncil.ca/grants/recipients/0l127245536828281250.htm; $423,815 from the Department of Canadian Heritage. See Canada, Department of Canadian Heritage, Canada Book Fund, Recipients, Canada Book Fund, Support to Publishers, Recipient List. www.pch.gc.ca/eng/1290024798857; and $43,502 from the Ontario Arts Council. As well, it would have received in the order of $300,000 to $400,000 in tax credits refunds through the Ontario Book Publishing Tax Credits program.
440 Telephone interview with Avie Bennett and Brad Martin, March 16, 2012, Brad Martin speaking.
441 Research-based statistic shared with the author on a not-for-attribution basis.
442 Telephone interview with Avie Bennett and Brad Martin, March 16, 2012, Brad Martin speaking.
443 Consider this. The M&S website lists 100 authors whose names begin with M, names such as David Macfarlane, Roy MacGregor, Hugh Maclennan, Alexander MacLeod, Alistair MacLeod, Roy Macskimming, Alberto Manguel, Preston Manning, Philip Marchand, Daphne Marlatt, Paul Martin, Elizabeth May, Rona Maynard, Seymour Mayne, Christina McCall, Bob McDonald, David Mcfadden, Ken McGoogan, Brian McKillop, Marshall McLuhan, Stephanie McLuhan, Anne Michaels, A.A. Milne, Rohinton Mistry, W.O. Mitchell, L.M. Montgomery, Susanne Moodie, Brian Moore, Christopher Moore, Shani Mootoo, Desmond Morton, Farley Mowat, Brian Mulroney, Alice Munro, Rex Murphy, and Susan Musgrave, a mere selection of M's.
444 Telephone interview with Avie Bennett and Brad Martin, March 16, 2012, Brad Martin speaking.
445 D&M Media Release. (March 2012) D&M Publishers Inc. "Canada's Leading Independent Publishing House, Addresses the Future."
446 Statistics Canada, "Book Publishing Industry," *The Daily*, July 10, 2008. According to the data reported, there are 311 Canadian-controlled firms. They originated 10,351 trade titles and reprinted 3,711 for a total of

14,062. Their similarly calculated average per-title earnings were $14,176.
447 Lorimer, Rowland, "Intellectual Property, Moral Rights and Trading Regimes." *Canadian Journal of Communication* 21 (2) (1996), 267–285.
448 Christopher Aide, "A More Comprehensive Soul: Romantic Conceptions of Authorship and the Copyright Doctrine of Moral Right," *University of Toronto: Faculty of Law Review*, 48, 2 (1990) 217–18.
449 Daniel Kahneman, *Thinking, Fast and Slow* (Toronto: Doubleday, 2011).
450 This is another reason why vertical niche publishers will not take over the industry.
451 Relative impact, for example, would suggest that if the expected sales for one book are in the tens of thousands of copies, and expected sales for another are in the thousands, an injection of marketing dollars will have greater effect on the first title than the second. An absolute effect would see an equal number of marketing dollars having an equal impact on both titles.
452 Chris Anderson, *Free: The Future of a Radical Price* (New York: Random House, 2009).
453 Seth Godin's Blog (June 30, 2009). "Malcolm is wrong." sethgodin.typepad.com/seths_blog/2009/06/malcolm-is-wrong.html.
454 BMW launched the new mini in North America using, in part, a free story book.
455 Alison Cairns, "An Analysis of the Operation of the University of British Columbia Press with an Emphasis on Scholarly Editing." (Simon Fraser University, Master of Publishing project report, 2005.)
456 Carol Tenopir and Donald W. King, Trends in scientific scholarly journal publishing in the United States. *Journal of Scholarly Publishing* (April 1997), 139
457 Anne Okerson, *Report of the ARL Serials Prices Project*. Washington: Association of Research Libraries, 1989.
458 J. Davidson Frame, F. Narin, and M.P. Carpenter, "World distribution of science," *Social Studies of Science* 7 (1997), 501–516.
459 A.M. Cetto, *Scientific Publications in the Developing World* (COSTED, Chennai, 1998).
460 V. Patel, S. Siribaddana, A. Sumathipala *Under-representation of developing countries in the research literature: ethical issues arising from a survey of five leading medical journals.* Biomed Central (2004): www.biomedcentral.com/content/pdf/1472-6939-5-5.pdf.
461 BioMed Central, "What Is BioMed Central?" www.biomedcentral.com/info/.

462 Rowland Lorimer, "Libraries, Scholars and Publishers in Digital Journal and Monograph Publishing," *Scholarly and Research Communication*. Forthcoming.

463 Anita de Waard and Joost Kircz. 2008. "Modeling scientific research articles — Shifting perspectives and persistent issues."

464 Telephone interview with Frances Pinter, January 2009.

465 There are those in the publishing industry, such as Anthony Watkinson, who question the sustainability of the open access model. See Anthony Watkinson, "Open Access: A Publisher's View," *Logos* 17 (2006), 47. It is certainly the case that for open access to be sustained, research agencies and, ultimately, governments must be on board. It is not difficult to imagine support for open access waning in times of cutbacks, especially if journals expand with any vigour, as they show signs of doing. Support could be further jeopardized by public access to scholarly journals that publish the seemingly irrelevant meanderings of clever but confused minds. Whatever happens, it is doubtful that commercial journal partners will yield the entire field to the not-for-profit sector.

466 See the Public Knowledge Project website at pkp.sfu.ca.

467 Raym Crow, "Campus-based Publishing Partnerships: A Guide to Critical Issues" (Washington, DC: SPARC, 2009), www.arl.org/sparc/partnering/guide/.

468 OAPEN, "Press release: A new service for Open Access Monographs: the Directory of Open Access Books." February 29, 2012.

469 A further development in the area of custom textbooks is CourseSmart (www.coursesmart.com/), a not-for-profit joint company that has content from such textbook publishers as Bedford, Freeman and Worth; Cengage; McGraw-Hill; Pearson; and Wiley in Canada and from CQ Press, Elsevier, F.A. Davis, Jones and Bartlett, Lippincott, Williams & Wilkins, Nelson Education, Princeton University Press, Sage, Sinauer, Taylor and Francis, and Wolters Kluwer Health. Instructors are free to source materials from the publications of any of these companies to build their own custom textbooks.

470 Ontario Arts Council, "Writers' Reserve," www.arts.on.ca/Page119.aspx.

471 Ontario Arts Council, "Writers' Works in Progress," www.arts.on.ca/Page118.aspx.

472 Maria Scala, "The New Face of Canadian Publishing: A Corporate History and Analysis of Alfred A. Knopf Canada" (Master of Publishing project report, Simon Fraser University, 2000); and Maureen Gillis, "Branding the Borzoi: Imprint Branding and the Knopf Canada List"

(Master of Publishing project report, Simon Fraser University, 2003).

473 Paul Shoebridge and Michael Simons, *Welcome to Pine Point*, Montreal: National Film Board, interactive.nfb.ca/#/pinepoint.

474 Al Gore. *Our Choice*, Push Pop Press, 2011. pushpoppress.com/ourchoice/.

475 Daniel Eran Dilger. "Push Pop Press drops iPad ebooks after being acquired by Facebook." Apple Insider: August 2, 2011. www.appleinsider.com/articles/11/08/02/push_pop_press_drops_ipad_ebooks_after_being_acquired_by_facebook.html.

476 An amusing illustration of the challenges faced by early adaptors of the socio-technical ensemble that the original incunabula represented can be found at www.youtube.com/watch?v-pQHX-SjgQvQ.

477 See Robin Perrin, *From Cambridge to Communication: McLuhan beyond McLuhanism* (Master's thesis, Simon Fraser University, 1991); and Rowland Lorimer, "Marshall McLuhan: Media Genius and Noted Author of Understanding Media: The Extensions of Man," *Logos: The Journal of the World Book Community* 12, 2 (2001), 78–85.

478 Evidently, the literary mash-up is already upon us. Jon Clinch has written a book called *Finn* in which he appropriates Mark Twain's character. According to Carole Goldberg (Carole Goldberg, "Huck Finn's Back Pages," *Vancouver Sun*, March 3, 2007, C10) other novelists have done likewise: Jane Smiley with *A Thousand Acres* (a modern retelling of *King Lear*) and *Ten Days in the Hills* (a "Hotel California version of Boccaccio's *The Decameron*"), and Chris Bohjalian with *The Double Bind*, in which he mixes characters from Fitzgerald's *Great Gatsby* with contemporary Vermonters.

BIBLIOGRAPHY

The vast majority of web citations derive from research done or checked during the preparation of penultimate and final drafts. I have added a date where I thought a site might change or disappear.

Adams, James. "Bookstores Reach Deal." *Globe and Mail*, January 13, 2007.
———. "Publishers' Plans Anger Bookstores." *Globe and Mail*, January 11, 2007.
———. "$2.5-Million Voyage." *Vancouver Sun*, August 9, 2008.
Aide, Christopher. "A More Comprehensive Soul: Romantic Conceptions of Authorship and the Copyright Doctrine of Moral Right." *University of Toronto: Faculty of Law Review* 48, 2 (1990): 217–26.
Alberta Foundation for the Arts. "Arts in Alberta." http://www.affta.ab.ca/arts-in-alberta.shtml.
———. *Moving Arts: 2006/2007 Review*, 16. http://www.affta.ab.ca/resources/AFA-06-07-Year-in-review.pdf.
Aldana, Patricia. *Canadian Publishing: An Industrial Strategy for its Preservation and Development in the Eighties*. Toronto: Association of Canadian Publishers, 1980.
Altbach, Philip and Edith Hishino. *International Book Publishing: An Encyclopedia*. New York: Garland, 1995. See section on Africa, 366–423.
Amazon Annual Report 2010. http://phx.corporate-ir.net/phoenix.zhtml?c=97664&p=irol-reportsannual
Anderson, Chris. *Free: The Future of a Radical Price*. New York: Random House, 2009.
———. *The Long Tail: Why the Future of Business Is Selling Less of More*. New York: Hyperion, 2006.
Anderson, Scott. "Heritage May Help GDS Clients Through Cash Crunch: But at Least One Publishing Executive is Calling for a Longer-term Solution." *Quill & Quire* 67, no. 4 (April 2001): 4.
Association of Canadian Publishers. *ACP Copyright Paper*. Toronto: ACP, 2008.

———. *Book Publishing and Canadian Culture: A National Strategy for the 1990s.* Toronto: ACP, 1991.

———. *A Mid-Decade Assessment.* Toronto: ACP, 1985.

———. *New Directions: Rethinking Public Policy for Canadian Books.* Toronto: ACP, 1995.

———. *Setting Priorities for Federal Book Publishing Policy.* Toronto: ACP, 1997.

Australian Council for the Arts. *Arts Funding Guide* 2009. http://www.australiacouncil.gov.au/grants/arts_funding_guide.

———. *Literature–July 2009.* http://www.australiacouncil.gov.au/grants/amr/literature/literature_-_july_2009.

Authors Guild. *Authors Guild v. Google Settlement Resources Page.* http://www.authorsguild.org/advocacy/articles/settlement-resources.html.

Baeker, Greg. "Sharpening the Lens: Recent Research on Cultural Policy, Cultural Diversity and Social Cohesion." *Canadian Journal of Communication* 27, 2 (2002): 179–96.

Baird, Jean. *Canadian Literature in High Schools: A Research Study.* Toronto: Writer's Trust, 2001.

Baker v. Sony, EMI, Universal Music, Warner, CMRRA, SODRAC. Ontario Superior Court File, number CV0800360651 00CP.

Barnes, Roger, and Rowland Lorimer. *Book Publishing: The Act.* Toronto: ACP, 1996.

———. *Book Purchasing in Canada: 1997.* A survey in a commissioned report. Toronto: Canadian Publishers Council, 1998.

———. *Children's Book Purchasing.* Toronto: Canadian Publishers Council, 1999.

———. *Merchandising in Independent Bookstores.* Toronto: Association of Canadian Publishers, 1998.

Barthes, Roland. *Camera Lucida: Reflections on Photography.* Translated by Richard Howard. New York: Farrar, Straus and Giroux, 1981.

Bijker, Wiebe. "Do Not Despair: There Is Life after Constructivism." *Science, Technology and Human Values* 18 (1993): 113–38.

BioMed Central. "What Is BioMed Central?" http://www.biomedcentral.com/info.

Book Publishers Association of Alberta. *Alberta Book Publishers Operating Support Initiative,* 2009. http://www.affta.ab.ca/cultural-industries-grant-program.aspx. Cited in "Alberta Bound: Support for Book Publishing in Alberta" by Kelsey Everton. Course paper, Simon Fraser University, 2010.

BookManager. Official website. http://www.bookmanager.ca.

BookNet Canada. Official website. http://www.booknetcanada.com.

———. "2008 in Review." *BNC Newsletter*, December 11, 2008.

BookSurge. Official website. http://www.booksurge.com.

Bourdieu, Pierre. "The Forms of Capital." In *Handbook of Theory and Research for the Sociology of Education*, edited by John G. Richardson. New York: Greenwood Press, 1986. Also available online at http://www.marxists.org/reference/subject/philosophy/works/fr/bourdieu-forms-capital.htm.

Bourdieu, Pierre, and Jean-Claude Passeron. "Cultural Reproduction and Social Reproduction." In *Knowledge, Education and Cultural Change*, edited by R. Brown. London: Tavistock, 1993.

Bowker. "Bowker Reports Traditional U.S. Book Production Flat in 2009." http://www.bowker.com/index.php/press-releases/616-bowker-reports-traditional-us-book-production-flat-in-2009.

British Columbia. "Book Publishing." August 2010. http://www.sbr.gov.bc.ca/business/income_taxes/corporation_income_tax/tax_credits/book_publishing.htm.

———. "Business." http://www.sbr.gov.bc.ca/business/income_taxes/corporation_income_tax/tax_credits/book_publishing.htm.

British Columbia Arts Council. Official website. http://www.bcartscouncil.ca.

British Columbia. Ministry of Finance. "British Columbia Book Publishing Tax Credit," *Tax Bulletin* CIT 008 (2009). http://www.sbr.gov.bc.ca/documents_library/bulletins/cit_008.pdf.

Bruner, Jerome S., Rose R. Olver, Patricia M. Greenfield, et. al. *Studies in Cognitive Growth: A Collaboration at the Center for Cognitive Studies.* New York: Wiley, 1966.

Budziak, Kimberly. "The Influence of Online Reading Communities on the Publishing Landscape." Student paper, Master of Publishing program, Simon Fraser University, December 2011.

Butalia, Urvashi. "English Textbook: Indian Publisher." *Media, Culture and Society*, 15 (1993): 217–32.

Cairns, Alison. "An Analysis of the Operation of the University of British Columbia Press with an Emphasis on Scholarly Editing." Project report, Master of Publishing program, Simon Fraser University, 2005.

Canada. The Broadcasting Act, S.C., c. 11, 1991. http://laws-lois.justice.gc.ca/eng/acts/B-9.01.

Canada. Canada Council for the Arts. "About the GGs." http://www.canadacouncil.ca/prizes/ggla/qz128686615675969592.htm.

———. "Book Publishing Support: Block Grants." www.canadacouncil.ca/grants/writing/ap127723094273982142.htm.

Canada. Canadian Conference of the Arts. "C-32 — An Act to Modernize or to Expropriate Copyright?" *Cultural Forum Series*, January 25, 2011. http://www.ccarts.ca/en/events/Thinking Culture/Forum3.htm.

Canada. Canadian Publishers Council. Official website. http://pubcouncil.ca/membership.php.

Canada. Canadian Research Knowledge Network. "History of CRKN." http://www.researchknowledge.ca/en/about/history.jsp.

Canada. Competition Bureau. "Competition Bureau Files Application with Consent of Indigo and Chapters." http//www.cb-bc.gc.ca/eic/site/cb-bc.nsf/eng/00802.html.

Canada. *Copyright Act. Book Importation Regulations*, SOR/99-324. *Copyright Act*, RSC (1985), c. C-42.

Canada. Department of Canadian Heritage. Book Policy and Programs Division. "Book Publishing Industry Development Program." http://www.pch.gc.ca/pgm/padie-bpidp/index-eng.cfm.

———. "Canada Book Fund" (formerly Book Publishing Industry Development Program), March 9, 2010. http://www.pch.gc.ca/eng/1268182505843/1268255450528.

———. "Canada Book Fund, Business Development Support." http://www.pch.gc.ca/pgm/flc-cbf/sae-sfp/guide/103-eng.cfm.

———. "Canada Book Fund, 2010–2011 — Application Guide." http://www.pch.gc.ca/pgm/flc-cbf/soa/guide-eng.cfm#a3.4.

———. "The Canadian Film Industry and Investment Canada." Fact sheet issued by Communications Canada, November 6, 2008. http//www.pch.gc.ca/invest/film-eng.cfm.

———. "Intersections: Reading the Cultural Landscape." http://www.pch.gc.ca/pc-ch/org/sectr/ac-ca/pblctns/anl-rpt/2007-2008/index-eng.cfm.

———. "Livres Canada Books" (previously Association for the Export of Canadian Books). Official website. http://www.livrescanadabooks.com/entry.

———. "Livres Canada Books: Export Marketing Assistance Program." http://www.livrescanadabooks.com/aecb/page/2009-2010_forms_-_final_report_-_emap/. (program cancelled)

———. "Livres Canada Books: Foreign Rights Marketing Assistance Program." http://www.livrescanadabooks.com/en/funding/frmap.

———. Official website. http://www.pch.gc.ca/eng/1266037002102/

———. *Printed Matters*. Ottawa: 2005. 1265993639778.

———. *Publishing Measures*. Ottawa: 2004, 2005, 2006, 2007, 2008, 2009.

———. *Review of the Revised Foreign Investment Policy in Book Publishing and Distribution*. http//www.pch.gc.ca/eng/1272486502392/1268255450528.

———. "Support for Organizations and Associations." http://www.pch.gc.ca/pgm/flc-cbf/soa/guide-eng.cfm#a3.1.

Canada. Department of Communications. *Vital Links: Canadian Cultural Industries*. Ottawa: Government of Canada, 1987.

Canada. Foreign Affairs and International Trade Canada. "New Strategies for Culture and Trade: Canadian Culture in a Global World." February 1999. http://www.international.gc.ca/trade-agreements-accords-commerciaux/fo/canculture.aspx?lang=en.

Canada. 40th Parliament, 3rd Session. Legislative Committee on Bill C-32. Presentation by the Council of Ministers of Education, March 24, 2011. http://www2.parl.gc.ca/HousePublications/Publication.aspx?DocId=5071024&Language=E&Mode=1&Parl=40&Ses=3#Int-3823706.

Canada. *Investment Canada Act, RSC*, 1985, c. 28 (1st Supp.), ss. 20–24. http://laws.justice.gc.ca/eng/I-21.8.

Canada. Ministry of Communications, Information Services. "Masse Announces New Book Publishing Policy Development Program and Additional Funding for the Canada Council." Press release, June 18, 1986.

Canada. *Royal Commission on National Development in the Arts, Letters and Sciences*. Ottawa: King's Printer, 1952.

Canada. Standing Committee on Cultural Heritage. "The Challenge of Change: A Consideration of the Canadian Book Industry." Ottawa: Government of Canada, 2000. http://www2.parl.gc.ca/HousePublications/Publication.aspx?DocId=1031737&Language=E&Mode=1&Parl=36&Ses=2.

CAREO Project. "Overview." http://www.ucalgary.ca/commons/careo/CAREOrepo.htm.

Carr, Nicholas. *The Big Switch: Rewiring the World, from Edison to Google*. New York, Norton: 2009.

Castledale Inc., and Nordicity. *A Strategic Study for the Book Publishing Industry in Ontario*. Toronto: Ontario Media Development Corporation, 2008.

Caves, Richard E. *Creative Industries: Contracts Between Art and Commerce*. Cambridge, MA: Harvard University Press, 2000.

Cavill, Pat. *Collections: How and Why Public Libraries Select and Buy Their Canadian Books*. Toronto, Association of Canadian Publishers, 1998.

Centre for Addiction and Mental Health. Official website. http://www.camh.net. See especially http://www.camh.net/hsrcu/v2/publications/reports/consulting/planning.htm.

Cetto, Ana Maria. *Scientific Publications in the Developing World*. Chennai: COSTED, 1998.

Chan, Vanessa, Cynara Geissler, and Ann-Marie Metten. "Notes Towards a B.C.-OMDC: New Media Partnership Potential in British Columbia." Student paper, Master of Publishing Program, Simon Fraser University, 2010.

Chapters/Indigo. Advertisement. *Globe and Mail*, June 12, 2010.

———. "Our Company: Management." http://www.chapters.indigo.ca/Our-Company-Management/oc_Management-artnb.html.

Charles, Sandra, Gilles McDougall, and Julie Tran. *The Importance of the Intellectual Property Industries in the Canadian Economy*. Ottawa: Industry Canada, 2001.

CK-12 Foundation. Official website. http://about.ck12.org.

Conference Board of Canada. *Valuing Culture: Compendium of Research Papers: The International Forum on the Creative Economy*. Ottawa, 2008. http://sso.conferenceboard.ca/e-Library/LayoutAbstract.asp?DID=2701.I

———. *Valuing culture: Measuring and Understanding Canada's Creative Economy*. Ottawa: 2008.

Cote, Marc. "Why's Everybody Always Picking On Us?" *Globe and Mail*, February 14, 2009.

CourseSmart. Official website. http://www.coursesmart.com.

Créatec+. *Reading and Buying Books for Pleasure*. Ottawa: Department of Canadian Heritage, 2006.

Creative Commons. "About." http://creativecommons.org/about.

———. "About Licenses." http://creativecommons.org/about/licenses.

Creighton, Donald. *The Commercial Empire of the St. Lawrence*. Toronto: Ryerson Press, 1937.

Cross River Publishing Consultants. *PMA White Papers: Book Industry Returns: An Analysis of the Problem; Opportunities for Improvement; With a Focus on Independent Publishers*. Manhattan Beach, CA: Publishers Marketing Association, 2001.

Crow, Raym. *Campus-based Publishing Partnerships: A Guide to Critical Issues*. Washington, DC: SPARC, 2009. Also available online at http://www.arl.org/sparc/partnering/guide.

D'Agostina, Giuseppina. "Healing Fair Dealing? A Comparative Copyright Analysis of Canadian Fair Dealing to U.K. Fair Dealing and U.S. Fair Use." *McGill Law Review* 53, 2 (2008): CLPE Research Paper 28/2007.

Darnton, Robert. *The Forbidden Best-Sellers of Pre-revolutionary France*. New York: Norton, 1995.

———. *The Literary Underground of the Old Regime*. Cambridge, MA: Harvard University Press, 1982.

de Waard, Anita, and Joost Kircz. *Modeling Scientific Research Articles: Shifting Perspectives and Persistent Issues.* Paper #13, 2008. http://www.utsc.utoronto.ca/~elpub2008. Also available at: http://www.informatik.uni-trier.de/~ley/db/conf/elpub/elpub2008.html#WaardK08.

Donner, Arthur, and Fred Lazar. *The Canadian Book Publishing Industry: Competitive Challenges and the Need for Restructuring.* Ottawa: Department of Canadian Heritage, Cultural Industries Branch, 2000.

———. *The Competitive Challenges Facing Book Publishers in Canada.* Ottawa: Department of Canadian Heritage, Cultural Industries Branch, 2000.

Duxbury, Nancy. "The Economic, Political, and Social Contexts of English-Language Book Title Production in Canada, 1973–1996." Ph.D. thesis, Simon Fraser University, 2000.

———. *The Reading and Purchasing Public: The English-Canadian Trade Book Market in Canada.* Toronto: Association of Canadian Publishers, 1995.

Edwards Brothers. Official website. http://edwardsbrothers.com.

Eisenstein, Elizabeth. *The Printing Press as an Agent of Change.* New York: Cambridge University Press, 1979.

English, John. *Citizen of the World: The Life of Pierre Elliott Trudeau,* Volume One, *1919–1968.* Toronto: Knopf, 2006.

———. *Just Watch Me: The Life of Pierre Trudeau,* Volume Two, *1968–2000.* Toronto: Knopf, 2009.

Europa. *Year of Creativity and Innovation 2009.* http://create2009.europa.eu.

European Office for Cultural Policy and Economy. Official website. http://www.european-creative-industries.eu/Home/tabid/79/Default.aspx.

Flatworld Knowledge. Official website. http://www.flatworldknowledge.com.

Fleming, Patricia L., Gallichan Gilles, and Yvan Lamonde, eds. *History of the Book in Canada: Volume One: Beginnings to 1840.* Toronto: University of Toronto Press, 2004.

Flood, Alison. "Stieg Larsson Becomes First Author to Sell 1m Books." *The Guardian,* July 28, 2010. http://www.guardian.co.uk/books/2010/jul/28/stieg-larsson-1m-ebooks-amazon.

Foreign Affairs and International Trade Canada. *New Strategies for Culture and Trade: Canadian Culture in a Global World.* February 1999. http://www.international.gc.ca/trade-agreements-accords-commerciaux/fo/canculture.aspx?lang=en.

Fox Jones. *Evaluation of the Book Publishing Industry Development Program: Economic Study* and *Financial and Market Impact Study.* Ottawa: Department of Communication, 1992.

Frame, J. Davidson, Francis Narin, and Mark P. Carpenter. "World Distribution of Science." *Social Studies of Science* 7 (1997): 501–16. http://www.jstor.org/stable/284718.

Galloway, Gloria. "Q&A. Peter Nicholson: The Man Paul Martin Chose to Craft the Cost-Cutting 1995 Budget Reflects with Gloria Galloway on Ottawa's Challenge This Time Around." *Globe and Mail*, March 1, 2010.

George, Ellen M. Review of *Level 26*, by Anthony Zuiker and Duane Swierczynski. *AuthorsDen.com*, September 28, 2009. http://www.authorsden.com/visit/viewarticle.asp?id=50960.

Gerson, Carole, and Jacques Michon. *History of the Book in Canada: Volume Three: 1918 to 1980.* Toronto: University of Toronto Press, 2007.

Ghazali, Abu Hamid Muhammad al-. *The Incoherence of the Philosophers.* Translated by Michael Marmura. Provo, UT: Brigham Young University Press, 1997.

Gillis, Maureen. "Branding the Borzoi: Imprint Branding and the Knopf Canada List." Project report, Master of Publishing program, Simon Fraser University, 2003.

Globe and Mail. "Remove This Barrier to Books: Readers, Through Bookstores, Should be Able to Buy at a Fair Price, Without a Largely Hidden Subsidy to Canadian-based Publishers." Editorial, May 11, 2010. http://www.theglobeandmail.com/news/opinions/editorials/remove-this-barrier-to-books/article1563936.

Godin, Seth. "Malcolm is Wrong." *Seth's Blog.* June 20, 2009. http://sethgodin.typepad.com/seths_blog/2009/06/malcolm-is-wrong.html.

Goldberg, Carole. "Huck Finn's Back Pages." *Vancouver Sun*, March 3, 2007.

Grant, Peter, and Chris Wood. *Blockbusters and Trade Wars.* Vancouver: Douglas & McIntyre, 2004.

Grescoe, Paul. *The Merchants of Venus: Inside Harlequin and the Empire of Romance.* Vancouver: Raincoast Books, 1996.

Gu, (Richard) Qianqiao. "The Establishment of Bertelsmann in China." Project report, Master of Publishing program, Simon Fraser University, 2006. An excerpt from this report was published as "Bertelsmann in China: Low Profile, Patient Growth." *Logos: The Journal of the World Book Community* 17, 4 (2006): 173–81.

Gundy, H.P. "The Development of the Book Trade in Canada." In *Background Papers, Ontario, Royal Commission on Book Publishing.* Toronto: Queen's Printer for Ontario, 1972.

Halpape, Jan. "Subsidies versus a Net Price System: A Comparison of the Canadian and German Book Industry Support Systems." Student paper,

Master of Publishing program, Simon Fraser University, 2008.

Hannigan, John. "Culture, Globalization and Social Cohesion: Towards a Deterritorialized Global Fluids Model." *Canadian Journal of Communication* 27, 2 (2002): 277–88.

Hartley, Matt. "New Chapter for Kobo as Firm Sold to Japan's Rakuten." *Financial Post*. November 8, 2011. http://business.financialpost.com/2011/11/08/new-chapter-for-kobo-as-firm-sold-to-japans-rakuten/

Heller, Michael. *The Gridlock Economy: How Too Much Ownership Wrecks Markets, Stops Innovation and Costs Lives*. New York: Basic Books, 2008.

Helliwell, John. *Globalization and Well-Being*. Vancouver: UBC Press, 2002.

Hesse, Carla. "Books in Time." In *The Future of the Book*, edited by Geoffrey Nunberg. Los Angeles: University of California Press, 1988.

———. "Enlightenment Epistemology and the Laws of Authorship in Revolutionary France, 1777–1793." In *Law and the Order of Culture*, edited by Robert Post. Berkeley: University of California Press, 1991. Also available online at http://content.cdlib.org.

———. "The Rise of Intellectual Property, 700 BC–AD 2000: An Idea in the Balance." *Daedalus* (Spring 2002): 34–43.

Hewlett Packard, *HP* (official website). http://h10088.www1.hp.com/cda/gap/display/main/index.jsp?zn=gap&cp=20000_4041_101__.

Hoskins, Colin, and Rolf Mirus. "Reasons for the U.S. Dominance of the International Trade in Television Programmes." *Media, Culture and Society*, 10 (1988): 499–515.

Hoskins, Colin, Stuart McFadyen, and Adam Finn. "Cultural Industries from an Economic/Business Research Perspective." *Canadian Journal of Communication*, 25, 41 (2000): 127–44. Also available online at http://www.cjc-online.ca/index.php/journal/article/view/1146/1065.

Huenefeld, John. *The Huenefeld Guide to Book Publishing*. Bedford, MA: The Huenefeld Company, 2001.

Hurtig, Mel. *The Betrayal of Canada*. Toronto: Stoddart, 1991.

Ibbitson, John. "Canada Risks Being Shut Out of Pacific Trade Pact, New Zealand Prime Minister Warns." *Globe and Mail*, April 15, 2010.

IDRC. Official website. http://www.idrc.ca/en/ev-66174-201-1-DO_TOPIC.html.

Indigo Books & Music Inc. *First Quarter Report*, p. 5. 2011. www.chapters.indigo.ca/investor-relations/corporate-documents/.

Innis, Harold. *The Bias of Communication*. Toronto: University of Toronto Press, 1951.

———. *The Cod Fisheries: The History of an International Economy*. Toronto:

University of Toronto Press, 1942.

———. *The Fur Trade in Canada: An Introduction to Canadian Economic History.* Toronto: University of Toronto Press, 1930.

Jenson, Jane. "Identifying the Links: Social Cohesion and Culture." *Canadian Journal of Communication* 27, 2 (2002): 141–52.

Jones, George. "Kobo Vox and Pulse: 'How Is Kindle Going to Catch Up With Us?'" *TabTimes*. October 28, 2011. http://tabtimes.com/news/media/2011/10/28/kobo-vox-and-pulse-how-kindle-going-catch-us.

Juby, Susan. "Editor to Author: Some Personal Reflections on Getting Published." In *Publishing Studies: Book Publishing 1*, edited by Rowland Lorimer, John Maxwell, and Jillian Shoichet. Vancouver: CCSP Press, 2005.

Kinders, Marsha. *Playing with Power in Movies, Television, and Video Games: From Muppet Babies to Teenage Mutant Ninja Turtles.* Los Angeles: University of California Press, 1991. Also available online at http://books.google.com/books?id=raDNuiIThHQC&lpg=PP1&pg=PP1#v=onepage&q=transmedia%20intertexuality&f=false.

Kindle Review. "Android's Impact on eBooks, Reading." October 27, 2009. http://ireaderreview.com/2009/10/27/androids-impact-on-ereading.

Laberge, Marc. "The Canadian Book Industry Supply Chain Initiative Business Plan: Prepared for the Canadian Book Industry Supply Chain Initiative Steering Committee." Ottawa: Department of Canadian Heritage, 2002.

Lamonde, Yvan, Patricia L. Fleming, and Fiona A. Black, eds. *History of the Book in Canada: Volume Two: 1840 to 1918*. Toronto: University of Toronto Press, 2005.

Laxer, James, and Robert Laxer. *The Liberal Idea of Canada: Pierre Trudeau and the Question of Canada's Survival*. Toronto: James Lorimer, 1977.

Leanpub. Official website. http://leanpub.com.

Levitin, Daniel. *This Is Your Brain on Music*. Toronto: Penguin, 2007.

Levitt, Kari. *Silent Surrender: The Multinational Corporation in Canada*. Toronto: Macmillan, 1970.

Lewis-Kraus, Gideon. "The Last Book Party: Publishing Drinks to a Life after Death." *Harper's* (March 2009): 41–51.

Lightning Source. Official website. www.lightningsource.com.

Litt, Paul. *The Muses, the Masses, and the Massey Commission*. Toronto: University of Toronto Press, 1992.

Lorimer, James, and Susan Shaw. *Book Reading in Canada*. Toronto: Association of Canadian Publishers, 1981.

Lorimer, Rowland. *Build It and They Will Flow: Book Distribution in English Canada.* Research report for Department of Canadian Heritage, Ottawa, 1997.

———. "A Canadian Social Studies for Canada." *The History Teacher* 16, no. 4 (1981): 45–55.

———. *Current Disruptions in Retail and Distribution and the State of British Columbia Book Publishers.* Victoria: Government of British Columbia, 2001.

———. "The Future of English-language Book Publishing." In *Beyond Quebec: Taking Stock of Canada*, edited by Kenneth McRoberts. Montreal: McGill-Queen's University Press, 1995.

———. "Intellectual Property, Moral Rights and Trading Regimes. *Canadian Journal of Communication.* 21, 2 (1996): 267–85.

———. "Libraries, Scholars and Publishers in Digital Journal and Monograph Publishing." *Scholarly and Research Communication* forthcoming, 2012 or 2013.

———. *Magazines Alberta: Vibrancy, Growth, Interactive Community Leadership.* Calgary: Alberta Magazine Publishers Association, 2008.

———. "Marshall McLuhan: Media Genius and Noted Author of Understanding Media: The Extensions of Man." *Logos: The Journal of the World Book Community* 12, 2 (2001): 78–85.

———. *The Nation in the Schools: Wanted: a Canadian Education.* Toronto: OISE Press, 1984.

———. "Scholarly and Research Communication: A Journal and Some Founding Ideas." *Scholarly and Research Communication* 1, 1 (2010). http://www.src-online.ca.

———. "Your Canadian Reader." *Lighthouse* 2, no. 3 (1978): 6–16.

Lorimer, Rowland, and Roger Barnes. "Book Reading, Purchasing, Marketing, and Title Production." In *Book Publishing 1*, edited by Rowland Lorimer, John Maxwell, and Jillian Shoichet. Vancouver: CCSP Press, 2005.

———. *Merchandising in Independent Bookstores.* Toronto: Association of Canadian Publishers, 1998.

Lorimer, Rowland, and Patrick Keeney. "Defining the Curriculum: The Role of the Multinational Textbook in Canada." In *Language, Authority and Criticism: Readings on the School Textbook*, edited by Suzanne De Castell, Allan Luke, and Carmen Luke. London: Falmer Press, 1988.

Lorimer, Rowland, and Lindsay Lynch. *The Latest Canadian National Reading Study, 2005: Publishers' Analysis.* Vancouver: CCSP Press, 2007.

Lorimer, Rowland, and Stephen Osborne. *A Financial, Circulation and Publications Analysis of 16 Canadian Literary and 17 Canadian Arts*

Magazines. Ottawa: Canada Council, 2000.

Lorimer, Rowland, Brian Owen, and Rea Devakos. *Digital Developments in Libraries, Journals, and Monograph Publishing: Emerging Pitfalls, Practices, and Possibilities.* (Please see Lorimer, Rowland. "Libraries, Scholars and Publishers in Digital Journal and Monograph Publishing" above.)

LSM Consulting. *Book Publishers: Training Gaps Analysis in Canada 2006.* Ottawa: Cultural Human Resources Council, 2006. http://www.culturalhrc.ca/research/CHRC_Book_Publisher_TGA-en.pdf.

MacDonald, Mary Lu. "Subscription Publishing." In *History of the Book in Canada: Volume One: Beginnings to 1840,* edited by Patricia L. Fleming, Gallichan Gilles, and Yvan Lamonde. Toronto: University of Toronto Press, 2004.

MacLean, Heather. "The Supply Chain Initiative: The Inception and Implementation of a New Funding Initiative for the Department of Canadian Heritage." Project report, Master of Publishing program, Simon Fraser University, 2009.

MacSkimming, Roy. "Baie-Comeau Gets Sealed." *Quill & Quire* 58, no. 10, 1992.

———. *Making Policy for Canadian Publishing.* Toronto: ACP, 2002.

———. *The Perilous Trade: Book Publishing in Canada 1946–2006.* Toronto: McClelland and Stewart, 2007.

———. *The Perilous Trade: Book Publishing in Canada, 1946–2006,* updated edition. Toronto: McClelland and Stewart, 2009.

Manitoba. Manitoba Culture, Heritage and Tourism. "Arts Branch: Services and Programs." http://www.gov.mb.ca/chc/artsbranch/services.html.

Manguel, Alberto. *A History of Reading.* Toronto: Vintage, 1998.

Marlow, Ian. "Bid for Nortel Patents Marks Google's New Push Into Mobile World." *Globe and Mail,* Apr. 04, 2011. http://www.theglobeandmail.com/news/technology/tech-news/google-bids-for-nortel-patents/article1969788.

Martell, George, ed. "From Pilgrim's Progress to Sesame Street: 125 years of Colonial Readers." In *The Politics of the Canadian Public School.* Toronto: James Lorimer, 1974.

Matas, Mike. "A Next-Generation Digital Book." 2011. http://www.ted.com/talks/mike_matas.html.

Mathews, Robin. *The Struggle for Canadian Universities: A Dossier.* Toronto: New Press, 1969.

Matarasso, François. "Use or Ornament: The Social Impact of Participation in the Arts." In *Valuing Culture,* Conference Board of Canada. London:

Comedia, 1997.

Matthew, Robyn. "Entering a New Market: Oxford University Press Canada's Foray into French-as-a-Second-Language Publishing." Project report, Master of Publishing program, Simon Fraser University, 2007.

Maurer, Rolf. "Brought to the Brink: It's Time for All Canadians to Rethink the Role of Literature in Our Lives." *Quill & Quire* 77, no. 7 (September 2011): 14.

McGuire, Brian Patrick. *Jean Gerson and the Last Medieval Reformation.* University Park: Pennsylvania State University Press, 2005.

McGuire, Hugh. "Sifting Through All These Books." *O'Reilly Tools of Change for Publishing.* June 14, 2010. http://toc.oreilly.com/2010/06/sifting-through-all-these-book.html.

McLuhan, Marshall. *The Gutenberg Galaxy: The Making of Typographic Man.* Toronto: University of Toronto Press, 1962.

———. *The Mechanical Bride: Folklore of Industrial Man.* New York: Vanguard Press, 1951.

———. *Understanding Media: The Extensions of Man.* New York: McGraw Hill, New American Library, 1964.

Menzies, Gavin. *1434: The Year a Magnificent Chinese Fleet Sailed to Italy and Ignited the Renaissance.* New York: William Morrow (HarperCollins), 2008.

Milroy, Peter. *Canadian Books in Review: A Quantitative Analysis of Coverage Given to Books in the Weekend Review Sections of Selected Canadian Newspapers.* Vancouver: CCSP Research Report #1, 1994.

MVB Marketing-und Verlagsservice des Buchhandels GmbH. "Verzeichnis Lieferbarer Bücher (VLB)." http://www.vlb.de.

NASA. "Earth at Night." http://apod.nasa.gov/apod/image/0011/earthlights2_dmsp_big.jpg.

Neatby, Hilda. *So Little for the Mind.* Toronto: Clarke Irwin, 1954.

Neil, Garry. "The Convention As a Response to the Cultural Challenges of Economic Globalisation." In *UNESCO's Convention on the Protection and Promotion of the Diversity of Cultural Expressions: Making It Work*, edited by Nina Obuljen and Joost Smiers. Zagreb: Institute for International Relations, 2006. http://www.culturelink.org/publics/joint/diversity01/Obuljen_Unesco_Diversity.pdf.

Nemmi, Max, and Monique Nemmi. *Young Trudeau: Son of Quebec, Father of Canada.* Translated by William Johnson. Toronto: McClelland & Stewart, 2006.

Newfoundland and Labrador. Department of Tourism, Culture and Recreation. "Newfoundland and Labrador Publishers Assistance Program." http://www.tcr.gov.nl.ca/tcr/artsculture/cedp/publishers_

assistance_program.html.

———. "Publishers' Assistance Program Provides $200,000 Investment." News release, August 4, 2009. http://www.releases.gov.nl.ca/releases/2009/tcr/0804n02.htm.

Newman, Jenna. "The Google Book Settlement: The Net Benefits of Private Contract Precedence in the Absence of Adequate Copyright Law." *Scholarly and Research Communication* 1, 3 (2010): 1–56.

Nova Scotia. Official website. http://www.gov.ns.ca.

OECD. *The Well-Being of Nations: The Role of Human and Social Capital*. Paris: OECD Centre for Educational Research and Innovation, 2001.

Okerson, Ann. *Report of the ARL Serials Prices Project*. Washington: Association of Research Libraries, 1989.

Olson. David R., and Nancy Torrance, eds. *The Making of Literate Societies*. Malden, MA: Blackwell, 2001.

Ontario. *The Business of Culture: The Report of the Advisory Committee on a Cultural Industries Sectoral Strategy (ACCISS)*. Toronto: Queen's Printer, 1994.

Ontario. Royal Commission on Book Publishing. *Canadian Publishers and Canadian Publishing*. Toronto: Queen's Printer, 1972.

Ontario Arts Council. Official website. http://www.arts.on.ca.

———. "Writers' Reserve." http://www.arts.on.ca/Page119.aspx.

———. "Writers' Works in Progress." http://www.arts.on.ca/Page118.aspx.

Ontario Media Development Corporation. "Ontario Book Publishing Tax Credit." http://www.omdc.on.ca/Page3397.aspx.

Ontario Ministry of Culture and Communications, Cultural Industries and Agencies Branch. *Canadian Book Publishing in Ontario*. Toronto, 1989.

O'Reilly Media. "About O'Reilly." http://www.oreilly.com/about.

Orpwood, Graham. "Canadian Content in School Texts and the Changing Goals of Education." *Education Canada* 20, 1 (1980): 16–19.

Park, Jane. "CC Talks with: CK-12 Foundation's Neeru Khosla on Open Textbooks." *Creative Commons Blog*. April 28, 2009. http://creativecommons.org/weblog/entry/14141.

Parker, Ellie, and Adrian Furnham. "Does Sex Sell? The Effect of Sexual Programme Content on the Recall of Sexual and Non-Sexual Advertisements." *Applied Cognitive Psychology* (2007): 1217–28.

Parker, George. *Beginnings of the Book Trade*. Toronto: University of Toronto Press, 1985.

Patel, Virkram, Sisira Siribaddana, and Athula Sumathipala. *Under-representation of Developing Countries in the Research Literature: Ethical Issues Arising from a Survey of Five Leading Medical Journals*. Biomed Central, 2004. http://

www.biomedcentral.com/content/pdf/1472-6939-5-5.pdf.

Paul Audley and Associates. *Book Publishing Policy: A Review of Background Information and Policy Options*. Ottawa: Department of Communications, 1990.

Pearson, Roger. "Introduction." *Candide and Other Stories*. Toronto: Oxford University Press, 1990.

Peat Marwick Consulting Group and Bill Roberts. *English-Language Book Publishing and Distribution in Canada: Issues and Trends, Final Report*. Ottawa: Department of Communications, October 1989.

Pelletier, Gérard. *Federal Aid for Publishing*. Text of speech, Montreal, Quebec, February 11, 1972, *Quill & Quire*, March 1972.

Pendakur, Manjunath. *Canadian Dreams and American Control: The Political Economy of the Canadian Film Industry*. Detroit: Wayne State University Press, 1990.

Perrin, Robin. "From Cambridge to Communication: McLuhan beyond McLuhanism." Master's thesis, Simon Fraser University, 1991.

PLAN Institute. Official website. http://www.planinstitute.ca.

Poulter, Sean. "Sex Sells: Mills and Boon Boom Sparked by Downloaded Romances." *Bookbee*. March 28, 2010. http://bookbee.posterous.com/sex-sells-mills-and-boon-boom-sparked-by-down.

Public Knowledge Project. Official website. http://pkp.sfu.ca.

Public Lending Right Commission. Official website. http://www.plr-dpp.ca.

Putnam, Robert. "Bowling Alone: America's Declining Social Capital." *Journal of Democracy* 6, 1 (1995): 65–78.

———. *Making Democracy Work: Civic Traditions in Modern Italy*. Princeton: Princeton University Press, 1993.

Quill & Quire. *The Book Trade in Canada, 2008 Edition*. Toronto: Quill & Quire, 2008.

Radway, Janice. *Reading the Romance: Women, Patriarchy and Popular Literature*. Chapel Hill: University of North Carolina Press, 1984.

Reza, Abeer, and Charles Saunders. *Economic Contribution of the Canadian Magazine Industry*. Toronto: Informetrica Ltd. for Magazines Canada, 2006.

Repo, Satu. "From Pilgrim's Progress to Sesame Street: 125 years of Colonial Readers." In *The Politics of the Canadian Public School*, edited by George Martell. Toronto: James Lorimer, 1974.

Roberge, Angela. "Adolescents' Reading in the Lower Mainland (of Vancouver)." Honours project, Simon Fraser University, 1995.

Robertson v. Globe and Mail; *Robertson v. Canwest Global, et al.* Toronto, 2009.

Rose, Jonathan. *The Intellectual Life of the British Working Classes*. New Haven,

CT: Yale University Press, 2001.

Ruiz, Janet. "A Literature Review of the Evidence Base for Culture, the Arts and Sport Policy." *Valuing Culture*, Conference Board of Canada. Edinburgh: Scottish Executive Education Department, 2004. Original paper available at http://www.scottishexecutive.gov.uk.Publications /2004/08/19784/41507.

Sanderson, Heather. "A PEXODyssey: Bibliographic Data Management and the Implementation of PEXOD at the Dundurn Group." Project report, Master of Publishing program, Simon Fraser University, 2004.

Sawchuk, Aaron, and Graham Henderson. "Intellectual Property: Engine for Growth and the Backbone of the Creative Economy." Conference Board of Canada, *Valuing Culture: Compendium of Research Papers*, Paper #22: 189–200.

Scala, Maria. "The New Face of Canadian Publishing: A Corporate History and Analysis of Alfred A. Knopf Canada." Project report, Master of Publishing program, Simon Fraser University, 2000.

Seller, Shyla. "Implications of Authorship: The Author/Editor Relationship from Proposal to Manuscript." Project report, Master of Publishing program, Simon Fraser University, 2005.

Shatzkin, Mike. "Digital Publishing in the U.S.: Driving the Industry to Vertical Niches?" *Logos: The Journal of the World Book Community* 19, 2 (2008): 56–60.

Shoebridge, Paul, and Michael Simons. *Welcome to Pine Point*. Montreal: National Film Board. http://interactive.nfb.ca/#/pinepoint.

Siemens, Raymond. "Raymond G. Siemens: Profile." http://web.uvic .ca/~siemens.

Simulating History. "Home Page." http://www.simulatinghistory.com.

Skelton, Caroline. "The Wal-Mart Level of Excellence." *Quillblog* (blog of *Quill & Quire*). September 2, 2005. http://www.quillandquire.com/blog/ index.php/2005/09/02/the-wal-mart-level-of-excellence.

Slocum, Marc. "Amazon Growth Fuels Online's Book Market Share." *O'Reilly Tools of Change for Publishing*. April 16, 2008. http://toc.oreilly .com/2008/04/amazon-growth-fuels-onlines-bo.html.

Statistics Canada. "Book Publishing Industry." *The Daily*. July 10, 2008. http://www.statcan.ca/Daily/English/080710/d080710a.htm.

———. *Book Publishers, 2008*. Cat. no. 87F0004X. Ottawa: 2010.

———. *Canadian Framework for Culture Statistics*. Cat. no. 81-595-MIE2004021. Ottawa: 2004. http://www.statcan.ca/english/research/81-595-MIE/81- 595-MIE2004021.pdf.

———. *Culture, Tourism and the Centre for Education Statistics.* Cat. no. 87F0004X. See also Statistics Canada, "Book Publishing Industry." *The Daily.* July 10, 2008. http://www.statcan.ca/Daily/English/080710/d080710a.htm.

———. Service Industries Division. *Book Publishers, 2006.* Cat. no. 87F0004X. Ottawa: Statistics Canada, 2009. http://www.statcan.gc.ca/pub/87f0004x/87f0004x2008001-eng.pdf.

———. Service Industries Division. *Book Publishers, 2007.* Cat. no. 87F0004X. Ottawa: 2009.

———. Service Industries Division. *Book Publishers, 2008.* Cat. no. 87F0004X. Ottawa: 2010.

Strauss, Marina. "Indigo Aims to Shake the Dust out of its Book Bins." *Globe and Mail*, November 24, 2011, B3.

———. "Less Tolstoy, More Toys." *Globe and Mail,* April 9, 2011.

———. "Nine Questions for Heather Reisman." *Globe and Mail,* May 27, 2010. http://www.theglobeandmail.com/report-on-business/rob-magazine/nine-questions-for-heather-reisman/article1577613/?cmpid=rss1.

Tawney, R.H. *Religion and the Rise of Capitalism.* London: John Murray, 1926.

Tenopir, Carol, and Donald W. King. "Trends in Scientific Scholarly Journal Publishing in the United States." *Journal of Scholarly Publishing,* April 1997:135–70.

Thakur, Neil. National Institutes for Health (USA). "Funding Policies and Research Access–Round Table." Panel discussion, International Conference on Electronic Publishing (Elpub), Toronto, June 27, 2008. Available at www.utsc.utoronto.ca/~elpub2008.

Tiessen, Paul. "From Literary Modernism to the Tantramar Marshes: Anticipating McLuhan in British and Canadian Media Theory and Practice." *Canadian Journal of Communication* 18, no. 4 (1993), 451–68.

Tomlinson, Nicole. "Canadian Books Now Mandatory in High Schools." *Vancouver Sun*, July 5, 2008.

Tong, Murray Chun-Kee. "Influences and Implications of Acquisitions Decisions at UBC Press." Project report, Master of Publishing program, Simon Fraser University, 2009.

Turku School of Economics and Business Administration and Rightscom. *Publishing Market Watch: Final Report.* Brussels: DG Enterprise of the European Commission, 2005. http://ec.europa.eu/information_society/media_taskforce/doc/pmw_20050127.pdf.

Turner-Riggs. *The Book Retail Sector in Canada.* Ottawa: Department of Canadian Heritage, 2007.

Uglow, Jenny. *The Lunar Men: Five Friends Whose Curiosity Changed the World.* New York: Farrar, Straus and Giroux, 2002.

United Nations. *Creative Economy: Report 2010, Creative Economy: A Feasible Development Option.* United Nations: UNCTAD, 2010. Also available online at http://www.unctad.org/Templates/WebFlyer.asp?intItemID=5763&lang=1.

UNESCO. MacBride Commission. *Many Voices: One World: Report of the International Commission on Communication Problems.* Paris: Unipub, 1980.

United Kingdom. Department of Culture, Media and Sport. *Creative Britain: New Talent for a New Economy.* London: British Government, 2008.

United States. United States District Court. Southern District Court of New York. The Author's Guild et al., plaintiffs against Google Inc. http://www.nysd.uscourts.gov/cases/show.php?db=special&id=115.

University of Delaware Library. Special Collections Department. "Seventy Years at the Hogarth Press: The Press of Virginia and Leonard Woolf." http://www.lib.udel.edu/ud/spec/exhibits/hogarth.

viHistory. "How It Works." http://vihistory.ca/content/about/intro.php.

Vygotsky, Lev. *Thought and Language.* Cambridge, MA: MIT Press, 1968.

Watkinson, Anthony. "Open Access: A Publisher's View." *Logos* 17 (2006): 47.

Webcom. Official website. http://www.webcomlink.com/tech/digital.htm.

White, Howard. *Books and Water.* Video portrait of Harbour Publishing, Vancouver: Simon Fraser University, Master of Publishing program, 1992.

Whittman-Hart, Divine. "Canadian Book Industry: Transition to the New Economy." Slide Presentation to the Association of Canadian Publishers, April 30, 2001.

Wigod, Rebecca. "New Rules Coming to Get Canadian Literature in Schools." *Vancouver Sun,* July 26, 2008.

Wikipedia contributors. "History of General Motors." *Wikipedia, The Free Encyclopedia.* http://en.wikipedia.org/w/index.php?title=History_of_General_Motors&oldid=484790844.

Williams, Erin. "The Chapters Effect on British Columbia-Based Literary Publishers." Project report, Master of Publishing program, Simon Fraser University, 2006.

Woll, Thomas. *Publishing for Profit.* Chicago: Chicago Review Press, 2006.

Woods, Stuart and Jason Spencer. "Publishers Brace for Uncertainty as Indigo Introduces New Returns Policy." *Quill & Quire.* July 7, 2011. www.quillandquire.com/google/article.cfm?article_id=11894

World Trade Organization. *Canada–Certain Measures Concerning Periodicals.* Dispute DS31, 1996, 1997.

Wright Investment Services. "Amazon.com, Inc. — Company Profile Snapshot, 2009." *Wrightreports.* 2009. http://wrightreports.ecnext.com/coms2/reportdesc_COMPANY_023135106.

Xiao, Dongfa. "Book Publishing in China (2011)." Guest presentation at Simon Fraser University, March 30, 2011.

Yarrow, Jay. "Dan Brown's E-Book Sales Aren't So Killer After All." *Business Insider,* September 22, 2009. http:// www.businessinsider.com/dan-browns-e-book-sales-arent-so-killer-after-all-2009-9.

Zifcak, Michael. "Australia Without Retail Price Maintenance." *Logos,* 2 (1991): 204–8.

Žiga, Turk. "Lisbon Strategy 2.1: Europe — The Most Creative Economy in the World." Presentation for a Pre-Presidency Conference, Ljubljana, December 3–4, 2007. http://www.slideshare.net/ziga.turk/lisbon-strategy-21-europe-the-most-creative-economy-in-the-world-full.

Zuiker, Anthony, and Duane Swierczynski. "Level 26: Dark Origins." *BSCReview,* August 26, 2009. http://www.bscreview.com/2009/08/level-26-dark-origins-by-anthony-zuiker-duane-swierczynski-reviewed.

INDEX

Page numbers in italics refer to information in tables.

AbeBooks, 212, 261, 284–285
Aboriginal culture, 14, 42, 85, 94, 176, *177*, 185, 294
Absolute Friends (Le Carré), 38
Access Copyright, 97, 154, 168–169, 243, 304, 331
access to information, free, 8, 244, 302–303, 305, 318
acquisition, 22–23, 26, 36. *See also* list building; origination
    and authors' agents, 236
    of scholarly works, 320, 322, 323
    in service publishing model, 319
    as socio-technical process, 5
Across the Universe, 293
Act Respecting Copyrights, 63
advances, 234, 236, 258
Advisory Committee on a Cultural Industries Sectoral Strategy (ACCISS), 127, 132. *See also The Business of Culture* (ACCISS)
Africa, 67
agencies represented by foreign-owned publishers, 232–233
agency publishing, 58, 63–66, 73, 210–211, 235
agents, authors', 25–26, 29, 236–237
Aid to Industry and Associations, 179
Aid to Publishers, 179
Aid to Scholarly Publishing Program (ASPP), 104, 174, 183, 322
Alberta, 188, 268–269, 275, 306
Alberta Foundation for the Arts, 188
Aldus Pagemaker, 256
alphabets, 21, 51, 254
Amazon, 13
    and AbeBooks, 285

    bargaining power of, 226, 270
BookSurge, 281, 285
CreateSpace, 281, *283*
    and databases, 261
    and digital browsing, 250
    distribution in Canada (Amazon.ca), 228, 229, 315
    and e-books, 15, 251, 333. *See also* Kindle
    inefficiencies of, 284
    Kindle, 252, 287, 332
    and the long tail, 227
    and selection of titles, 261
    and Trafford, 282
    user data, 286
American Books in Print, 209
Anderson, Chris, 227, 284
Anderson, Scott, 239
Android, 251, 287, 288
Anglo-American Reciprocal Copyright Agreement, 63
Anne of Green Gables (Montgomery), 69
Apple, 252, 256, 287, 288, 335
apps, 286–288
Arouet, François-Marie. *See* Voltaire
Artists and Community Collaboration Fund, 186
Association for the Export of Canadian Books (AECB)/Livres Canada Books, 82, 92, 105, 174, 178–179
Association of American Publishers, 250
Association of Canadian Publishers (ACP), 98, 126, 137–138, 244
    Book Publishing and Canadian Culture, 110–112
    Canadian Publishing, 101–104
    Collections, 198

eBOUND project, 334–335
A Mid-Decade Assessment, 104–107
New Directions, 116–119
and NIICD, 146
unrealistic goals of, 105, 106
Association of Manitoba Book Publishers, 189
Association of Universities and Colleges of Canada, 85
Athabasca University Press, 323
Atlantic Books Today, 200
Atwood, Margaret, 85, 237
Audley, Paul, 113–114
Austen, Jane, 196, 197
Australia, 16, 44, 151, 209, 210
Australia Council for the Arts, 16
AuthorHouse, 318
authors, Canadian. *See* Canadian authors
authors, general
    as central to publishing, 37
    commissioned, 36
    contracted, 94, 95, 335
    creative process of, 272
    cultivation of, 30–31
    and databases, 261
    as factor influencing book purchase, 216–218
    interactivity with readers, 272, 274
    and relationships with publishers, 5, 23, 142–143, 305
    as representatives of communities of interest, 275
    and rights sales, 29
    as risk takers, 335
    scholarly, 321
    unestablished, 22–23, 26, 30, 316
Authors Guild, 250
authorship, history of, 264
Author Solutions, 318
awards. *See also specific awards*
    Canada Council, 83, 186
    and cultural partners to the book industry, 4, 92, 201–202, 307
    and marketing, 25
    Ontario Royal Commission on Book Publishing on, 90
    provincial, 188, 189

baby boom, 84, 293
backlist, 90, 249, 260, 313, 317
Bacque, James, 85
Baie-Comeau Agreement, 82, 106, 106–109, 113–114, 119, 233. *See also* Investment Canada Act
Baird, Jean, 198–199
Barnes & Noble, 143, 251, 333
Barthes, Roland, 265
BC Bookworld, 200
Bédard, Pierre, 57
The Beginnings of the Book Trade in Canada (Parker), 55, 58, 59, 61, 62
Belford Brothers, 58
Belgium, 44
Béliveau, Jean, 308
Bennett, Avie, 308, 309, 312, 313–314
Berne Convention, 63, 166, 278, 303
Berners-Lee, Tim, 336
Bertelsmann, 12, 138, 233
Berton, Pierre, 73
Besse, Ron, 74
Bevington, Stan, 85
Bible, 38, 41, 255, 318
The Big Switch (Carr), 262
Bijker, Wiebe, 251, 263
Bill C-11, 169, 243–244, 302, 303
Bill C-32, 169, 243–244, 265, 302, 303
Bill of Rights (Britain, 1689), 8
binding, 5
BioMed Central, 320–321
Birds of series (Lone Pine), 213
BitTorrent, 287
BlackBerry, 271, 287
Black Rock (Connor), 64
Blockbusters and Trade Wars (Grant, Wood), 140–145
Block Grants (Canada Council), 98, 181, *183*, 326
Bloomsbury Academic, 323
Bloomsbury Group, 32, 34, 36, 69, 272, 293
BMO financial group, 202
book clubs, 39, 85, 223, 272, 301
book fairs. *See* rights fairs
"the book habit," 48
Book Importation Regulations, 46, 166

Book Manager, 92
BookNet Canada, 171–172
    efficiencies due to, 207
    establishment of, 83, 151, 152
    sales data, 92, 234, 242, 261
    and the Supply Chain Initiative (SCI), 128, 151
Book Policy and Programs Division, 179
Book Publishers Operating Support Initiative (BPOSI), 188
Book Publishers Project, 189
Book Publishing and Canadian Culture (ACP), 110–112
Book Publishing Board, Canadian, 90, 98
Book Publishing Development Program, 82, 100, 104, 105–106. *See also* Book Publishing Industry Development Program (BPIDP)/Canada Book Fund (CBF)
Book Publishing Industry Development Program (BPIDP)/Canada Book Fund (CBF), 174–176, 186
    establishment of, 83, 107
    evaluation of, by Fox Jones, 115–116
    renewal of, 115–116, 118, 307
    in service publishing model, 326
    university presses as, 321
book purchasers. *See also* readers
    and databases, 261
    demand structure of, 215–224
    factors influencing, 143, 216–218, 217
    women as, 37, 39, 184, 216, 222, 294
The Book Report (DCH), 179–181
book retailing, 13, 149, 220–224. *See also* booksellers; *specific bookstores*
books
    as cultural objects, 3–6, 10, 21, 39, 88, 249
    definition of, 21
    as a distinctive medium, 39, 266
    as physical objects, 21
    as preservers of ideas, 21, 38
    as repositories of thought, 21
    and social change, 41–43
booksellers, 4, 25. *See also* book retailing; *specific bookstores*

    as advisors, 223
    associations of, 66
    and BookNet Canada, 171
    Confederation to 1950, 58, 63
    foreign ownership of, 228
    lobbying for Foreign Reprints Act, 57–58
    in pre-Confederation Canada, 56
    as printers, 65
    as publishers, 65
    resistance of, to change, 125
    as uneconomical for publishers, 260
Books in Print, 209–210
bookstores. *See* booksellers; chain bookstores; independent bookstores; university bookstores; *specific bookstores*
BookSurge, 281, 285
Bordieu, Pierre, 9, 147
Boulton, Matthew, 34
branch plants. *See also* foreign-owned publishers
    in the early twentieth century, 65
    in educational publishing, 68, 72–73, 84, 95
    and run-ons, 210, 211
    and title origination, 235
Brave New Words, 189
Brave New World (Huxley), 42
Brazil, 10, 333
A Brief History of Time (Hawking), 39
Briggs, William, 64
British Books in Print, 209
British Colonial Office, 56
British Columbia
    arts funding cuts in, 307
    booksellers in, 172
    and GDS collapse, 240
    Harbour Publishing on, 32
    libraries in, 91
    schools and Canadian literature, 199
    support for book publishing in, 202, 306
    tax credits, 137, 174, 186–187, 307
British Columbia Arts Council, 187
British Columbia National Award for Canadian Non-Fiction, 201
British North America (BNA), 57, 59–60, 61.

See also colonial history of Canada
British North America Act, 62
British Parliament, and copyright in Canada, 63
Broadcasting Act (1968), 88, 315
Broadview Press, 198
Brock, Isaac, 61
Brown, Dan, 251
Bruce, Phyllis, 329
Bruner, Jerome, 40
Budziak, Kimberly, 253
Bunyan, John, 38
Burns, Robbie, 197
The Business of Culture (ACCISS) and the creative economy, 127–129, 131–136, 153–154, 191, 195
 release of, 83, 127
 and Toronto as cultural centre, 131–132, 336–337
"buying around." *See* parallel importation

Callaghan, Morley, 73
Camera Lucida (Barthes), 265
Camp, Dalton, 86, 88
Campus Alberta Repository of Educational Objects (CAREO), 268–269
Canada
 colonial history of, 46, 52, 55–62, 67–68, 294, 328
 cultural identity of. *See* Canadian cultural identity
 government of. *See* federal government
 publishing infrastructure in, lack of, 44
Canada Book Fund. *See* Book Publishing Industry Development Program (BPIDP)/ Canada Book Fund (CBF)
Canada Council for the Arts
 ACP on, 112
 Block Grants, 98, 181, *183*, 326
 Emerging Publisher Grants, 181–182, 183, *183*, 193
 establishment of, 70–71
 grants to university presses, 322
 and literary magazines, 199–200
 not addressed in 1984 ACP report, 104
 in service publishing model, 325–326
 support for book publishers, 82, 98–99, 116, 174, 186, 306, 322. *See also specific grants*
 support for writers, 185–186, 199
Canada Foundation for Innovation, 249
Canada-Japan Literary Awards, 186
Canada New Media Fund, 185
Canada West. *See also* Ontario; Upper Canada
Canadian authors
 ACP on, 117
 associations of, 66
 and BPIDP eligibility for publishers, 108, 175
 and choice of publisher, 237
 and copyright, 63
 of educational materials, 93, 94, 335
 emerging, 12, 160
 as factor influencing book purchase, 217–218
 future of, 329–330
 government support for, 79, 80
 interaction of, with Canadian educators, 93
 international recognition of, 9, 46, 80, 184, 246, 315, 329
 in libraries, 197–198
 and McClelland & Stewart, 84, 86, 89
 numbers of, 117
 Ontario Royal Commission on Book Publishing on, 87–88
 published by foreign/foreign-owned publishers, 69, 73, 137, 307, 312
 readers' interest in, *218–219, 219*
 recognition of, within Canada, 202–203
 royalty insurance for, 90, 91
Canadian Bank Act, 87
Canadian book market
 ACP on, 102, 117
 competitive disadvantage of, 43–44, 80, 130, 131–132
 in First World War, 65
 size of, 46, 213
 structure of, 6, 208–215
Canadian Book Publishing Development Program (CBPDP), 82, 100, 104, 105–106. *See also* Book Publishing Industry

Development Program (BPIDP)/Canada
  Book Fund (CBF)
Canadian Bookseller, 200
Canadian Booksellers Association (CBA), 150,
  235
Canadian Books in Print, 209
Canadian Copyright Act (1872), 62–63
Canadian Copyright Licensing Agency, 168.
  See also Access Copyright
Canadian cultural identity, 22, 45, 52, 55, 69,
  70, 88, 120, 123, 124, 125, 153, 329, 333
  vs. educational theory, 93
Canadian Cultural Observatory, 185
Canadian Culture Online, 185
The Canadian Encyclopedia, 308, 309
Canadian Federation of the Humanities and
  Social Sciences, 183
Canadian Institutes for Health Research
  (CIHR), 305
Canadian Knowledge Research Network, 333
Canadian Learning Materials Centre, 91
Canadian Literary Review, 200
Canadian Materials, 200
Canadian-owned publishers. *See also specific
  Canadian-owned publishers*
  access to grants, 235
  and Canadian educators, 93
  competitive disadvantage of, 235, 245
  financial status of, 49, 114, 127
  future of, 329–330
  and lack of access to loans, 50
  and lack of earnings, 50
  as local industry focused on domestic
    market, 69
  market share of, 12, 111, 129, 160
  and non-mainstream products, 49
  and own titles, 11, 12, 161, 162, *163*, *164*,
    165
  resilience of, 245
  support for. *See* support for book
    publishing
Canadian ownership
  and BPIDP/Canada Book Fund, 175
  Marcel Masse on, 106
  regulations, 307, 310, 314–315. *See
    also* Baie-Comeau Agreement;
  Investment Canada Act
  statistics on, *165*
Canadian Publishers' Council (CPC), 105, 150
Canadian publishing. *See also* publishing
  1960s to 1980s, 70, 79
  barriers to entry, 192–193
  beginnings of, 21–22
  Canadian culture as focus of, 74, 336
  Confederation to 1950, 62–67
  cuts to, 135
  emergence of, 70, 294
  and Foreign Reprints Act, 58
  as foundation of Canadian culture, 88
  history of, 6, 17, 293, 333
  vs. international publishing, 67–68
  not addressed in Massey Commission,
    72
  pre-Confederation, 55–58, 59–62
  technological effects on, 289
Canadian Publishing (ACP), 101–104
Canadian Publishing Development
  Corporation, 98
Canadian Radio-television and
  Telecommunications Commission
  (CRTC), 315
Canadian Research Knowledge Network
  (CRKN), 249
Canadian Studies, 93, 184
Canadian Television Fund, 326–327
Le Canadien, 57
Cancopy. *See* Access Copyright
Candide (Voltaire), 35–36
Carr, Nicolas, 262
Carrier, Roch, 150
Carroll, Lewis, 254
Carson, Rachel, 294
Catholicism, 60
Cats (Webber), 37
CBC, 92
  Literary Awards, 186
  Radio, 195
Cengage, 12, 96, 159
Centennial College, 200
Centre for Culture Industries and Technology,
  134
chain bookstores, 6, 13, 105, 143, 149, 221, 223.

See also specific chain bookstores
Challen, Paul, 27
Chapters, 128, 136, 149–152, 239. *See also* Chapters/Indigo
Chapters/Indigo, 13, 224–231. *See also* Chapters; Kobo
   advertisements, 195, 228
   an alternative model, 226–229
   bargaining power of, 270
   and databases, 261
   influence of, on demand patterns, 207
   marketing and, 48
   as publisher (Prospero Books), 25
   purchasing choices of, 143
   and relationship with foreign-owned firms, 238
   terms of sale, 242
Charles Taylor Award, 201
Charter of Rights, Canadian, 123
Chateau Clique, 57
Chatelaine, 143
Chen, Steve, 336
children's books
   awards, 202
   bookseller knowledge about, 222, 224
   Canada Council funding for, 98–99, 182
   displays of, in bookstores, 269
   sales of, 11
children's publishers, 26, 222, 311. *See also specific children's publishers*
Chin, Judge Denny, 250
China, 10, 39, 41, 274, 333
Chrétien, Jean, 120, 135
Christian Guardian, 328
Churchill, Winston, 73
Cineplex, 224
Circular 14 (Trillium List), 94
Clarke Irwin, 74, 84
Clarkson, Adrienne, 308
Clarkson Gordon and Company, 86
Classics (bookstore), 13
Clemens, Samuel Langhorne (Mark Twain), 58
cloud computing, 253, 262–263, 276
Coach House Press/Books, 84, 85
codex, 21
Coles, 13

Collections (ACP), 198
Colonial Advocate, 57
colonial history of Canada, 46, 52, 55–62, 67–68, 294, 328
colonialism, general, 67–70
Commissions of National Education (British), 59
Common Sense Revolution, 244
Commonwealth countries, 48, 56, 209, 210. *See also specific Commonwealth countries*
Competition Bureau, 225
compilations (course packs), 260, 324
concentration of ownership
   ACP on, 111
   and costs to the consumer, 228
   as detriment to culture, 81, 315, 316
   Donner and Lazar on, 137–138
Condorcet, Marquis de, 7
Confederation, Canadian, 58
Conference Board of Canada, 295, 296, 300
Connor, Ralph (Rev. Charles W. Gordon), 64, 69
Conseil des Arts et Lettres du Québec, 188
Conservatives (Canada), 106, 120, 170, 228, 307
Conservatives (Ontario), 244
Constitution, Canadian, 123
"consumption" of cultural products, 141, 142
content management systems (CMS), 253, 258
content quotas, 132, 135, 144, 299
contracts, publishing, 26, 73, 201, 258, 313
Convention on Cultural Diversity (CCD)
   as Canadian achievement, 145, 154
   creation of, 83, 145
   and the creative economy, 159, 172, 191, 296, 337
   effects of, 146–147, 170
Convention on the Protection and Promotion of the Diversity of Cultural Expressions. *See* Convention on Cultural Diversity (CCD)
Cooke Agency, 237, 310
Copp Clark, 73, 84
Copps, Sheila, 118, 120, 140, 146
co-publication, 98, 102, 161, 234
copy editing, 23
copyright, 5. *See also specific copyright laws/*

agreements
   in Canada, 62–63, 96–97, 166–169, 172.
      See also Copyright Act (Canada)
   and Creative Commons, 306
   early history of, 264, 271
   and education, 243–244, 288, 303–304, 331
   infringement, 55, 57
   international agreements, 97, 294. *See also* Berne Convention
   laws, as structural support, 46, 166, 302–305
   and rights sales, 25–26
   in the U.K., 56
   in the U.S., 58, 63, 64
Copyright Act (Canada). *See also* Bill C-11
   of 1872, 62–63
   and distribution right, 47, 81, 83, 102, 113, 114, 118, 154, 166, 235
   and educational copying, 287
   Ontario Royal Commission on Book Publishing on, 96–97
   today, 243, 302
Copyright Board of Canada, 169
Costco, 13, 207, 283
cost of sales, 117, 188, 208
Council of Ministers of Education of Canada, 244, 304
course packs, 260, 324
Cramer, Gabriel and Philibert, 35
CreateSpace, 281, *283*
Creative Britain, 298
Creative Commons, 305–306
creative economy
   The Business of Culture on, 128, 135
   as a concept, 9–10, 81, 159, 246
   infrastructure, in Canada, 14, 148, 334, 337
   and macro-economics of publishing, 315
   Valuing Culture on, 295–301
Creative Economy Report (UN), 297–298, 299, 301
Creighton, Donald, 42
criticism, literary, 4, 39, 80, 88, 182, 243
Crosbie, John, 308
crowd-sourcing, 274, 336

cultural discount, 141, 207
cultural diversity
   Book Publishing and Canadian Culture on, 111
   CCD on, 146. *See also* Convention on Cultural Diversity (CCD)
   as central to Canadian publishing, 9, 336, 337
   and the creative economy, 299
   Grant and Wood on, 144
   and NIICD, 139
   and technology, 333
cultural environment necessary for publishing, 44–45, 327–330
cultural exemption, 3, 138, 139
Cultural Human Resources Council (CHRC), 200
cultural industries
   as assets, 148, 154, 300. *See also* creative economy
   in Canada, 55, 125–126, 127–129, 133, 330
   economics of, 141–142, 145, 207
   policy on, 9, 144, 190
   SAGIT on, 139
Cultural Industries Advisory Council, 134
Cultural Industries Development Fund (Canada), 112, 114
Cultural Industries Development Fund (Saskatchewan), 188–189
Cultural Industries Guarantee Fund, 188
Cultural Industries Sectorial Advisory Group on International Trade (SAGIT), 138
cultural partners of the book industry, 193–201
Cultural Personalities Exchange Program, 184
cultural practice, publishing as, 36–37
cultural support for publishing, 79, 80, 82, 181–183, 202, 306–307. *See also specific cultural support program*
cultural tool kit (Grant, Wood), 144
Culture.ca, 185
Culturescope.ca, 185
curatorial function of publishers, 263, 319, 323
custom publishing, 260, 324. *See also* service publishing
Cutler, May, 308

415

Index

D&M Publishers, 27, 138, 160, 234, 314, 315
D'Agostina, Giuseppina, 304
Darnton, Robert, 34–35
Darwin, Erasmus, 34
databases and publishing, 253–254, 257–263, 276, 280, 285–286, 289
David, Jack, 31
Davidson, Andrew, 236
Davies, Robertson, 308
The Da Vinci Code (Brown), 251
Defoe, Daniel, 38
Dennys, Louise, 329
Department of Canadian Heritage (DCH). *See also specific DCH programs*
   and AECB/Livres Canada Books, 178
   Book Policy and Programs Division, 179
   on book publishing industry weaknesses, 131
   and BPIDP/Canada Book Fund, 174, 326
   Canadian Culture Online, 185
   Canadian Studies program, 93
   cuts to publishing, in 1995, 116
   and magazine policy, 136
   and NIICD, 140, 145, 146
   Publishing Measures, 92
   Reading and Buying Books for Pleasure, 216
   recovery plan, in 1997, 118
   Standing Committee on Canadian Heritage on, 150
   statistics on support for book publishing, 179–181
   study, in 2000, of book publishing, 136–138
   support for book publishing, 174, 306
   Valuing Culture, 295
Department of Foreign Affairs and International Trade (DFAIT), 183, 306
design, book, 4, 5, 24, 66, 208, 256–257
desktop publishing, 256
developmental editing, 23, 27–28
De Waard, Anita, 322
Dewey, John, 72
Diderot, Denis, 271–272
digi-novels, 273

digital books, 176, 267, 332. *See also* e-books
digital rights management (DRM), 303
Digital Transition Fund, 188
digitization, book, 250, 333, 334
Directory of Open Access Books (DOAB), 324
discounts, publishers', 15, 149, 226, 241
Distican, 239
distribution, 25
   ACP on, 103, 111
   and agency publishing. *See* agency publishing
   Canadian distributors, 161
   consolidation of, 240
   cost structure of, 208
   Egerton Ryerson and, 61
   in Europe, 15
   as foundation of publishing in Canada, 44, 104
   infrastructure, 168
   and market functioning, 237–241
   and small publishers, 240
   and the Supply Chain Initiative (SCI). *See* Supply Chain Initiative (SCI)
distribution right
   abolishment of, potential, 235
   ACP on, 102, 104, 112, 118
   establishment of, 9, 47, 83, 97, 114
   as structural support, 81, 154, 166, 302
La Dolce Vita (Fellini), 265–266
Donner, Arthur, 136–138
Doubleday, 108, 160
Douglas & McIntyre. *See* D&M Publishers
Douglas Gibson Books, 310
Drupal, 274
Dryden, Ken, 28
The Duchovny Files, 31
Duthie Books, 231, 244
Dutton, 273

Eayrs, Hugh, 65
e-books, 285–286, 332–333. *See also* digital books
   and apps, 287
   and Chapters/Indigo, 229, 230
   and decline of print, 230
   distribution of, 270, 315

enhanced, 51
formats of, 252
free, 318
market for, 249–252
Ontario Media Development
    Corporation on, 188
pricing of, 15
production of, 270
sales of, 231
supply chain for, 270
eBOUND project, 334–335
economics
    "curious," of cultural industries,
        141–142, 145, 207
    general, and book consumption, 48
    of publishing, 37, 46, 48–49, 100,
        102, 117, 126. *See also* macro-
        economics of publishing;
        micro-economics of publishing
The Ecstasy of Rita Joe (Ryga), 42
ECW Press, 31, 329
editing, 4, 5, 23–24, 27–28, 208. *See also*
    proofreading
editors, 192, 329
education. *See also* educational publishing;
    educators
    Canadian content in, 49, 72, 85, 91, 94,
        331
    Canadian literature in, 198–199
    colonial, 67
    as cultural partner of book industry,
        198–199
    and fair dealing (education exemption),
        243–244, 288, 303–304, 311
    and literacy, 172–173
    post-secondary, 85, 198, 331. *See also*
        universities
    pre-Confederation, 59–62
    and the provinces, 49–50, 93
educational environment necessary for
    publishing, 49–50, 51–52, 330–331
educational publishing
    ACP on, 106, 113
    and apps, 287
    in Canada today, 330–331
    Canadian content in, 94

in the colonial era, 59–60
and copyright, 330
custom publishing in, 260, 324
and data on digital consumption, 286
foreign influence on, 50
foreign ownership of, 12, 66, 68, 72–73,
    84, 91, 119, 159
interactivity in, 275
multimedia in, 335
Ontario Royal Commission on Book
    Publishing on, 88, 92–96
separation of, from trade publishing, 66,
    68, 95, 115, 120
Statistics Canada on, 113
educators. *See also* education
    and Canadian content in curriculum,
        72, 91, 331
    custom publishing for, 324–325
    and data on digital consumption, 286
    post-secondary, 269
    resistance of, to change, 125
    training of, 92, 93, 94
Edwards Books and Art, 244
Edwards Brothers, 280
Egypt, 274
Eisenstein, Elizabeth, 269
electronic data interchange (EDI), 105, 151, 171
Eliot, T.S., 37
Elsevier, 160, 322
Emergency Committee of Canadian
    Publishers, 85. *See also* Association of
    Canadian Publishers (ACP)
Emerging Publisher Grants (Canada Council),
    181–182, 183, *183*, 193
Encyclopedia of British Columbia, 32
England, 15, 35, 44. *See also* Great Britain;
    United Kingdom
Enlightenment, 7, 264
entitlement, culture of, 191, 326
environmental concerns, 4, 50, 294
environments, necessary, for publishing, 6,
    43–50, 51–52, 69–70, 104, 327–333
e-readers, 5, 222, 332. *See also specific e-readers*
Essays on Canadian Writing, 31
Europe. *See also specific European countries*
    books about, in First World War, 65

and British copyright law, 57
and the creative economy, 299
emigration from, 59
public lending right in, 46
publishing support in, 14–15
run-on copies in, 210
Voltaire in, 34–36
European Year of Creativity and Innovation, 299
exporting. *See also* Association for the Export of Canadian Books (AECB)/Livres Canada Books
   ACCISS on, 133
   ACP on, 111, 112, 117
   and authors' agents, 237
   marketing for, 98, 105
   Ontario Royal Commission on Book Publishing on, 90, 92
   stifling of, due to U.S. Copyright Act, 64
Export Marketing Assistance Program (EMAP), 178

Faber & Faber, 37
Facebook, 252, 253, 271, 335, 336
fact-checking, in editorial process, 23
fair dealing, 169, 243, 287, 303–304
Famous Players, 224
fan fiction, 274
federal government. *See also specific federal departments and agencies*
   ACP on, 112
   and international agreement for trade of culture products, 83
   as leader in publishing support, 15–16
   and magazine policy, 9, 136, 142
   and Ontario Royal Commission on Book Publishing recommendations, 86
   and Public Lending Right (PLR), 169
   publishing, 96
   publishing policies, review of, 113
   support for book publishing, 13–14, 22, 46, 98, 104, 119, 148, 153, 179–181. *See also specific support programs*
   support for culture, 45, 99

and tariff on importation, 63
view on social capital, 147
Fellini, Federico, 265–266
festivals, literary, 92, 185
fiction
   awards for, 201–202
   Canada Council funding for, 182
   by Canadian authors, 64
   Canadian content in, 73
   editing, 23
   in education, 94
   low sales of, 317
   Massey Commission on, 71
   and McClelland & Stewart, 30, 86, 308, 312
   and Methodist Book and Publishing House, 64
   and social change, 42
Fiere, Paulo, 42
film industry
   vs. book industry, 265–266, 330
   in Canada, 132, 134, 167, 328
   Capital Cost Allowance, 132
   and corporate concentration, 316
   distribution in, 167
film phototypesetting, 256
Finkelstein, Jesse, 314
Finova, 239
Firefly Books, 232
First World War, 65
Foreign Investment Review Act, 82, 99
foreign-owned publishers, 12, 84. *See also* branch plants; *specific foreign-owned publishers*
   and agencies, 232–233
   and authors' agents, 236
   and Canadian authors, 22, 235, 307
   competitive advantage of, 234, 238
   in educational publishing, 12, 66, 68, 72–73, 84, 91, 119, 159
   market share of, 165
   Ontario Royal Commission on Book Publishing on, 86
   proposed patriation of Canadian branches of, 90, 105
   size of, 12, 138

Vital Links on, 110
foreign ownership
    ACP on, 103, 105
    and the Baie-Comeau Agreement, 82, 106–109, 110, 113–114, 119, 233. *See also* Investment Canada Act
    Donner and Lazar on, 138
    dynamics of, 232–233
    Grant and Wood on, 144
    loans to deter, 87
    Ontario Royal Commission on Book Publishing on, 87
    policy on, 8
Foreign Reprints Act, 57–58
Foreign Rights Marketing Assistance Program (FRMAP), 178
Fox, Francis, 100, 108
Fox Jones, 115
France
    arts and culture in, 88
    as colonial power, 42, 43, 328
    and the creative economy, 106
    literature in, about book publishing, 36
    and NIICD, 145
    price maintenance in, 14
    publishing in, 34, 44
    VAT reductions in, 14
    Voltaire's social circle in, 35
Franco-Ontarians, 94
freedom of speech, 8–9, 10, 45
free trade, 9, 108, 109, 128, 140, 145. *See also* Free Trade Agreement (FTA); North American Free Trade Agreement (NAFTA)
Free Trade Agreement (FTA), 83, 109, 113–114, 138–139
French-language publishing in Canada, 16, 160–161, *180*, 184
French Revolution, 7
Freud, Sigmund, 33
Friesens, 280
fulfillment, 25, 92, 208–210, 221, 322

Gage, 85
Gallant, Mavis, 73, 237
Gardens of Shame (Vine, Challen), 27–28

gender equality, as a principle, 8–9
General Distribution Services (GDS), 224, 238–239
General Motors, 103
General Publishing, 138, 239. *See also* General Distribution Services (GDS)
geography, and technology, 268
George Brown College, 200
George Gershwin (Pollack), 143
Germany, 14, 15, 35, 44, 65
Gibson, Douglas, 224, 308, 309–310
Giller Prize, Scotiabank, 201, 308, 310
globalization, 109, 112, 115, 134
Globe and Mail, 195–197, 225, 230
Godfrey, Dave, 85
Golding, William, 42
Good, Cynthia, 329
Google, 250, 251, 262, 273, 288, 333
Google Books Amended Settlement Agreement (ASA), 250
Gordon, Rev. Charles W. *See* Connor, Ralph (Rev. Charles W. Gordon)
Gore, Al, 335
government printers, 55, 57, 59
government publishing, 96, 98
government reports and social change, 42
government support. *See* support for book publishing
Governor General's Literary Awards, 186, 202
Grant, Peter, 140–145
grants. *See also specific grant programs*
    ACP on, 103
    in Australia, 16
    Grant and Wood on, 144
    in Great Britain, 15
    vs. GST, 14
    industrial dependence on, 116
    in Manitoba, 189
    Massey Commission on, 71
    to McClelland & Stewart, 309
    postal subsidies, 116
    sales-based, 104
    as temporary, 126–127
    title, 90, 110, 126, 326
    to university presses, 320, 321–322
    to writers, 325

Great Britain, 34, 38, 63. *See also* United
	Kingdom
The Great Democracies, 73
Great Depression, 28, 33, 66, 293
Greystone Books, 27, 314. *See also* D&M
	Publishers
The Gridlock Economy (Heller), 304
Griffin Poetry Prize, 201
Grosset & Dunlop, 69
GST, 14, 112, 115, 118, 120, 148
guidebook publishing, 287
Gulf + Western, 106
Gundy, H. Pearson, 58–59, 60, 64
Gutenberg, Johannes, 21, 41, 263
The Gutenberg Galaxy (McLuhan), 38
Gzowski, Peter, 39, 223, 227

Halpenny, Francess, 91, 197
HanWang reader, 251
Harbour Publishing, 31–32, 124
Harlequin, 12, 16–17, 37, 161, 162, 251, 305
Harper, Stephen, 170. *See also* Harper
	government
HarperCollins Canada
	and agencies, 232
	and Chapters/Indigo, 150, 238
	and D&M Publishers, 314
	as foreign-owned mega-corporation, 12,
		138, 160, 234
	Phyllis Bruce at, 329
Harper government, 172, 228, 307, 310
Harris, Mike, 244
Harry Potter series, 323
Haunted series (Lone Pine), 32
Hawking, Stephen, 39
H.B. Fenn, 232, 233, 314
health information, 268, 303, 305
Heather's Picks, 227
Heller, Michael, 304
Helliwell, John, 147, 148
Here, There and Everywhere (Emerick), 143
Hesse, Carla, 7
heterogeneity
	of Canadian book publishing, 9, 81, 100,
		110–111, 152, 245, 316, 336
	of cultural industries, 144

Historica, 309
history of ideas, 32–36
Hogarth Press, 32–34, 69, 272
House of Anansi, 84, 85, 124
HP, 288
Humber College, 200
Hurley, Chad, 336
Hurtig, Mel, 308
Hushion, Jackie, 150
Huxley, Aldous, 42

ideas, history of, 32–36
identity, publishing, 29–31
immigration, 59, 61, 62, 64, 94, 124
Imperial Copyright Law (1842), 56, 57, 58
importation
	ACP on, 103, 111
	in agency publishing. *See* agency
		publishing
	from Britain, pre-Confederation, 55–56,
		57
	dominance of, in Canadian book
		market, 14, 207
	as foundation of publishing in Canada,
		44, 104
	as market challenge in Canada, 46, 48,
		80–81, 207
	parallel. *See* parallel importation
	policy supporting, 125
	pre-Confederation, 55
	of single copies, 97
	statistics on, 113, 162
	tariff on, 63, 167
	from the U.S., 57, 63, 66
Impressionists, 34
independent bookstores
	ACP on, 105
	advertisements of, 196
	and BookNet Canada, 172
	collapse of, 222, 230–231
	and delivery/fulfillment problems, 48,
		149
	and demand structure, 13, 220–221
	inventory of, 209, 220–221
	and price maintenance, 15
	and title recommendations, 223

Independent Publishers Association, 85. *See also* Association of Canadian Publishers (ACP)
India, 10, 39, 67, 209, 210, 333
Indian Charter Act, 67
Industrial Revolution, 33, 34, 253–254, 264
industrial support for publishing, 79, 80, 82, 112, 116, 117, 119, 174–181, 202, 306–307. *See also specific industrial support programs*
Industry-Wide Assistance Program, 189
information revolution, 7, 254
information technology (IT), 254, 261, 266
Ingram, 168, 281, 285
ink, 4, 21, 51
Innis, Harold, 42, 264
intellectual property. *See* copyright
interactivity, 9, 253, 272–276, 332
International Development Research Agency, 96
International Marketing Assistance, 179
International Network on Cultural Diversity, 146
Internet
    and bookselling, 212, 242. *See also* online bookstores
    and cloud computing, 263
    as competition to other media, 196, 249, 284, 330
    and copyright, 303
    and crowd-sourcing, 274
    free content on, 302
    potentials of, for publishing, 336
    and scholarly publishing, 200
    and social networking, 271
    and title availability, 210
internships, 178, 185
Investment Canada, 12, 310–311, 314
Investment Canada Act, 47, 106, 137, 144, 233, 235. *See also* Baie-Comeau Agreement
iPad, 252, 315, 332, 335
iPhone, 287, 332
Ireland, 14, 15, 44, 59–60, 151
Irish National School Books, 59, 60
Irish potato famine, 59
Islamic cultures, books in, 39
Italy, 14, 44

"Jabberwocky" (Carroll), 254
Japan, 39, 44, 88
Japan-Canada Fund, 186
Jeanneret, Marsh, 86, 88
Johnston, Wayne, 38
Joseph S. Stauffer Prizes, 186
journalism programs, 200–201
Journey Prize, 311
Juby, Susan, 236

Kahneman, Daniel, 317
Kain, Karen, 308
Karim, Jawed, 336
Keats, John, 272, 316
Kew Gardens (Woolf), 33
Key Porter, 108, 138, 314
Kids Can Press, 124
Kinders, Marsha, 268
Kindle, 252, 287, 332
Knopf, 160
Knopf, Alfred, 30
Kobo, 229–230, 251, 252–253, 286, 315, 332
Kogawa, Joy, 42
Kruze, Martin, 27–28

lang, k.d., 31
Langara College, 200
Larson, Steig, 251
The Latest Canadian National Reading Study (Lorimer, Lynch), 216, *217*, 222–223
laws, as structural support for publishing. *See* copyright
Lazar, Fred, 136–138
L.C. Page, 69
Leanpub.com, 262, 282
Le Carré, John, 38
Lee, Dennis, 85
Lessig, Lawrence, 305–306
Level 26 (Zuiker, Swierczynski), 273–274
LexisNexis, 160
Liberals (Canada), 106, 116, 120, 135, 151
librarians, 4, 31, 91, 200, 244, 321
libraries. *See also* Public Lending Right (PLR)
    ACP on, 112
    Canada Foundation for Innovation grant to, 249

　　　　as cultural partners to publishing, 21,
　　　　　　125, 197–198, 332
　　　　digital, 250, 277
　　　　and e-books, 332
　　　　funding for, 173
　　　　and Google digitization program, 250
　　　　and literacy, 173, 264
　　　　marketing to, 80, 91, 112
　　　　and parallel importation, 110, 167
　　　　and readership, 283
　　　　and scholarly works, 250, 320, 321, 324
Library and Archives Canada, 71, 211
Life of Title, 280
Lightning Source, 281, 285
LinkedIn, 271
linotype, 256
list building, 29–31
literacy
　　　　and "the book habit," 48
　　　　and education, 52, 172–173
　　　　political and social, 42
　　　　the printed book and, 264
　　　　relationship with knowledge, 40–41
　　　　value of, 39–40
Literary Copyright Act (1709), 56
literary criticism, 4, 39, 80, 88, 182, 243
literary magazines, 185, 199–200, 325
literature, societal importance of, 39–40
Livres Canada Books. *See* Association for the
　　　　Export of Canadian Books (AECB)/
　　　　Livres Canada Books
loans
　　　　ACP on, 103, 112
　　　　and Cultural Industries Development
　　　　　　Fund, 112, 114
　　　　decreasing profitability, 49
　　　　lack of access to, 50
　　　　loan guarantees, 87, 89, 106, 118, 188,
　　　　　　244
　　　　Paul Audley on, 113
　　　　provincial government, 87, 89, 99, 188
lobbying, 18, 57, 101, 106, 109, 307. *See also*
　　　　Association of Canadian Publishers
　　　　(ACP)
Local Initiatives Projects, 124
Lolita (Nabakov), 42

Lone Pine Publishing, 32, 213
long tail, 227, 260, 284
The Long Tail (Anderson), 227
Lord of the Flies (Golding), 42
Lorimer, Rowland, 216, *217*
The Lost Symbol (Brown), 251
Lovell, John, 58, 59, 63, 328
Lovell's Library, 59
Lower Canada, 57, 61
low wages in publishing, 8, 114, 127, 192
LSD Leacock (Rosenblatt), 85
Lulu, 281, *283*, 285
Lunatics, 34
Lynch, Lindsay, 216, *217*

Macaulay, Thomas, 67
MacBride Commission, 140
MacDonald, Flora, 109
Macdonald, John A., 168
Macintosh computer, 256
Mackenzie, William Lyon, 57, 295
Mackinlay Brothers, 60
MacLean, Heather, 151
Maclean's, 73, 142
MacLennan, Hugh, 73
MacLeod, Alistair, 308
Macmillan of Canada, 65, 73, 104
macro-economics of publishing, 8, 269, 297,
　　　　301, 315, 326. *See also* creative economy
MacSkimming, Roy, 85, 128, 234, 237–238
magazines
　　　　advertisers in, 143
　　　　and audience building, 134
　　　　Canadian content in, 328
　　　　as cultural partners of the book industry,
　　　　　　199–200
　　　　DCH policy on, U.S. challenge to, 9,
　　　　　　136, 142
　　　　and interactivity with readership, 275
　　　　non-fiction writers for, 73
　　　　and Royal Commission on Publications,
　　　　　　88–89
Maillet, Antonine, 42
Make magazine, 276
Man Booker Prize, 202
The Man from Glengarry (Connor), 64, 69

Manguel, Alberto, 249
Manitoba, 189
Manitoba Arts Council, 189
Manitoba Book Publishers Marketing Assistance Program, 189
Manitoba Book Week, 189
Manitoba Ministry of Culture, Heritage and Tourism, 189
Manning, Preston, 116, 118, 135
Manutius, Aldus, 21, 41, 264, 269
Maple Leaf Gardens, 27–28
Maritimes, 60, 93, 227. *See also specific Maritime provinces*
marketing and publicity, 5, 25, 80
    ACP on, 103, 106
    and the Canada Book Fund, 176
    in Canadian media, 91–92
    and Chapters/Indigo, 225, 231
    costs of, 208
    and cover design/cover copy, 24
    for e-books, 332
    and educational publishing, 95
    for export, 98. *See also* Association for the Export of Canadian Books (AECB)/Livres Canada Books
    impact of, as given in BookNet Canada, 171
    international, 183–184
    Manitoba assistance for, 189
    and market functioning, 237–241
    Ontario Royal Commission on Book Publishing on, 91
    of *Seabiscuit* (Hillenbrand), 28–29
    for self-publishers, 282
    at university presses, 320
    word of mouth, 223
Mark Twain. *See* Clemens, Samuel Langhorne (Mark Twain)
Martin, Brad, 310, 311, 312, 313
Martin, Paul, 116, 118, 135
Martin, Peter and Carol, 84, 85
Martin Kruze Memorial Fund, 28
Marx, Karl, 41
Masse, Marcel, 106–107, 111, 114, 115, 120
Massey, Vincent, 45, 55, 70–71, 72
Massey Commission. *See* Royal Commission on National Development in the Arts, Letters and Sciences (Massey Commission)
Master of Publishing, 311, 314
Matas, Mike, 286
Mathews, Robin, 85
Maurer, Rolf, 240–241
McArthur, Kim, 224, 225
McArthur & Company, 224
McClelland, Jack, 30, 84, 86, 295, 308, 328
McClelland, John, 30, 60, 65
McClelland & Stewart (M&S)
    bailout, 87, 89
    and bestsellers, 160
    business model of, 84, 312–313
    and Chapters/Indigo, 224
    and consolidation of ownership, 138. *See also* McClelland & Stewart (M&S): sale to Random House Canada
    as early Canadian publisher, 65, 73
    Giller winners at, 308, 310
    history of, 308–312
    impact of Baie-Comeau/Investment Canada Act on, 234
    lecture series, 311
    modern identity of, 30
    as publisher of Canadian literature, 30, 74, 86, 89
    sale to Random House Canada, 234, 237, 307, 309–311, 314
McGraw-Hill, 74, 138
McGraw-Hill Ryerson, 74, 96, 160–161, 324
McGuire, Hugh, 282
McLuhan, Marshall, 38, 40, 264, 265, 269, 336
McNally Robinson, 261
The Mechanical Bible (McLuhan), 38
Mechanics' Institutes, 264
metadata, 151, 171, 253–254, 258, 259, 269
Methodist Book and Publishing House, 60, 64, 65. *See also* McGraw-Hill Ryerson; Ryerson Press
micro-economics of publishing, 8, 48–49, 117, 297, 315–317, 334
Microsoft, 287
A Mid-Decade Assessment (ACP), 104–107

Minister's Advisory Council on Arts and Culture, 134
Mistry, Rohinton, 237, 308
modernism, 32–33, 64, 69
Molson Prizes, 186
Monkey Beach (Robinson), 42
Montgomery, L.M., 69
Montreal Review of Books, 200
Moore, G.E., 33
Moore, James, 235
motivators of publishing, 36–37
Mount Royal College, 200
movable type, 41, 51, 254–256, 269
Mubarek, Hosni, 274
Mulroney, Brian, 106, 108
multimedia, 134, 261, 267, 286–288
Munro, Alice, 237, 308
Murdoch, Rupert, 12
museums and archives, 268
music, as art form, 265
music industry, 134, 151, 287, 305, 328, 330. *See also* radio

Nabakov, Vladimir, 42
Napster, 287
NASA, 269
National Archives, 71. *See also* Library and Archives Canada
National Book Week, 82, 99
National Institutes of Health (NIH), 305
nationalism, 65, 74, 85, 120, 295, 309. *See also* Canadian cultural identity
National Library, 70. *See also* Library and Archives Canada
The Navigator of New York (Johnston), 38
Neale, John, 309
Neatby, Hilda, 72
Neil, Garry, 146–147
Nelson Education, 12, 96, 159, 161
net pricing, 14–15, 242
New Brunswick, 189
New Canadian Library, 308
New Democratic Party (NDP) (Ontario), 131, 132
New Directions (ACP), 116
Newfoundland and Labrador, 189, 299

New International Instrument for Cultural Diversity (NIICD), 140, 145, 191, 296
Newman, Peter C., 150
New Press, 84, 85, 124
News Corp, 12, 131, 138
newspapers, 56, 57, 92, 195–197
New Star Books, 240
New Strategies for Culture and Trade (SAGIT), 138, 145, 296
New Zealand, 44, 151, 209, 210
niche publishing, 13, 96, 231, 276, 300, 330
1984 (Orwell), 42
Nobel Prize for Literature, 202
non-fiction
    acquisition and editing, 23
    awards for, 201–202
    Canadian content in, 73
    in education, 93, 94
    literary, 182
    and social change, 41–42
non-traditional retailers, 13, 25, 179, 242
Nortel, 250
North American Free Trade Agreement (NAFTA), 83, 139, 146
Northwest rebellion, 61
Nova Scotia, 56, 189, 269, 299

Obama, Barack, 196, 214
Obasan (Kogawa), 42
octavo, 4, 21, 269
Old Possum's Book of Practical Cats (Eliot), 37
Ondaatje, Michael, 85
Onex, 224
online bookstores, 13, 212, 223, 229, 230, 277, 283–285. *See also* AbeBooks; Amazon
online publishing, 96, 200
Ontario. *See also* Canada West; Upper Canada
    bailout of M&S, 87
    and the Book Publishing Board, 90
    book publishing policy in, 113, 187–188
    booksellers in, 172
    and *The Business of Culture*, 83, 135, 295
    and the creative economy, 147, 299
    cultural industries in, 133, 134
    and government publishing, 96
    loans in, for cultural production, 82

misalignment with federal government, 129
and the Ontario Royal Commission on Book Publishing. *See* Ontario Royal Commission on Book Publishing
publishers in, 87, 133, 325. *See also specific Ontario-based publishers*
publishing in, 86, 129–131
support for book publishing in, 202, 306. *See also* Ontario Arts Council
tax credits, 137, 174, 187
Ontario Arts Council, 99, 187, 325
Ontario Book Publishers' Assistance Program, 89
Ontario Book Publishing Tax Credit (OBPTC), 187
Ontario Development Corporation, 87
Ontario Institute for Studies in Education, 93
Ontario Media Development Corporation, 134, 187–188, 333
Ontario Ministry of Culture and Communications, 129–131
Ontario Ministry of Education, 94
Ontario Municipal Employees pension fund, 12
Ontario Royal Commission on Book Publishing, 86–98
as basis for publishing policy, 6, 45, 74–75, 79–80, 89, 119
establishment of, 6, 86
on publishing as cultural, 82, 126, 245–246, 328
success of, 97–98
On the Principles of Political Economy and Taxation (Ricardo), 41
open access, 305, 321, 323–324
Open Access Publishing in European Networks (OAPEN), 324
Open Journal Systems, 324
Open Monograph Press (OMP), 324
Opportunities for Youth, 124
Orange Prize, 202
O'Reilly Media, 276–278, 334
The Organization Man, 73
origination. *See also* own titles
abandonment of, post-war, 66
by Canadian-owned firms, 11, 12, 117, 153, 160, 235
as central to cultural aspect of publishing, 4
cost structure of, *208*
financed by agency publishing, 65, 234
statistics on, 161, 165, *165*, 315, 316
U.S. domination of, 46
Orwell, George, 42
outline, book, 23, 27–28
own titles
as defining products of publishers, 11
versus imports, 129–130
sales of, 161, 162, *163*, *164*, 165
Oxford University Press (Canada), 65, 96

Pagemaker, 256
paper, 5, 21, 22–26, 24, 188, 255, 278
parallel importation, 110, 113, 114, 167
Paris Corporation of Printers and Publishers, 34
Parker, George, 55, 58, 59, 60, 62
Partner Program (Google), 250
Pearson
as educational publisher, 96, 260, 288, 324, 325
as foreign-owned mega-corporation, 12, 138, 160, 232
Pearson, Roger, 35–36
Peat Marwick, 112, 114
The Pedagogy of the Oppressed (Friere), 42
peer review, 274, 320–321, 322, 323
Pélagie-la-Charrette (Maillet), 42
Pelletier, Gérard, 98
Penguin, 30
Penguin Canada, 12, 138, 160, 232, 234, 238, 314, 329
Pepper, Doug, 310
Peter Martin, 84
Peter Rabbit (Potter), 335
photocopying, 97, 168–169, 243, 278, 294
photography, 265
Piaget, Jean, 41–42
Pierce, Lorne, 65–66, 295, 328
The Pilgrim's Progress (Bunyan), 38

Pinter, Frances, 323
PlayBook, 287
poetry, 65, 182, 201, 202, 214, 311
policy, book publishing, 6, 8–10, 153. *See also* structural support for publishing; *specific policies*
    landmarks of, 82–83
    Ontario Royal Commission on Book Publishing as foundation for, 6, 45, 74–75, 79–80, 89, 119
    Paul Audley on, 113
political environment necessary for publishing, 45–47, 295
Porter, Anna, 314
Prairie Books Now, 189, 200
price maintenance, 14–15, 242
pricing, book
    ACP on, 102, 111
    and copyright acts of 1800s, 63
    and distribution right, 236
    Donner and Lazar on, 136
    and e-books, 270
    of imported titles, 81
    as influence on book choice, 143
    as market challenge in Canada, 48, 80, 207, 215
    net pricing/price maintenance, 14–15, 242
    and remainders, 212
Priestley, Joseph, 34
Primeau, Claude, 150
Prince Edward Island, 189
Printed Matters (DCH), 179–181
printers, 24
    and Berne Convention, 63
    and databases, 261
    and Foreign Reprints Act, 58
    government, 55, 57
    print-on-demand, 278–281
    as publishers, 55, 57, 58, 60, 65, 281
    U.S., 55, 58
printing, 4, 24
    digital, 278–281
    and economies of scale, 278
    history of, 4, 254–256
    magazine, 89
    offset, 278, 279
Print Measurement Bureau, 143
print on demand (POD), 259, 278–283, *283*, 319, 323
privacy concerns, 229, 286
Prix Goncourt, 42, 202
production, 4, 24
    centralization of, 253–254
    costs of, 208, 326–327
    efficiencies, 48, 131
    scheduling, 24, 258
professional development, 177, 192, 201, 326, 333. *See also* training
proofreading, 24
Prospero Books, 25
Protestantism, 59, 264
Province of Canada, 56
provinces
    and education, 49–50, 93
    support for book publishing from, 46, 116, 119, 148, 186–189, 306. *See also* provincial arts councils
provincial arts councils, 80, 99, 202, 306, 325–326
Publications Distribution Assistance Program (PDAP), 92, 149
Public Lending Right (PLR), 47, 97, 154, 169–170, 172–173, 243, 294
Public Lending Right Commission, 198
public ownership of media, 144
publishers, Canadian. *See* Canadian-owned publishers
publishers, general
    as cultural actors, 51
    as curators of content, 263, 319, 323
    as factor influencing book purchase, 218
    and hubris, 319, 327
    as originator of ideas, 36
    and POD, 280–281
    and relationships with authors, 5, 23, 142–143, 305
    and relationships with booksellers, 143
    vs. self-publishers, 282
Publishers Assistance Program, 189
publishing
    in Canada. *See* Canadian publishing

as cultural pursuit, 79, 81–82, 86, 111,
    115, 117, 119, 120, 126, 131, 137,
    246, 333, 336
database-driven, 257–263
desktop, 256
feudal system of, 316
as high-risk pursuit, 316, 317, 318
identity, 29–31
process, 4, 10, 22–26, 103, 269
professionals, 5, 117–178, 326–327, 329,
    332. *See also specific publishing
    professionals*
Publishing Measures (DCH), 92
Pulitzer Prize, 202
Pulse (news feed), 252
purpose and mission, publisher's, 22, 26, 29–31
Push Pop Press, 335
Putnam, Robert, 9, 147, 148

Quebec
    copyright act in, 56
    and the creative economy, 299
    educational materials in, 120
    electronic data interchange (EDI) in, 105
    policies, review of, 113
    printers in, 60
    and social capital, 147
    support for publishing in, 174, 188, 306
Quill & Quire, 200, 239

radio, 66, 92, 195, 328
Radiohead, 304–305
Radway, Janice, 294
Rae, Bob, 132
Raincoast Books, 161, 232, 233, 238
Raincoast Chronicles, 31–32
Rakuten, 230
Random House (New York), 28–29
Random House Canada
    and Canadian authors, 312
    and Cooke Agency, 237
    as foreign-owned mega-corporation, 12,
        138, 160, 314
    importation and distribution by, 232
    imprints of, 238
    Louise Dennys at, 329

    and McClelland & Stewart, 234,
        309–311
readers. *See also* book purchasers
    ACP on, 116–117
    Canadian, 79, 87–88, 111, 114, 213, 331
    and choice of books, 143
    as contributors to the world of ideas, 37
    growth and development of, 271
    interaction of, with books, 38, 39
    interactivity with authors, 272
    interest of, in Canadian authors,
        218–219, *219*
    numbers of, 117
Reading and Buying Books for Pleasure (DCH),
    216
Reading Life, 252
recommendations, and book purchasing,
    222–223, 228–229
Red River rebellion, 61
Reformation, 264
Reform Party, 116, 135
regional publishers, 26, 32, 105, 172, 231. *See
    also* small publishers
Reisman, Heather, 224, 227, 230
remainders, 167, 211–212
Renaissance, 264
Repo, Satu, 59
returns, 25, 92, 260
    and BookNet Canada, 171
    and chain bookstores, 221
    and Chapters/Indigo, 149, 225, 230,
        239–241
    digital, 332
    and General Distribution Services,
        238–240
    and remainders, 212
    Woll study on, 241–242
review publications, 88, 90, 194, 200
reviews, 194, 195–197, 198, 200, 223
Ricardo, David, 41, 141, 142
Richler, Mordecai, 73
rights
    and authors' agents, 236
    lack of training about, 201
rights acquisition, 61. *See also* agency
    publishing

rights fairs, 105, 179, 184, 188
rights sales, 4, 25–26, 29
   and the AECB, 178–179
   and distribution right, 235–236
   as market challenge in Canada, 48
   statistics on, 162, 178
   subsidiary, 96, 102
RIM, 287, 288
Roberts, Charles G.D., 64
Robertson, John Ross, 58
Robinson, Eden, 42
Robinson Crusoe (Defoe), 38
Rohmer, Richard, 86, 88
romance reading, 37, 39
Rose, Jonathan, 38
Rosenblatt, Joe, 85
Rough Cuts, 277
Rowling, J.K., 323
Rowman and Littlefield, 280
Royal Commission on National Development in the Arts, Letters and Sciences (Massey Commission), 45, 55, 70–71, 74, 80, 328
Royal Commission on Publications, 88
Royal Society, 34
royalties
   authors' acceptance of low, 114
   and authors' agents, 236
   and foreign-owned publishers, 234
   insurance, 90, 91
   in the music industry, 304–305
   and publishers' eligibility for funding, 175
   within a title's budget, 258
run-ons, 48, 66, 207, 210–211, 212. *See also* importation
Russell, Bertrand, 33
Russia, 333. *See also* Soviet Union
Ryerson, Egerton, 60–62, 173, 295, 328
Ryerson Press, 65, 73, 74, 85. *See also* McGraw-Hill Ryerson; Methodist Book and Publishing House
Ryerson University, 200
Ryga, George, 42

Safari (online bookstore), 277
SAGIT. *See* Cultural Industries Sectorial Advisory Group on International Trade (SAGIT)
sales, 25, 237–241
   of Canadian-originated titles, 315
   data, 92, 151, 171, 242. *See also* BookNet Canada
   international, 92, 237. *See also* exporting
   relation to marketing, 28–29
   of scholarly works, 320
   statistics on, 11, *163–164*, *181*
Sanders, Rob, 314
Sapir-Whorf hypothesis, 40
Saskatchewan, 188–189, 199, 306
Saskatchewan Arts Board, 188–189
Saturday Night, 73
Schedule C, 167
scholarship and scholarly publishing. *See also* university presses
   and BPIDP/Canada Book Fund, 175
   electronic publishing in, 323–324
   journals, 200, 320, 320–321, 324
   monographs, 140, 200, 260, 319–324
   multimedia in, 267–268
   Ontario Royal Commission on Book Publishing on, 88, 90
   open access to, 302–303
   peer review, 274, 320–321, 322, 323
Scholastic, 96
School Act (1864), 60
Schultz, Howard, 230
Schwartz, Gerry, 224
science textbooks, 50, 275, 331
"scorched earth policy," 106
Scotiabank Giller Prize, 201, 308, 310
Scotland, 15, 44
Scott Foresman, 74
Scott, Mark, 314
scrolls, 21
Seabiscuit (Hillenbrand), 28–29
Second World War, 42, 66
selection of books in Canada, 209, 315
self-publishing. *See* service publishing
Seligman, Ellen, 308
Seller, Shyla, 27–28
sell-through, 143, 171
service publishing, 281–283, 318–327, 332, 334

Setting Priorities for Federal Book Publishing Policy (ACP), 118
Shelley, Percy B., 316
Shorebridge, Paul, 332
Shortcovers, 229. *See also* Kobo
Short Cuts, 277
Siemens, Ray, 269
Simon & Schuster, 12, 96, 138, 160, 161, 232, 238, 314
Simon Fraser University, 200, 311, 314
Simons, Michael, 332
The Sky Pilot (Connor), 64
small publishers. *See also* regional publishers
 and AbeBooks, 261
 acquisition at, 237
 and BookNet Canada, 172
 and CBPDP, 100
 culture of poverty at, 191
 and distribution and retailing, 240–241
 market share of, 12
 and non-traditional retailers, 13, 242
 and price maintenance, 15
 and print on demand, 281
Smith, Adam, 41
social capital, 81, 147–149, 154, 159, 190–191, 246, 297, 315, 318
social change and books, 41–43
social media/networking, 191, 253, 271–276, 328, 336
Social Sciences and Humanities Research Council (SSHRC), 70–71, 112, 174, 183
Société de développement des entreprises culturelles (SODEC), 188
Société typographique de Neuchâtel (STN), 35
socio-technical ensembles, 4, 251, 263–264, 336
So Little for the Mind (Neatby), 72
Sony Reader, 251, 332, 333
Soros Foundation, 323
South Africa, 151, 209, 210
Soviet Union, 88. *See also* Russia
Spain, 14, 44
specificity, title, 212–213
spin-offs and community building, 9, 301
Standing Committee on Canadian Heritage, 128, 150
Staples, 285

Starbucks, 230
Statistics Canada
 on book publishing, 92, 113, 160–162, *163–164*, 165
 on book sales, 11, 315
 on Canadian book market, 12
 definition of cultural sector, 296
 on distributors, 165
 on number of new titles, 211
 on ownership of publishers in Canada, 11, 165, 233
Steele, James, 85
Stein, Gertrude, 33
Stevenson, Larry, 222, 224
Stoddart, Jack, 224, 239–240
Stoddart, Susan, 239
structural support for publishing, 46, 81, 112, 119, 125, 166–173, 202. *See also specific structural support programs*
The Struggle for Canadian Universities (Mathews, Steele), 85
styles of publishing, 31–32
stylistic editing, 23
subsidies. *See* grants
substantive edit, 23, 27–28
succession, 137, 260, 309, 314, 326
Summer Publishing Workshops, 311
Supply Chain Initiative (SCI), 50, 92, 128, 151, 152, 154, 179, 185, 296
support for book publishing. *See also specific support programs*
 in Australia, 16
 in Canada vs. internationally, 13–16
 cultural. *See* cultural support for book publishing
 drawbacks of, 189–193
 versus free trade, 108
 in Great Britain, 15
 industrial. *See* industrial support for publishing
 as laid out in the Ontario Royal Commission on Book Publishing, 79
 permanence of, in Canada, 148
 SAGIT on, 139
 structural. *See* structural support for

publishing
    as temporary, 126
Support for Organizations and Associations, 176
Suzuki, David, 85
Sweden, 14, 88
Swierczynski, Duane, 273
Switzerland, 35–36

tablets, 50, 252
tags, in database-driven publishing, 256
Tamblyn, Michael, 229
tax credits
    ACP on, 112, 118–119
    in British Columbia, 137, 174, 186–187, 307
    Donner and Lazar on, 136–137
    in Ontario, 137, 174, 187
    Paul Audley on, 113
    provincial, 188
technological environment necessary for publishing, 50, 52, 332–333
technological protection measures (TPM), 303
technology
    The Business of Culture on, 134
    and copyright, 243
    and economics of publishing, 318
    funding for, 326
    -inspired programs, 185
    Ontario Media Development Corporation on, 188
    training in, 185, 277, 333
    as transformative to publishing, 6–7, 249–276, 328, 334
Technology Strategy Board (Britain), 298
Tecumseh, 61
Teenage Mutant Ninja Turtles, 268
television, 92, 194, 195, 328
terms of sale, 25, 225, 242, 278
Teron, Jacques-Benjamin, 35
textbooks. *See also* education; educational publishing
    in British colonies, 59–60, 67
    Canadian content in, 94, 331
    from Canadian-owned publishers, 93
    development of, 66, 88, 95
    Egerton Ryerson and, 60–61
    from foreign-owned publishers, 66, 91
    interactive, 275
    in pre-Confederation Canada, 61
    returns of, 241
    U.S., in Canada, 49, 61, 62
Thomas Allen, 73, 232, 233
Thomson Corporation, 12
Thoreau, Henry David, 294
Thought and Language (Vygotsky), 40
Time, 142
title specificity, 212–213
Title Value Management, 280
Tools of Change, 277
topic, as factor influencing book purchase, 216–218, 222
Toronto, as cultural centre, 132, 133, 135, 336–337
Torstar, 12
tourism, 132, 135, 301
trade publishing
    Confederation to 1950, 62–67
    pre-Confederation, 55–59
    separation of, from educational publishing, 66, 68, 95, 115, 120
    service model of, 325–327
Trafford Publishing, 281–282, 285, 318–319
training. *See also* internships; professional development
    for educators, 92, 93, 94
    funding for, 178
    not addressed in ACP report, 103
    Ontario Royal Commission on Book Publishing on, 91
    publishers' attitudes towards, 192
    in publishing, 200–201
    in technology, 185, 277, 333
translation, 99, 176, 183, 185, 202
Trans-Pacific Partnership, 170
Trillium List, 94
Trilogy Retail Enterprises, 224
Trudeau, Pierre Elliott, 123–125, 153, 293, 308
Tundra Books, 308, 311
Twitter, 271, 274

Understanding Media (McLuhan), 38

UNESCO, 140, 145–146
United Church of Canada, 74
United Kingdom. *See also* England; Great Britain
  arts and culture in, 88
  as colonial power, 21–22, 42, 43, 328
  copyright in, 58–59
  and *Creative Britain* report, 298
  cultural dominance of, 10
  cultural identity of, 44
  publishers in, 48, 56, 57, 69
  remainders from, 211
  run-on copies from, 66, 111, 210
  supply chain data system in, 151–152
  VAT reductions in, 14
United Nations (UN), 44, 297–298, 301, 337
United Nations Development Programme, 297
United States
  arts and culture in, 88
  Civil War, 61
  content in Canadian education, 49–50
  copyright in, 58, 63, 64
  cultural influence of, 10, 14, 22, 42, 43, 52, 61, 328
  entertainment industries in, 126
  as exporting nation, 140
  film industry in, 167
  intrinsic publishing infrastructure of, 44
  libraries and scholarly works in, 324
  literature about book publishing, 36
  and magazine policy, 9, 136, 142
  origination of titles in, 46
  and price maintenance, 15
  printers in, 55, 58
  publishers, 58, 63, 66, 84, 167, 211. *See also specific U.S.-based publishers*
  reaction to Baie-Comeau, 108, 109, 119
  returns in, 241–242
  revolution, 46, 61
  rights sales, 48
  run-on copies from, 66, 111, 210–211
  supply chain data system in, 151
  textbooks from, 49–50
Universal Copyright Convention, 278
Universal Declaration of Human Rights, 146

Universal Declaration on Cultural Diversity, 83, 146
universities, 85, 95, 304, 319, 331. *See also* education: post-secondary; *specific universities*
university bookstores, 277
University of Chicago Press, 280
University of Illinois, 277
University of Toronto, 60, 309, 310, 311
University of Toronto Press, 65, 74, 198
university presses, 26, 178, 183, 319–323. *See also specific university presses*
Upper Canada, 57, 60, 61. *See also* Canada West; Ontario
Urquhart, Jane, 237
U.S. Copyright Act, 63, 64
used-book sellers, 210, 284. *See also* AbeBooks
utility-based computing. *See* cloud computing
UTP Higher Education, 198

value-added tax (VAT), 14
Valuing Culture (Conference Board of Canada), 295, 337
Victoria, B.C., 269
Victorian era, 32–33, 69
Vietnam War, 124, 293
Vine, Cathy, 27
Virtual Museum of Canada, 185
Vital Links (Government of Canada), 109–110, 129, 130, 131
Vogue, 143
Voltaire, 35–36
Vygotsky, Lev, 40

wages in publishing, low, 8, 114, 127, 192
Wales, 15
Walmart, 13, 207, 227, 229
The Walrus, 143
warehousing, 25, 61, 171, 239–241, 245, 280
War of 1812, 61
Watt, James, 34
The Wealth of Nations (Smith), 41
Webber, Andrew Lloyd, 37
Wedgwood, Josiah, 34
Welcome to Pine Point, 332
Wells, H.G., 33

White, Trena, 314
W.H. Smith, 13
Whyte, William, 73
Wikipedia, 274
Wiley-Blackwell, 96, 138, 314
Williams, Erin, 238
Wilson, Charles, 103
Winfrey, Oprah, 39, 223, 227, 272
W.J. Gage, 74
women, as book purchasers and readers, 37, 39, 184, 216, 222, 294
Wood, Chris, 140–145
Woolf, Leonard, 32, 69
Woolf, Virginia, 32–33, 69
WordPress, 262, 274
Wordsworth, William, 316
World Creative Business Conference, 298
World Intellectual Property Organization (WIPO), 303
World Trade Organization (WTO), 136, 139, 142, 145, 146
World War I, 65
World War II, 42, 66
writers-in-schools program, 199
Writers' Reserve, 325
Writers' Trust of Canada, 198
Writers' Works in Progress, 325
WYSIWYG, 256–257

Xerox PARC, 256

YouTube, 336

Zuckerberg, Mark, 336
Zuiker, Anthony, 273–274

get the eBook FREE!

Get the eBook free!

At ECW Press, we want you to enjoy this book in whatever format you like, whenever you like. Leave your print book at home and take the eBook to go! Purchase the print edition and receive the eBook free. Just send an email to ebook@ecwpress.com and include:

- the book title
- the name of the store where you purchased it
- your receipt number
- your preference of file type: PDF or ePub?

A real person will respond to your email with your eBook attached. Thank you for supporting an independently owned Canadian publisher with your purchase!